METHODS IN MOLECULAR BIOLOGY

For further volumes:
http://www.springer.com/series/7651

For over 35 years, biological scientists have come to rely on the research protocols and methodologies in the critically acclaimed *Methods in Molecular Biology* series. The series was the first to introduce the step-by-step protocols approach that has become the standard in all biomedical protocol publishing. Each protocol is provided in readily-reproducible step-by-step fashion, opening with an introductory overview, a list of the materials and reagents needed to complete the experiment, and followed by a detailed procedure that is supported with a helpful notes section offering tips and tricks of the trade as well as troubleshooting advice. These hallmark features were introduced by series editor Dr. John Walker and constitute the key ingredient in each and every volume of the *Methods in Molecular Biology* series. Tested and trusted, comprehensive and reliable, all protocols from the series are indexed in PubMed.

Protein-Protein Interaction Networks

Methods and Protocols

Edited by

Stefan Canzar and Francisca Rojas Ringeling

Gene Center, Ludwig-Maximilians-Universität München, München, Germany

Humana Press

Editors
Stefan Canzar
Gene Center
Ludwig-Maximilians-
Universität München
München, Germany

Francisca Rojas Ringeling
Gene Center
Ludwig-Maximilians-
Universität München
München, Germany

ISSN 1064-3745 ISSN 1940-6029 (electronic)
Methods in Molecular Biology
ISBN 978-1-4939-9872-2 ISBN 978-1-4939-9873-9 (eBook)
https://doi.org/10.1007/978-1-4939-9873-9

This Humana imprint is published by the registered company Springer Science+Business Media, LLC, part of Springer Nature.
The registered company address is: 233 Spring Street, New York, NY 10013, U.S.A.

Preface

Proteins interact in complex networks to drive cellular function. These networks can be rewired in different biological contexts or disrupted in genetic diseases. Comprehensive and well-annotated protein–protein interaction (PPI) networks are thus indispensable for a systems-level interpretation of the massive amounts of molecular data produced by high-throughput experimental technologies. Experimental approaches to detect interactions between proteins are challenged by a wide range of binding affinities and different assays show different strengths and weaknesses in coping with them. This book presents step-by-step protocols for computational methods that can reduce the gaps left by experimental approaches to detect interactions. It also includes methods that can help to transfer our knowledge of interactions between different species, integrate our knowledge scattered across databases that cover different types of interactions, combine them with context-specific data, provide functional insights into the discoveries made from the analysis of omics datasets, and assemble individual interactions into higher order semantic units, i.e., protein complexes and functional modules.

Computational methods can learn from existing data on interactions obtained through, e.g., high-throughput experimental approaches such as yeast two-hybrid and affinity purification coupled to mass spectrometry, or exploit indirect evidence to expand our knowledge of interacting proteins. *SPRINT* (Chapter 1), for example, allows to computationally lift existing PPI networks to the species' interactome level by learning from the sequences of proteins that are known to interact. Chapter 2 describes a text-mining protocol that extracts and visualizes PPI networks from the rich source of biomedical literature in PubMed. In contrast, the method by Shim and Lee (Chapter 3) relies on protein domain annotations to infer functional PPI from shared functional units, that is, from similar compositions of domains.

Alternatively, homology between protein sequences can help to reveal or refine novel interactions. *Path2PPI* (Chapter 4) uses homology search to transfer interactions from a collection of well-annotated reference species to a target organism for which little to no data on protein–protein interactions are available. In Chapter 5, the authors present a protocol on how to exploit co-evolution of contact residues to increase the resolution of interactions from protein families to the precise pair of interacting paralogs.

Given a pair of interacting proteins, computational methods can further zoom in to the residue level and shed light on the 3D structure of a protein–protein interaction or a protein complex to enhance our understanding of their function. In Chapter 6, Jing et al. describe step by step how to employ the *RaptorX-ComplexContact* web server to predict inter-protein residue–residue contacts. The underlying algorithm uses a deep-learning model based on sequential features and co-evolution signals derived from the alignment of multiple homologous sequences. The identified residue–residue contacts can guide the prediction of PPI, as well as their structural modeling (protein docking). In fact, the step-by-step protocol provided in Chapter 7 uses residue–residue contacts in the docking step, before using the *COZOID* tool to guide the selection of a protein docking model based on its similarity to an evolutionary conserved complex structure of homologous proteins. Zooming in one more level, Chapter 8 describes the use of the *Path-LZerD* software to predict the assembly steps of a multimeric protein complex. Knowing how such a complex is formed can inform the artificial design of protein complexes as well as the design of drugs that target critical interactions in a complex. Finally, Halakou et al. unify the network view and the structural perspective of interactions and show in Chapter 9 how a PPI network can be refined by docking alternative confirmations of proteins participating in binary interactions.

Our knowledge about protein–protein interactions is scattered over multiple PPI databases that cover various types of interactions across several species. Being able to systematically query such databases is crucial for the meaningful integration of PPIs in downstream analysis methods. In Chapter 10, the authors illustrate how tissue or disease-specific interactions from major curated databases can be retrieved and combined through the *Integrated Interactions Database* (IID). The *HIPPIE* web tool (Chapter 11) similarly integrates various (human) interaction databases but additionally overlays gene expression information and other annotation resources to build protein networks specific to a tissue, a disease, or to subcellular localization. Access to a cancer-specific PPI network is provided through the *OncoPPi* Portal. The OncoPPI network was previously constructed in high-throughput screening experiments and in Chapter 12 Ivanov describes its integration with cancer genomics and pharmacological data through the web portal.

PPI networks can guide the interpretation of high-throughput molecular profiling datasets. The function of a set of differentially expressed genes detected through RNA-seq, for example, can be elucidated by identifying enriched biological pathways or by considering their interactions in a larger network context. *ReactomeFIViz* (Chapter 13) is a Cytoscape app that supports the pathway- and network-based analysis of RNA-seq and other omics datasets based on the Reactome pathway database. *KeyPathwayMiner* (Chapter 14) solves the reverse problem. It combines interaction networks and omics datasets to detect novel functional modules. In Chapter 15, the authors describe how *PathClass* can use such de novo pathways to classify breast cancer subtypes.

Much can be learned from the topology of a PPI network. In Chapter 16, Biedermann et al. provide a practical guide to the use of their software *VieClus*. It detects functional modules or protein complexes by searching for sets of proteins that exhibit dense interactions within the sets but sparse interactions between the sets. *TD-WGcluster* (Chapter 17) generalizes the notion of modules by combining the topology with a dynamics component obtained through time series data. Even more, a comparative analysis of PPI networks can uncover evolutionary relationships between different species, transfer knowledge between them, and identify conserved pathways. In Chapter 18, Hayes instructs the users of *SANA* on how to perform an alignment-based comparison of PPI networks.

München, Germany

Stefan Canzar
Francisca Rojas Ringeling

Contents

Contributors

GREGORIO ALANIS-LOBATO • *Human Embryo and Stem Cell Laboratory, The Francis Crick Institute, London, UK*

NICOLAS ALCARAZ • *Department of Biology, The Bioinformatics Centre, University of Copenhagen, Copenhagen, Denmark*

CARLO BALDASSI • *Bocconi Institute for Data Science and Analytics, Bocconi University, Milan, Italy; INFN, Sezione di Torino, Torino, Italy*

RICHA BATRA • *Helmholtz Zentrum München, Research Center for Environmental Health, Institute of Computational Biology, Neuherberg, Germany; Department of Dermatology and Allergy, Technical University of Munich, München, Germany*

SONJA BIEDERMANN • *Faculty of Computer Science, University of Vienna, Vienna, Austria*

JAN BYSKA • *Department of Informatics, University of Bergen, Bergen, Norway; Faculty of Informatics, Masaryk University, Brno, Czech Republic*

CHARLES CHRISTOFFER • *Department of Computer Science, Purdue University, West Lafayette, IN, USA*

KATARINA FURMANOVA • *Faculty of Informatics, Masaryk University, Brno, Czech Republic*

THOMAS GUEUDRÉ • *Italian Institute for Genomic Medicine, Turin, Italy*

ATTILA GURSOY • *Computer Science and Engineering Department, Koc University, Istanbul, Turkey*

FARIDEH HALAKOU • *Computer Science and Engineering Department, Koc University, Istanbul, Turkey*

ANDREA HAMANN • *Department of Biosciences, Molecular Developmental Biology, Institute of Molecular Biosciences and Cluster of Excellence Frankfurt, "Macromolecular Complexes", Johann Wolfgang Goethe-University, Frankfurt am Main, Germany*

ANNE HARTEBRODT • *TUM School of Life Sciences, Technical University of Munich, Freising, Germany*

ROBIN HAW • *Informatics and Biocomputing Program, Ontario Institute for Cancer Research, Toronto, ON, Canada*

WAYNE B. HAYES • *Department of Computer Science, University of California, Irvine, CA, USA*

YONGQUN HE • *University of Michigan Medical School, Ann Arbor, MI, USA*

MONIKA HENZINGER • *Faculty of Computer Science, University of Vienna, Vienna, Austria*

LUCIAN ILIE • *Department of Computer Science, The University of Western Ontario, London, ON, Canada*

ANDREY A. IVANOV • *Department of Pharmacology and Chemical Biology, Emory University, Atlanta, GA, USA; Emory Chemical Biology Discovery Center, Emory University, Atlanta, GA, USA; Winship Cancer Institute, Emory University, Atlanta, GA, USA*

XIAOYANG JING • *Toyota Technological Institute at Chicago, Chicago, IL, USA; School of Computer Science, Fudan University, Shanghai, China*

ADAM JURCIK • *Faculty of Informatics, Masaryk University, Brno, Czech Republic*

IGOR JURISICA • *Princess Margaret Cancer Centre, University Health Network, Toronto, ON, Canada; Krembil Research Institute, University Health Network, Toronto, ON, Canada; Department of Medical Biophysics, University of Toronto, Toronto, ON, Canada;*

Department of Computer Science, University of Toronto, Toronto, ON, Canada; Institute of Neuroimmunology, Slovak Academy of Sciences, Bratislava, Slovakia

OZLEM KESKIN • *Chemical and Biological Engineering Department, Koc University, Istanbul, Turkey*

DAISUKE KIHARA • *Department of Biological Sciences, Purdue University, West Lafayette, IN, USA; Department of Computer Science, Purdue University, West Lafayette, IN, USA*

INA KOCH • *Institute of Computer Science, Johann Wolfgang Goethe-University, Frankfurt am Main, Germany*

MAX KOTLYAR • *Princess Margaret Cancer Centre, University Health Network, Toronto, ON, Canada; Krembil Research Institute, University Health Network, Toronto, ON, Canada*

BARBORA KOZLIKOVA • *Faculty of Informatics, Masaryk University, Brno, Czech Republic*

FINN KUUSISTO • *Regenerative Biology, Morgridge Institute for Research, Madison, WI, USA*

PAOLA LECCA • *Department of Mathematics, University of Trento, Trento, Italy; Faculty of Computer Science, Free University of Bolzano-Bozen, Bolzano, Italy*

INSUK LEE • *Department of Biotechnology, College of Life Science and Biotechnology, Yonsei University, Seoul, South Korea*

YIWEI LI • *Department of Computer Science, The University of Western Ontario, London, ON, Canada*

MARKUS LIST • *TUM School of Life Sciences, Technical University of Munich, Freising, Germany*

FRED LONEY • *Knight Cancer Institute, Oregon Health & Science University, Portland, OR, USA*

JEYAKUMAR NATARAJAN • *Data Mining and Text Mining Laboratory, Department of Bioinformatics, School of Life Sciences, Bharathiar University, Coimbatore, Tamil Nadu, India*

EDISON ONG • *University of Michigan Medical School, Ann Arbor, MI, USA*

HEINZ OSIEWACZ • *Department of Biosciences, Molecular Developmental Biology, Institute of Molecular Biosciences and Cluster of Excellence Frankfurt, "Macromolecular Complexes", Johann Wolfgang Goethe-University, Frankfurt am Main, Germany*

ANDREA PAGNANI • *Italian Institute for Genomic Medicine, Turin, Italy; INFN, Sezione di Torino, Torino, Italy; Dipartimento di Scienza Applicata e Tecnologia, Politecnico di Torino, Torino, Italy*

JAN J. PALECEK • *Faculty of Science, National Centre for Biomolecular Research, Masaryk University, Brno, Czech Republic; Mendel Centre for Plant Genomics and Proteomics, Central European Institute of Technology, Masaryk University, Brno, Czech Republic*

CHIARA PASTRELLO • *Princess Margaret Cancer Centre, University Health Network, Toronto, ON, Canada; Krembil Research Institute, University Health Network, Toronto, ON, Canada*

OLIVER PHILIPP • *Institute of Computer Science, Johann Wolfgang Goethe-University, Frankfurt am Main, Germany*

KALPANA RAJA • *Regenerative Biology, Morgridge Institute for Research, Madison, WI, USA; Data Mining and Text Mining Laboratory, Department of Bioinformatics, School of Life Sciences, Bharathiar University, Coimbatore, Tamil Nadu, India*

ANGELA RE • *Systems and Synthetic Biology Laboratory, Centre for Sustainable Future Technologies, Fondazione Istituto Italiano di Tecnologia, Torino, Italy*

IAN ROSS • *Computer Sciences Department, Center for High Throughput Computing, University of Wisconsin, Madison, WI, USA*

MARTIN H. SCHAEFER • *Department of Experimental Oncology, European Institute of Oncology, Milan, Italy*
CHRISTIAN SCHULZ • *Faculty of Computer Science, University of Vienna, Vienna, Austria*
BERNHARD SCHUSTER • *Faculty of Computer Science, University of Vienna, Vienna, Austria*
JUNG EUN SHIM • *Yonsei Biomedical Research Institute, Yonsei University College of Medicine, Seoul, South Korea*
JOHN STEILL • *Regenerative Biology, Morgridge Institute for Research, Madison, WI, USA*
RON STEWART • *Regenerative Biology, Morgridge Institute for Research, Madison, WI, USA*
GENKI TERASHI • *Department of Biological Sciences, Purdue University, West Lafayette, IN, USA*
JAMES THOMSON • *Regenerative Biology, Morgridge Institute for Research, Madison, WI, USA; Department of Cell and Regenerative Biology, University of Wisconsin, Madison, WI, USA*
SHENG WANG • *Toyota Technological Institute at Chicago, Chicago, IL, USA*
MARTIN WEIGT • *CNRS, Institut de Biologie Paris Seine, Biologie Computationnelle et Quantitative—LCQB, Sorbonne Université, Paris, France*
GUANMING WU • *Department of Medical Informatics and Clinical Epidemiology, Oregon Health & Science University, Portland, OR, USA*
JINBO XU • *Toyota Technological Institute at Chicago, Chicago, IL, USA*
HONG ZENG • *School of Computer Science and Technology, Hangzhou Dianzi University, Hangzhou, China*

Chapter 1

Predicting Protein–Protein Interactions Using SPRINT

Yiwei Li and Lucian Ilie

Abstract

Understanding protein–protein interactions (PPIs) is vital to reveal the function mechanisms in cells. Thus, predicting and identifying PPIs is one of the fundamental problems in system biology. Various high-throughput experimental and computation methods have been developed to predict PPIs. Here, we provide a straightforward guide of using the program "SPRINT" to predict the PPIs on an interactome level in an organism. First, some installation guides and input file formats are described. Then, the commands and options to run SPRINT are discussed with examples. In addition, some notes on possible extended installation and usage of SPRINT are given.

Key words Protein–protein interaction (PPI), PPI prediction, Human interactome

1 Introduction

Proteins are essential molecules in organisms. It is quite often that proteins perform their functions by interacting with other proteins, binding into a stable or temporary complex. Protein–protein interaction (PPI) is a vital process in molecular biology. Identifying PPIs is one of the fundamental problems in system biology [1].

Many experimental methods have been developed including the widely used approaches such as yeast two-hybrid (Y2H) [6] and tandem affinity purification (TAP) [20]. Such in vivo methods require intensive time and labor, and therefore, it is infeasible to perform interactome PPI identification. Various computation methods [5, 7–9, 14–17, 19, 22–26] have been proposed to help predicting PPIs. Such in silico methods could be useful screening tools for biologists to identify PPIs. However, many computational PPI predicting programs are slow, inaccurate, or hard to use.

In order to alleviate the above issues, we proposed the SPRINT (Scoring PRotein INTeractions) algorithm and program [14] that predicts PPIs fast and accurately. SPRINT requires only the protein sequences and the boolean PPI information of an organism. Unlike 3D structure or domain information, such knowledge is available

Stefan Canzar and Francisca Rojas Ringeling (eds.), *Protein-Protein Interaction Networks: Methods and Protocols*, Methods in Molecular Biology, vol. 2074, https://doi.org/10.1007/978-1-4939-9873-9_1,

on proteomes level for most of the well-studied organisms. Predicting the entire human interactome using SPRINT takes between 15 and 100 min on a 12-core machine depending on the size of the known PPIs. SPRINT is written in C++ and can be run under any UNIX-like system.

2 Materials

2.1 The SPRINT Source Code

The SPRINT algorithm is written in C++ and has been described and evaluated in detail previously [14]. The program is freely available at https://github.com/lucian-ilie/SPRINT under the GNU General Public License v3.0. The source code is maintained regularly on Github. SPRINT can be obtained using the "clone or download" button on GitHub or using the command "git clone https://github.com/lucian-ilie/SPRINT.git" in a UNIX-like environment with git installed. The git environment can be installed from https://git-scm.com/.

2.2 The Computing Environment

After downloading the source code, SPRINT can be installed on Mac OS, Linux, or UNIX machines. To install SPRINT, GCC compiler, boost library, and OpenMP API are needed.

The GCC compiler can be obtained from https://gcc.gnu.org/. SPIRNT is developed under GCC version 4.8.2.

The boost library is needed for the container `unordered_map` when compiling `compute_HSP` (Subheading 3.1), and the library can be obtained from https://www.boost.org/. SPIRNT is developed under boost version 1.53.

The OpenMP API is used in the parallel version of SPRINT. OpenMP is an application program interface that enables programmers to parallelize the code from high level. The OpenMP API can be obtained from https://www.openmp.org/, and SPRINT is developed under the OpenMP version 1.8.3. Compiling the serial version of SPRINT does not require OpenMP (*see* Subheading 2.3).

2.3 Installing SPRINT

The SPRINT program consists of two parts, computing the high-scoring segment pairs (HSPs) and predicting protein–protein interactions. The executable file names are `compute_HSP` and `predict_interactions`, respectively.

The first component, `compute_HSP`, takes a set of protein names and their sequences as its input and produces the similar subsequence pairs as its output. Since the output is a set of high-scoring segment pairs, its extension name is .hsp.

The second component, `predict_interactions`, takes a set of proteins, a set of known PPIs among them, and an hsp file as its input. Its output is a set of scores of any given pair of potential PPIs among the input proteins. The higher the score, the more confident SPRINT predicts the corresponding pair to interact.

Both `compute_HSP` and `predict_interactions` have a serial version and a parallel version. The implementation difference is that the most time-consuming for loop is parallelized using OpenMP in the parallel version, so multiple CPU cores will be used to achieve higher speed.

The installation is done through GNU make. After entering the SPRINT directory, the following commands are used for the installation.

- make `compute_HSPs_serial` (the serial version of `compute_HSPs`)
- make `compute_HSPs_parallel` (the parallel version of `compute_HSPs`)
- make `predict_interactions_serial` (the serial version of `predict_interactions`)
- make `predict_interactions_parallel` (the parallel version of `predict_interactions`).

2.4 The Input Datasets

To predict the PPIs of an organism, SPRINT requires two types of data. First, a set of protein names and sequences in FASTA format. Second, a set of known PPIs.

The FASTA file uses single-letter codes to represent peptide sequences of each protein. Each protein has two lines. The first line is the protein name, and the second line is its amino acid sequence. The line containing protein names starts with a ">" sign and is followed by the protein name. In the sequence line, each letter encodes an amino acid. An example is given in Fig. 1, the amino acid sequence of protein P32479 is MKVVKFPWLAHREESRKYE IYTVDVSHDGKRLA.

The raw FASTA file with the protein sequences of most organism can be downloaded from Uniprot [4] at https://www.uniprot.org/. For example, manually annotated all human proteins and their sequences can be downloaded by clicking on "Swiss-Prot" and then "Human" and then "Download." For details on using SPRINT on other organisms *see* **Note 1**. Uniprot contains additional information such as function and gene ontology information

```
>P32479
MKVVKFPWLAHREESRKYEIYTVDVSHDGKRLA
>P43571
MGIRRLVSVITRPIINKVNSSGQYSRVLATREDQDKASPKYMNNDKIAKKPYTYR
```

Fig. 1 A FASTA file example. Each protein consists of the protein name which starts with ">" and the amino acid sequence

```
P41812 P38142
Q3E755 P23900
P37265 P34216
Q99195 P25649
P10961 Q02772
```

Fig. 2 An input PPI file example. Each line contains a pair of protein names

which can be filtered by editing "Columns" before downloading. Further dataset processing is probably needed. An in-house C++ program SPRINT/`file_formater/fasta_formatter.cpp` is provided to convert the default FASTA file in Uniprot to a version that is accepted by SPRINT.

The PPI file needed by SPRINT contains the boolean information about known PPIs. Each line is a pair of proteins which are known to interact. For example, as shown in Fig. 2, protein P41812 and P38142 are known to interact. Note that the protein names in the PPI file should match the ones in the input FASTA file.

The PPI dataset is usually obtained through major protein–protein interaction databases [21] such as Biogrid [18], MINT [3], InnateDB [2], IntAct [12], and HPRD [13]. PPI databases normally contain more information than SPRINT needs, so further data processing is needed based on the raw data format. The pre-processed human dataset used in the original paper [14] can be downloaded at http://www.csd.uwo.ca/~ilie/SPRINT/.

3 Methods

As briefly discussed in Subheading 2.3, there are two components in SPRINT to predict PPIs, computing HSPs and predicting PPIs. Figure 3 shows the work flow of SPRINT. To predict the PPIs of an organism, its FASTA file is first fed to the component Compute HSP. The output of this component is the HSP file of the organism. The HSP file is then fed to the second component, Predict Interactions, along with the known PPI file of the organism. The known PPIs train the SPRINT algorithm to detect novel interactions. In the end, new interactions with confidence scores are output.

The following sections first describe the two major steps in SPRINT. Then example commands of using SPRINT are given.

3.1 Computing HSPs

High-scoring segment pairs (HSPs) are similar subsequences among all input sequences. SPRINT first detects these HSPs from the input protein sequences. Figure 4 shows the idea of computing HSPs. Four proteins, P1–P4, are input to the Compute HSPs program. After computation, three HSP pairs are found, that is subsequences A1, B1, C1 are similar to A2, B2, C2, respectively.

Similar subsequences are detected and marked by SPRINT and are stored in an hsp file. For each HSP pair, the first line describes

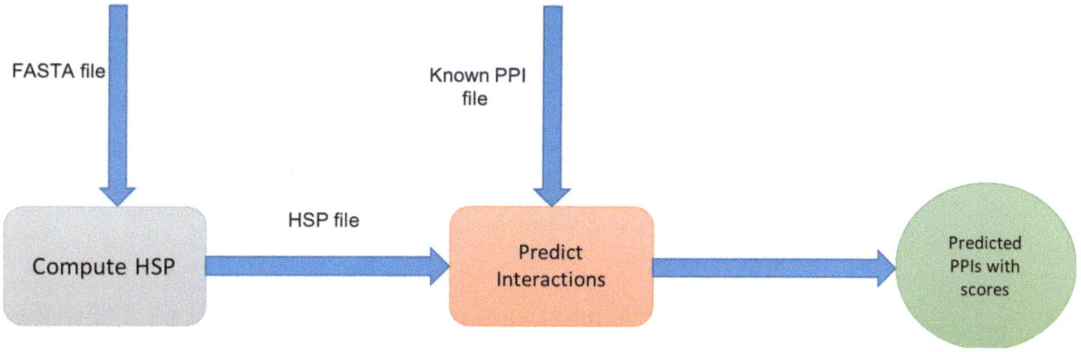

Fig. 3 The work flow of SPRINT. SPRINT has two components: Compute HSP and Predict Interactions. FASTA files and known PPI files are the inputs. Predicted PPIs with scores are the output

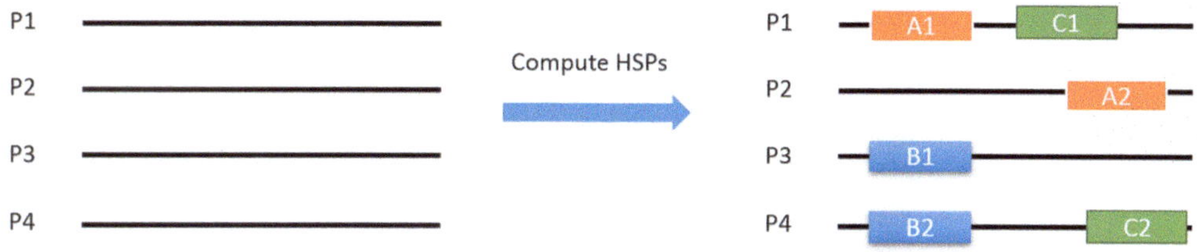

Fig. 4 The idea of computing HSPs. P1–P4 represent four protein sequences. (A1, A2), (B1, B2), (C1, C2) represent three pairs of HSPs. Blocks with the same color indicate a pair of HSP

```
> Q05934 and Q12734
668 1079 20
674 835 24
740 209 20
742 211 22
```

Fig. 5 An hsp file example of all the hsps between the two proteins Q05934 and Q12734

the two proteins. It starts with a ">" sign and is followed by "[protein 1] and [protein 2]." The rest of the lines are the HSP pairs between the two proteins. Each line indicates one HSP pair. The format is "[start position in protein 1] [start position in protein 2] [the length of this HSP]." An example of hsp file is shown in Fig. 5. Protein Q05934 and Q12734 have four HSP pairs. The first pair is of length 20. The starting positions are 668 and 1079 in each protein. This indicates the subsequence from position 668–687 in Q05934 is similar to the subsequence from position 1079–1098 in Q12734.

After installing SPRINT (described in Subheading 2.3), the executable file `compute_HSPs` appears in the directory "SPRINT/bin/". The command to run `compute_HSPs` is "bin/`compute_HSPs` [options]" where [options] is as follows:

-p <input protein file name, FASTA format>(required)

-h <output hsp file name>(required)

-Thit <an integer, the threshold Thit>(optional, default: 15)

-Tsim <an integer, the threshold Tsim>(optional, default: 35)

-M <an integer, Scoring matrix. 1: PAM120, 2: BLOSUM80, 3: BLOSUM62>(optional, default: 1)

-add <new protein file><existing hsp file name>(optional)

The two required options are -p and -h. They are followed by the input protein file name and output hsp file name. The tuneable parameters -Thit, -Tsim, and -M are discussed in **Note 2**. Computing all HSPs requires a relatively long time. For example, computing all HSPs among 20,102 human protein sequences takes about 29 h on a 12-core machine. HSPs need to be computed once for one organism as long as no new proteins are added. The pre-processed human protein sequences and their HSPs can be downloaded at http://www.csd.uwo.ca/~ilie/SPRINT/. The -add option is a time-saving feature when pre-computed HSPs are already available but a few proteins are added to the proteome and only the HSPs involving the new proteins need to be appended to the existing HSP file. If the number of added proteins is much lower than the original ones, computing HSPs for newly added proteins only is much faster than computing all HSPs from scratch. For example, if 100 human sequences are added to the original 20,102 protein set, computing only HSPs involving new proteins takes 2500 s on the same 12-core machine.

3.2 Predicting Interactions

Predicting interactions requires two sets of information for a given set of proteins: their HSPs and the known PPIs. The idea of predicting interactions in SPRINT is shown in Fig. 6. A, B, and C are HSP segments, and P1-3 and Q1-3 are proteins. Known PPIs (P1, Q1) and (P2, Q2) are fed to SPRINT along with the HSPs they contain. SPRINT assumes that in a known interaction, all HSPs are possible binding sites that cause the interaction. For example, (A, B) and (A, C) are assumed to be the binding sites in the interactions (P1, Q1) and (P2, Q2), repetitively, and they are expected to behave the same in a novel interaction (P3, Q3). Thus, the interaction score of (P3, Q3) is increased based on the possible biding between (A, B) and (A, C). The higher the score, the more confidently SPRINT claims the interaction.

The executable file `predict_interactions` is created under the directory "SPRINT/bin/" after the installation (*see* Subheading 2.3). The command to run `predict_interactions` is "bin/`predict_interactions` [options] where [options] is as follows:

-p <input protein file name, FASTA format>(required)

-h <input HSP file name>(required)

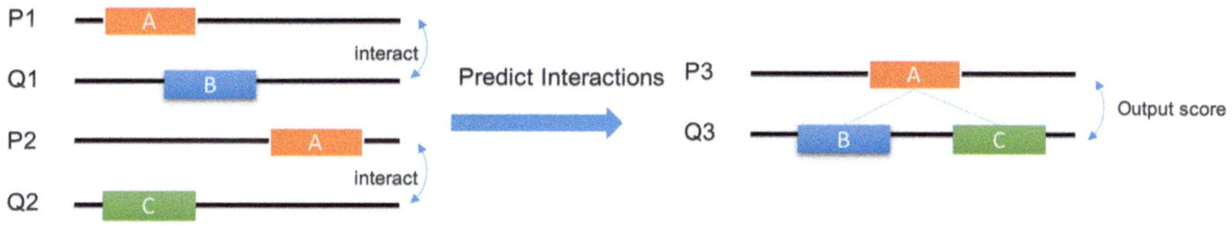

Fig. 6 The idea of predicting PPIs. A, B, C are HSPs. The blocks with the same color indicate HSP pairs. (P1, Q1) and (P2, Q2) are known interactions. The interaction score of (P3, Q3) is calculated

-Thc <an integer, the threshold to be considered a high count> (optional, default: 40)

-tr <training PPI file (known interactions)>(required)

-pos <positive testing PPI file (classified as true interaction)> (optional)

-neg <negative testing PPI file (classified as false interaction)> (optional)

-o <output file name>(required)

-e (instruct SPRINT to perform the entire interactome prediction) (optional)

The required options are -p, -h, -tr, and -o, which are followed by input FASTA file name, input HSP file name, input known PPIs file name, and output file name, respectively. The -Thc option is discussed in **Note 2**. The -pos and -neg options are followed by testing files. The main purpose of having these two options is for the convenience of performance evaluation. Quite often, to compare different predicting tools, statistics measurements such as sensitivity, precision, and F1-score need to be calculated. If testing files are specified, only the scores of PPIs in those files will be calculated and labeled. The output file name is passed using the -o option. If the -e option is passed to SPRINT, the program will perform the entire proteome prediction, that is, the score for every possible protein pair in the given protein set is calculated. The output files are appended with extension names .pos, .neg, or .txt for positive testing, negative testing, or interactome results, respectively. Figure 7a–c shows examples of three output types of SPRINT. The first two outputs are the results of positive and negative testing PPIs. They have the same format. Each line contains <[score] [label]>. In line *i*, the [score] is the prediction value that corresponds to the *i*-th protein pair in testing input, and the [label] is denoted with 1 (positive) and 0 (negative). This is convenient for future developers to further process the data. The output file format for the interactome prediction is different. Each line contains <[Protein1] [Protein2] [Score]>, which represent the predicted score for the pair ([Protein1], [Protein2]).

```
1.1172 1
0 1
0.193279 1
0 1
0.145658 1
0.605425 1
0.0288128 1
0.00742424 1
```

(a) Example result of positive testing PPIs. The two columns contain scores and labels repestively.

```
2.16309 0
0.00027667 0
1.221 0
0.430376 0
0.000987133 0
0.000332767 0
0.0139698 0
0.00134454 0
```

(b) Example result of negative testing PPIs. The two columns contain scores and labels repestively.

```
P32479 P32479 0.00080924
P32479 P43571 0.00793651
P32479 Q12130 0.0019099
P32479 P40961 0.656197
P32479 P38843 0.00106992
P32479 Q06315 0.00173814
P32479 P53885 0.004274
P32479 P40035 0.329972
```

(c) Example result of predicted interactome PPIs with scores. The first two columns are protein names and the third column is the predicted scores.

Fig. 7 Examples of the three types of SPRINT output

3.3 Example Commands to Run SPRINT

For illustration purposes, the directory `toy_example/` contains four example files, `protein_sequences.fasta`, `train_ppi.txt`, `test_positive_ppi.txt`, and `test_negative_ppi.txt`. Some example commands are described in this section.

The following commands assume that the current working directory is in SPRINT/. To create the HSP file HSP/`hsps_toy_example.hsp` for the proteins in `protein_sequences.fasta`, the following command is used.

```
> bin/compute_HSPs - p  toy_example / protein_sequences . fasta
- h  HSP / hsps_toy_example . hsp
```

Once the HSPs are computed, the entire interactome prediction can be executed by the command below. The PPIs in `train_ppi.txt` are used as training, and the output file is written to `result_interactome.txt`. Notice that no negative PPIs are needed for SPRINT as training dataset because SPRINT does not employ machine learning algorithms.

```
> bin/ predict_interactions  - p  toy_example / protein_sequences . fasta
- h  HSP / hsps_toy_example . hsp - tr  toy_example / train_ppi . txt  - e
- o  toy_example / result_interactome . txt
```

If only the scores for some testing pairs that are stored in `test_positive_ppi.txt` and `test_negative_ppi.txt` are needed, the following command is used.

```
> bin/ predict_interactions - p toy_example / protein_sequences . seq
- h HSP / hsps_toy_example . hsp - tr toy_example / train_ppi . txt
- pos toy_example / t e st_positive_ppi . txt - neg toy_example / test_negative_ppi . txt
- o toy_example / r e s ul t _ t e s t . txt
```

4 Notes

1. Predicting PPIs for Organisms Other than Human: SPRINT can perform the interactome prediction on any given organisms. The default parameters in SPRINT are optimized experimentally using human datasets. If the desired prediction is not on human dataset, it is recommended to tune the parameters first. The general method of tuning the parameters is to write scripts with parameters as variables and then cross validate on known datasets.
2. Parameter Tuning: There are some tuneable parameters (*see* Table 1) in SPRINT. In general, the default values are tuned experimentally and set based on the human datasets used in [14], so it is recommended to use the default values if PPI prediction is performed on human datasets. The detailed descriptions on each parameter can be seen in [14]. How each parameter affects the behavior of SPRINT is discussed here.

   When computing HSPs, a similarity matrix M is used to quantify the similarity between two amino acids. The available matrices in SPRINT are PAM120, BLOSUM80, and BLOSUM62. The -M option is to set the matrix, and the performance of these three is nearly identical.

Table 1
Tuneable parameters in SPRINT

Parameter	Component	Description
-Thit	Computer HSP	The similarity threshold to form a hit
-Tsim	Computer HSP	The similarity threshold to form a length-20 HSP
-M	Computer HSP	The scoring matrix used to compute similarity between two subsequences
-Thc	Predict Interaction	The threshold to consider a position high count

Multiple spaced seeds are pre-computed by [10, 11] and used in SPRINT to compute HSPs. The -Thit and -Tsim are the related options to this computation. In general, for a given weight and length of seed, the smaller the -Thit and -Tsim are, the more similarities SPRINT will detect. At the same time, the detected similarities will be less significant.

Some amino acid positions are involved in too many HSPs and believed just repeats in protein sequences with no interaction relevance. The -Thc option eliminates positions that appear too often in the HSP set. The smaller this value is, the fewer HSPs will contribute to the interactions prediction process.

Acknowledgements

L.I. has been partially supported by a Discovery Grant and a Research Tools and Instruments Grant from the Natural Sciences and Engineering Research Council of Canada (NSERC). We would like to thank the support from Robert and Ruth Lumsden Graduate Awards that is awarded to Y.L.

References

1. Bonetta L (2010) Protein–protein interactions: interactome under construction. Nature 468 (7325):851
2. Breuer K, Foroushani AK, Laird MR, Chen C, Sribnaia A, Lo R, Winsor GL, Hancock RE, Brinkman FS, Lynn DJ (2012) InnateDB: systems biology of innate immunity and beyond–recent updates and continuing curation. Nucleic Acids Res 41(D1):D1228–D1233
3. Chatr-Aryamontri A, Ceol A, Palazzi LM, Nardelli G, Schneider MV, Castagnoli L, Cesareni G (2006) Mint: the molecular interaction database. Nucleic Acids Res 35 (suppl_1):D572–D574
4. Consortium U *et al* (2014) Uniprot: a hub for protein information. Nucleic Acids Res p. gku989. https://doi.org/10.1093/nar/gku989
5. Ding Y, Tang J, Guo F (2016) Predicting protein-protein interactions via multivariate mutual information of protein sequences. BMC Bioinf 17(1):398
6. Fields S, Song OK (1989) A novel genetic system to detect protein protein interactions. Nature 340(6230):245–246
7. Guo Y, Yu L, Wen Z, Li M (2008) Using support vector machine combined with auto covariance to predict protein–protein interactions from protein sequences. Nucleic Acids Res 36(9):3025–3030
8. Hamp T, Rost B (2015) Evolutionary profiles improve protein–protein interaction prediction from sequence. Bioinformatics 31 (12):1945–1950
9. Huang YA, You ZH, Chen X, Chan K, Luo X (2016) Sequence-based prediction of protein-protein interactions using weighted sparse representation model combined with global encoding. BMC Bioinf 17(1):184
10. Ilie L, Ilie S (2007) Multiple spaced seeds for homology search. Bioinformatics 23 (22):2969–2977
11. Ilie L, Ilie S, Bigvand AM (2011) Speed: fast computation of sensitive spaced seeds. Bioinformatics 27(17):2433–2434
12. Kerrien S, Alam-Faruque Y, Aranda B, Bancarz I, Bridge A, Derow C, Dimmer E, Feuermann M, Friedrichsen A, Huntley R *et al* (2006) Intact–open source resource for molecular interaction data. Nucleic Acids Res 35(Suppl_1):D561–D565
13. Keshava Prasad T, Goel R, Kandasamy K, Keerthikumar S, Kumar S, Mathivanan S, Telikicherla D, Raju R, Shafreen B, Venugopal A *et al* (2008) Human protein reference

database–2009 update. Nucleic Acids Res 37 (Suppl_1):D767–D772

14. Li Y, Ilie L (2017) Sprint: ultrafast protein-protein interaction prediction of the entire human interactome. BMC Bioinf 18(1):485
15. Martin S, Roe D, Faulon JL (2005) Predicting protein–protein interactions using signature products. Bioinformatics 21(2):218–226
16. Schoenrock A, Dehne F, Green JR, Golshani A, Pitre S (2011) Mp-pipe: a massively parallel protein-protein interaction prediction engine. In: Proceedings of the international conference on supercomputing. ACM, New York, pp 327–337
17. Shen J, Zhang J, Luo X, Zhu W, Yu K, Chen K, Li Y, Jiang H (2007) Predicting protein–protein interactions based only on sequences information. Proc Natl Acad Sci 104 (11):4337–4341
18. Stark C, Breitkreutz BJ, Chatr-Aryamontri A, Boucher L, Oughtred R, Livstone MS, Nixon J, Van Auken K, Wang X, Shi X *et al* (2010) The biogrid interaction database: 2011 update. Nucleic Acids Res 39(Suppl_1): D698–D704
19. Sun T, Zhou B, Lai L, Pei J (2017) Sequence-based prediction of protein protein interaction using a deep-learning algorithm. BMC Bioinf 18(1):277
20. Suter B, Kittanakom S, Stagljar I (2008) Two-hybrid technologies in proteomics research. Curr Opin Biotechnol 19(4), 316–323
21. Szklarczyk D, Jensen LJ (2015) Protein-protein interaction databases. In: Protein-protein interactions. Springer, Berlin, pp 39–56
22. You ZH, Lei YK, Zhu L, Xia J, Wang B (2013) Prediction of protein-protein interactions from amino acid sequences with ensemble extreme learning machines and principal component analysis. BMC Bioinf 14:S10. BioMed Central
23. You ZH, Zhu L, Zheng CH, Yu HJ, Deng SP, Ji Z (2014) Prediction of protein-protein interactions from amino acid sequences using a novel multi-scale continuous and discontinuous feature set. BMC Bioinf 15:S9. BioMed Central
24. Yu CY, Chou LC, Chang DTH (2010) Predicting protein-protein interactions in unbalanced data using the primary structure of proteins. BMC Bioinf 11(1):167
25. Yu J, Guo M, Needham CJ, Huang Y, Cai L, Westhead DR (2010) Simple sequence-based kernels do not predict protein–protein interactions. Bioinformatics 26(20):2610–2614
26. Zahiri J, Yaghoubi O, Mohammad-Noori M, Ebrahimpour R, Masoudi-Nejad A (2013) Ppievo: protein–protein interaction prediction from PSSM based evolutionary information. Genomics 102(4):237–242

Chapter 2

Automated Extraction and Visualization of Protein–Protein Interaction Networks and Beyond: A Text-Mining Protocol

Kalpana Raja, Jeyakumar Natarajan, Finn Kuusisto, John Steill, Ian Ross, James Thomson, and Ron Stewart

Abstract

Proteins perform their functions by interacting with other proteins. Protein–protein interaction (PPI) is critical for understanding the functions of individual proteins, the mechanisms of biological processes, and the disease mechanisms. High-throughput experiments accumulated a huge number of PPIs in PubMed articles, and their extraction is possible only through automated approaches. The standard text-mining protocol includes four major tasks, namely, recognizing protein mentions, normalizing protein names and aliases to unique identifiers such as gene symbol, extracting PPIs, and visualizing the PPI network using Cytoscape or other visualization tools. Each task is challenging and has been revised over several years to improve the performance. We present a protocol based on our hybrid approaches and show the possibility of presenting each task as an independent web-based tool, NAGGNER for protein name recognition, ProNormz for protein name normalization, PPInterFinder for PPI extraction, and HPIminer for PPI network visualization. The protocol is specific to human but can be generalized to other organisms. We include KinderMiner, our most recent text-mining tool that predicts PPIs by retrieving significant co-occurring protein pairs. The algorithm is simple, easy to implement, and generalizable to other biological challenges.

Key words Protein–protein interaction, Network visualization, Information extraction, NAGGNER, ProNormz, PPInterFinder, HPIminer, KinderMiner

1 Introduction

The published literature is the best source of knowledge to understand the importance of proteins, their interactions, and the role in cellular functions through biological processes such as phosphorylation. Advances in high-throughput experiments and availability of bioinformatics tools to answer important biological questions result in the exponential growth of the biomedical literature [1]. PubMed contains ~28 million articles, and nearly ten million articles have been added in the last 10 years. Comprehensive manual extraction of information from this overwhelming dataset in a

Stefan Canzar and Francisca Rojas Ringeling (eds.), *Protein-Protein Interaction Networks: Methods and Protocols*, Methods in Molecular Biology, vol. 2074, https://doi.org/10.1007/978-1-4939-9873-9_2,

reasonable period of time is impossible. Text-mining approaches receive much interest within biology and biomedicine by facilitating automated extraction of information in a structured format and visualization of the relationships between the biomedical entities by building networks [2, 3]. Text mining helps the researchers deal with the information overload and summarizes the knowledge hidden within the published literature. For example, our recent text-mining tool, KinderMiner, identified 56,252 drugs, 68,847 diseases, 41,496 human genes, and 16,854 organisms mentioned in ~28 million published articles [4].

Text-mining approaches for extracting relation information include co-occurrence [5], pattern-based [6], rule-based [2], machine learning (ML) [7], and hybrid approaches [8]. Co-occurrence assumes a relation between two biological entities within a sentence or an abstract. These approaches tend to produce many false positives (FP) and report high recall but low precision [9]. The pattern-based approaches define a set of pattern templates related to the syntactic pattern of entities and give high precision but low recall. The pattern templates derived for a specific training dataset are not always applicable to other data not included in the training set [10, 11]. Rule-based approaches use rules derived by experts with specific domain knowledge. The rules are specific to the text from which they are derived and generally do not perform well on new text. ML approaches achieve better performance than co-occurrence, pattern-based, and rule-based approaches by automatically learning the relations from the annotated training data. The relationship between the biomedical entities is represented as a vector or matrix of features, and the extraction task is mostly treated as a binary classification problem with the suitable ML approach. Though ML approaches tend to decrease FP rate, they generally require a large volume of annotated training data for learning a wide range of possible relations. It is impossible to annotate training data for various domains of interest. Hybrid approaches are the combinations of one or more approaches described above and always tend to give better performance [12].

Text mining in biology and the biomedical domain is a hybrid discipline which emerged from three major fields including information science, bioinformatics, and computational linguistics [13]. It came into existence in 1999 when first applied for extracting protein–protein interaction (PPI) [14] and gene expression profiling [15]. Cellular information and energy are transported through proteins, lipids, metabolites, and other small molecules [16, 17]. Proteins play a critical role by interacting with other proteins. The knowledge on PPI is critical for understanding the function of individual proteins, the mechanisms of biological processes, and the diseases [18]. A wide range of interaction databases such as IntAct Molecular Interaction Database from European Molecular Biology Laboratory–European Bioinformatics Institute

(EMBL-EBI) [19], Molecular INTeraction database (MINT) from UniProt [20], Biomolecular Interaction Network Database (BIND) from the National Center for Biotechnology Information (NCBI) [21], and Database of Interacting Proteins (DIP) from the University of California, Los Angeles–Department of Energy (UCLA-DOE) [22] are developed by manually curating PPI information from multiple resources. Alternatively, databases such as STRING [23] and PolySearch [5] use text mining to extract PPI information from published literature. Despite the existing databases and text-mining tools, extracting and visualizing PPI information still remain an open challenge to text-mining researchers to develop more sophisticated approaches.

We present a text-mining protocol to extract and visualize PPI. This protocol is derived from our recent publications [2, 3, 12, 24]. We describe the development of web-based text-mining systems from each task. Other text-mining systems built using different approaches exist, but a discussion of these systems is not within the current scope. We discuss the extraction of PPIs scattered across multiple sentences and apply the approach to >28 million PubMed abstracts. Our approach is utilized for human-specific PPIs, and we show how this protocol can be generalized to other organisms.

2 Materials and Methods

2.1 Overview

We present a protocol for extracting and visualizing PPI information from the published literature (Fig. 1). The pipeline consists of four major tasks. First, protein mentions are identified by a rule-based post-processing approach and conditional random field (CRF), a ML approach widely used for entity recognition [12]. Next, protein names and aliases are normalized to unique identifiers [24]. Then, PPI is extracted from sentences with at least two proteins using Tregex [25], a tree query language from the Stanford Natural Language Processing (NLP) group. Finally, PPI from literature and other existing resources is visualized by building the network [3]. The pipeline can be extended to extract posttranscriptional modifications (PTM) of proteins such as protein phosphorylation with minor modification [26].

2.2 Protein Name Recognition

Automatic recognition of protein mentions in published literature is a prerequisite to PPI extraction [27]. The task is commonly known as named entity recognition (NER). The protein names

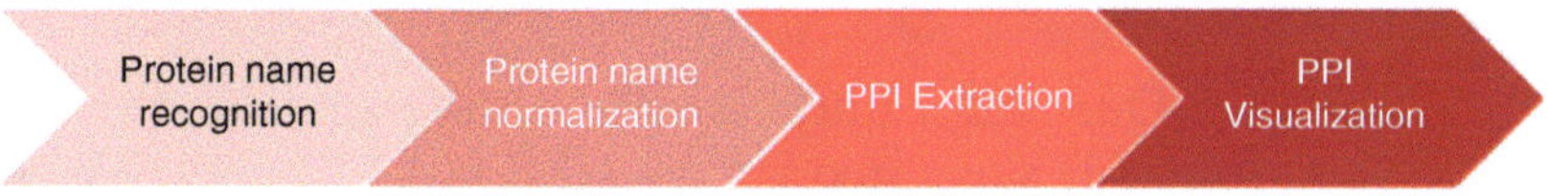

Fig. 1 Major tasks

are often chunks of text, and their recognition requires the identification of left and right boundaries. The ambiguity of natural language as well as the constant changes and advances in biomedical nomenclature make this a challenging task [28]. The major challenges are descriptive naming conventions where a protein name includes a premodifier (e.g., soluble form of growth factor receptor), abbreviations that are highly ambiguous, and nonstandardized naming where a protein name is mentioned in multiple forms across the literature (e.g., *N*-acetyltransferase/*N*-acetyl-transferase/*N*-acetyl transferase). Our hybrid approach to recognize protein names in literature consists of three processing steps: a two-stage abbreviation expansion algorithm to update the abbreviations with their original term, a CRF that uses a variety of ML features for initial learning and labeling, and a set of post-processing rules for addressing the tagging errors by the CRF.

Abbreviations are common in the published literature, and expanding them to their original term is mandatory to improve the performance of any text-mining system. Abbreviations across the literature map to different entities, and a search using abbreviation leads to many false positives. For example, abbreviation "HF" defines hepatic fibrosis, a disease [29], as well as H-ferritin, a protein [30], in two different articles. Among various approaches available for expanding the abbreviations to original terms, we combined the features of two algorithms that are simple to implement and have short execution times [31, 32]. We assumed the first appearance of an abbreviation within a pair of braces next to the original term and obtained a range of preceding words matching the number of characters in the abbreviation. The original term is then extracted by mapping the characters in the same order such that the first character of the original term and the abbreviation must match. Our abbreviation expansion algorithm is generalized to expand the abbreviation of any entity to its original term [33].

CRFs are probabilistic frameworks for labeling and segmenting sequential data. Like all supervised ML approaches, a CRF is trained on labeled data to maximize the conditional probability, $p(y|x)$. Here x represents the input sequence, and y represents the sequence of output labels. The partition function $Z(x)$ is obtained by summing y and gives an accurate conditional probability.

$$p(y|x) = \frac{1}{Z(x)} \exp \sum_{(x,y)_c} \left(-E\left((x,y)_c\right)\right)$$

$$Z(x) = \sum_{y'} \exp \sum_{(x,y')_c} \left(-E\left((x,y')_c\right)\right)$$

Our ML features are the orthographic and semantic features primarily suggested for the NER task [12, 34]. We used the standard implementations JLEX (http://www.cs.princeton.edu/

~appel/modern/java/JLex/) for text tokenization and MALLET (http://mallet.cs.umass.edu/) for generating a CRF model. We observed common tagging errors by NER approaches developed prior to ours, i.e., GENNIA tagger [35], NLProt [36], ABNER [34], and BANNER [37], by evaluating their performance on 1000 randomly selected PubMed abstracts related to human. We also noticed that the performance of ABNER [34] and BANNER [37] built on CRF is much better than the rest. This motivated us to use CRF for initial labeling and to develop a set of 14 rules to overcome the common errors of existing approaches [12]. Our approach is applicable to other organisms (*see* **Note 1**).

We implemented the above-discussed NER methodology as a web-based tool called "NAGGNER" (http://www.biominingbu.org:8080/NAGGNER/). It is written using Java servlets, and the web interface is deployed in an Apache Tomcat server (Fig. 2). NAGGNER is free to obtain and use. The input text is one or more PubMed abstracts in plain text, MEDLINE or XML format. The initial preprocessing converts the input text in MEDLINE or XML format into NAGGNER's format where the article ID and text (i.e., title and/or abstract) are tagged for further processing (i.e., <ID>16043634</ID> <TEXT> Cloning and ... </TEXT>). While most of the abstracts contain both title and abstract, a few of them contain only title. Some PubMed abstracts are segregated into subsections such as background, results, and conclusion. The retrieval module of NAGGNER is designed in such a way to collect the text spread across various subsections.

2.3 Protein Name Normalization

Protein mentions across the literature might include various synonyms, and their identification as the same entity is important for building a PPI network. The process is known as normalization. In humans, a large number of biological processes are regulated by protein kinases through PTM of protein [26]. Thus, differentiating protein kinases from other proteins in a PPI network can explain many underlying information on the regulatory mechanism [24]. To our knowledge, we are the first to normalize and differentiate protein kinases to facilitate both PPI and protein–protein kinase interaction (PPKI) networks. We differentiate human proteins and protein kinases from proteins specific to other organisms, thus facilitating the construction of PPI and PPKI networks between human proteins, protein kinases, and proteins specific to other organisms.

We compiled a dictionary of genes and proteins specific to human from three resources, namely, the synonyms dictionary from BioCreAtIvE-II challenge task on Critical Assessment for Information Extraction in Biology containing 32,975 entries [38], the human protein/gene name dictionary from NCBI containing 47,177 entries [39], and the human protein synonyms from UniProt containing 14,893 entries [40]. We combined the

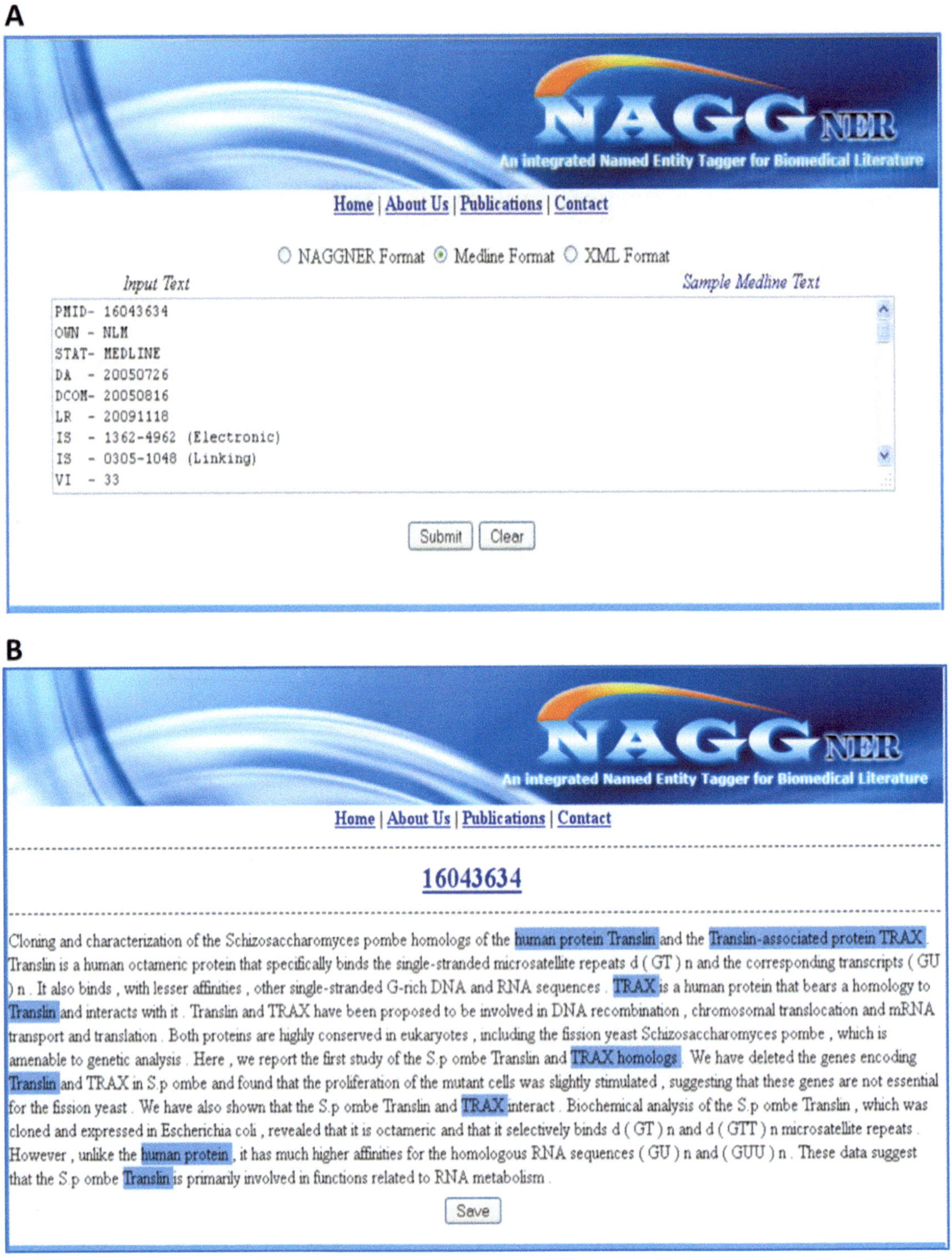

Fig. 2 NAGGNER—a hybrid named entity tagger

synonyms across the resources and filtered the ones approved by HUGO Gene Nomenclature Committee (HGNC) [41] from Entrez Gene. We compiled a second dictionary on human protein kinases from the existing resources, KinBase [42], with 516 protein kinases, and Kinweb [43] with 518 protein kinases. Both the dictionaries were combined to get a specialized dictionary for human proteins/genes and protein kinases that consists of 33,063 official symbols, 33,063 gene IDs, 201,496 synonyms, and annotation for 32,546 proteins and 517 protein kinases.

The protein mentions in the literature are recognized and tagged by NAGGNER or other NER systems such as BANNER. Our normalization system is a rule-based approach to map the protein/gene mentions in the literature to the dictionary for human proteins/genes and protein kinases. The rules were generated based on the protein mentions in the standard corpus, BioCreative-II gene normalization [44]: nine rules to dictionary terms and tagged proteins to resolve the morphological and syntactic variations and six rules to tagged proteins to resolve all possible ways of using "/" within the protein mentions [24]. The rules facilitate the replacement of protein/gene mentions with gene ID and official symbol in the literature. However, overlapping synonyms between proteins/genes make the task challenging to decide the appropriate gene ID and official symbol. We used the existing disambiguation method [45] that consists of three processing steps: (1) identify a list of ambiguous synonyms and the corresponding gene IDs, (2) retrieve all possible gene IDs for any ambiguous synonym in the literature, and (3) get the gene ID frequency in the literature to finalize the gene ID that is repeated many times in the literature. In addition to normalization, our approach classifies the normalized protein names as "HUMAN PROTEIN" or "HUMAN PROTEIN KINASE" by referring to the dictionary for human proteins/genes and protein kinases.

We implemented the above-discussed normalization methodology as a web-based tool called "ProNormz" (http://www.biominingbu.org/pronormz/). The tool is listed in OMICtools [46], an informative didactic resource for bioinformaticians, experimental researchers, and clinicians (https://omictools.com/pronormz-tool). We integrated NAGGNER (http://www.biominingbu.org:8080/NAGGNER/) and BANNER [37] to achieve NER task prior to normalization. While NAGGNER is specific for recognizing human proteins/genes, BANNER is capable of recognizing proteins/genes of any organism. The entire package is written using Java, Perl, and CGI. ProNormz is free to use and facilitates the option to process PubMed abstracts in plain text, MEDLINE format, XML format, or abstracts tagged with protein/gene mentions (Fig. 3).

A

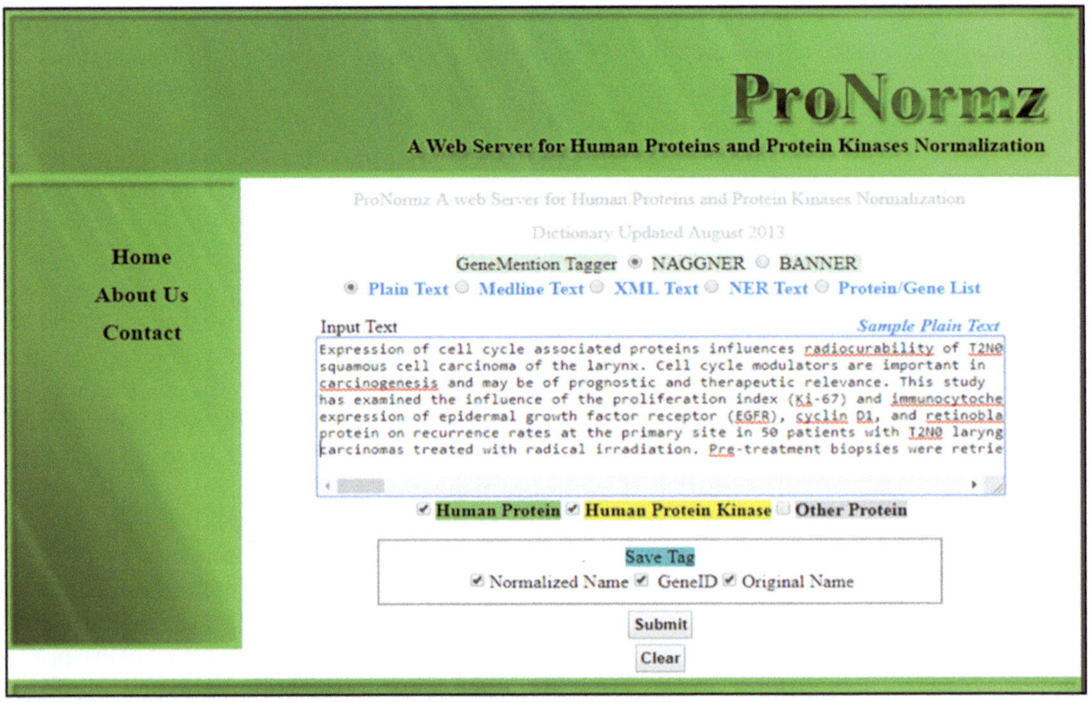

B

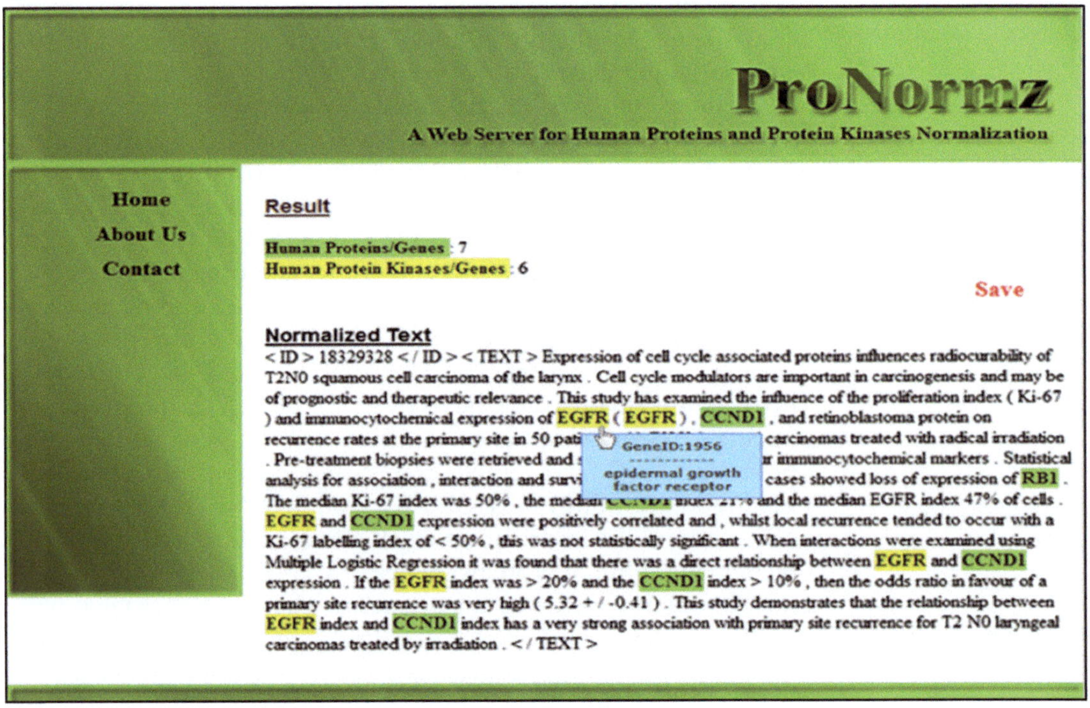

Fig. 3 ProNormz—a tool for normalizing protein/gene names

2.4 Mining Causal Relations of Human Proteins

Mining PPI information or similar relations (e.g., gene-gene interaction, gene-disease association, gene-drug interaction) from the published literature is commonly known as information extraction. It is achieved with a wide range of approaches such as co-occurrence of proteins in the same sentence or abstract, NLP approaches, and ML methods. One of the NLP approaches parses the input sentence to generate constituent tree or dependency tree and query PPI information using syntactic patterns [8]. The approach is capable of identifying a protein pair and a keyword that confirms their interaction.

We use NAGGNER for recognizing the protein names and ProNormz for normalizing the protein names prior to mining PPI information. Our approach consists of the following consequent processing steps: text preprocessing, recognizing interaction-related keyword, categorizing positive and negative relations, identifying candidate PPI pair, pattern matching, and PPI extraction. The text preprocessing module segments the input abstracts into individual sentences and assigns PubMed ID (PMID) to all sentences. It filters sentences with at least two protein mentions for further processing. The keywords describing the interaction between a pair of proteins (e.g., interact, bind, activate) are recognized by a dictionary lookup approach. Each keyword is expected to have various morphological forms based on their context in the sentence. For example, the keywords "interacts," "interacting," "interacted," "interaction," and "interactions" belong to the same root keyword "interact." We compiled a dictionary of 354 interaction related keywords from various approaches [6, 8, 47, 48] and databases [19, 20, 22]. We grouped the keywords into 88 subtypes by assigning a common root keyword to each subtype. The keyword in a sentence is either a verb or a noun. It is identified by parsing the sentence using Stanford lexical parser with grammar settings to englishPCFG module to generate the constituent tree and further querying the tree using Tregex expressions and dictionary lookup. Tregex is a tree query language, and the expressions are very similar to regex expressions to query the relationships between the tree nodes [25]. Identifying negative relations is mandatory to avoid false PPI extraction. Our approach looked for the presence of negation keyword (i.e., "not," "no," and "neither/nor") within the same noun/verb phrase of interaction related keyword. We used three types of "abstract forms" to define the position of the interaction-related keyword with protein pairs and defined seven rules for identifying candidate PPI pairs. Our rules are capable of extracting the candidate PPI pairs from sentences with two or more protein mentions with or without negation keyword. The candidate PPI pairs themselves do not always convey the true PPI information. We constructed 11 patterns using

Tregex expression to identify candidate PPI pairs with true interaction information [2].

The above-discussed methodology is implemented as a web-based tool called PPInterFinder (http://www.biominingbu.org/ppinterfinder/) and listed in OMICTOOLS [46] (https://omictools.com/ppinterfinder-tool). The input text can be a plain text, MEDLINE, XML, or text with pre-tagged protein/gene names (NER format). Similar to NAGGNER, an initial processing is carried out to extract and tag the PMID within a pair of ID tags and title/abstract within a pair of TEXT tags. The tool is free to obtain and use. Figure 4 shows the input and the extracted PPI output with highlighted human protein/gene names and interaction-related keyword.

2.5 Visualizing PPI Networks and Pathways

PPIs have attracted biologists for being useful in identifying various pathways within biological systems [3]. Text-mining efforts have extracted a large number of PPIs from the published literature and presented them as open resources [23, 49]. In addition, many research groups have dedicated their time and effort to constructing pathway databases such as Kyoto Encyclopedia of Genes and Genomes (KEGG) [50], Reactome [51], and BioCyc [52]. Updating PPI and pathway resources to keep up with the speed of articles being published is very challenging. In addition, PPI resources and pathway databases still remain independent of one another. Integrating them can uncover many hidden facts on biological processes and cellular functions. We made an effort to integrate PPI and pathway information by combining curated PPIs from Human Protein Reference Database (HPRD) [53], PPIs extracted from literature using PPInterFinder [2], and pathways related to human from KEGG [50].

We constructed two types of interaction databases: PPI and protein–nonprotein (i.e., DNA, RNA) interactions. The latter is obtained by extracting the information from HPRD. The proteins in PPI from HPRD are assigned with HPRD ID. Merging PPIs from HPRD with PPIs extracted by PPInterFinder from literature is not possible because HPRD ID does not match with HGNC gene symbols. We used ProNormz to assign Entrez Gene ID and official symbol to proteins in HPRD. Our approach identified duplicate PPIs (based on Entrez Gene ID and official symbol) and subsequently removed the duplicates. The interaction databases included 39,376 PPIs and 479 protein–nonprotein interactions. We constructed the human pathway database by mapping the proteins in PPIs to KEGG pathways using Gene ID. For every PPI (between protein A and protein B) extracted with PPInterFinder, we obtained the interactions of protein A and protein B with other proteins or nonproteins from the curated human PPI database. For every PPI, we obtained the related pathways by mapping the interacting proteins to the pathways. Our approach obtained pathways

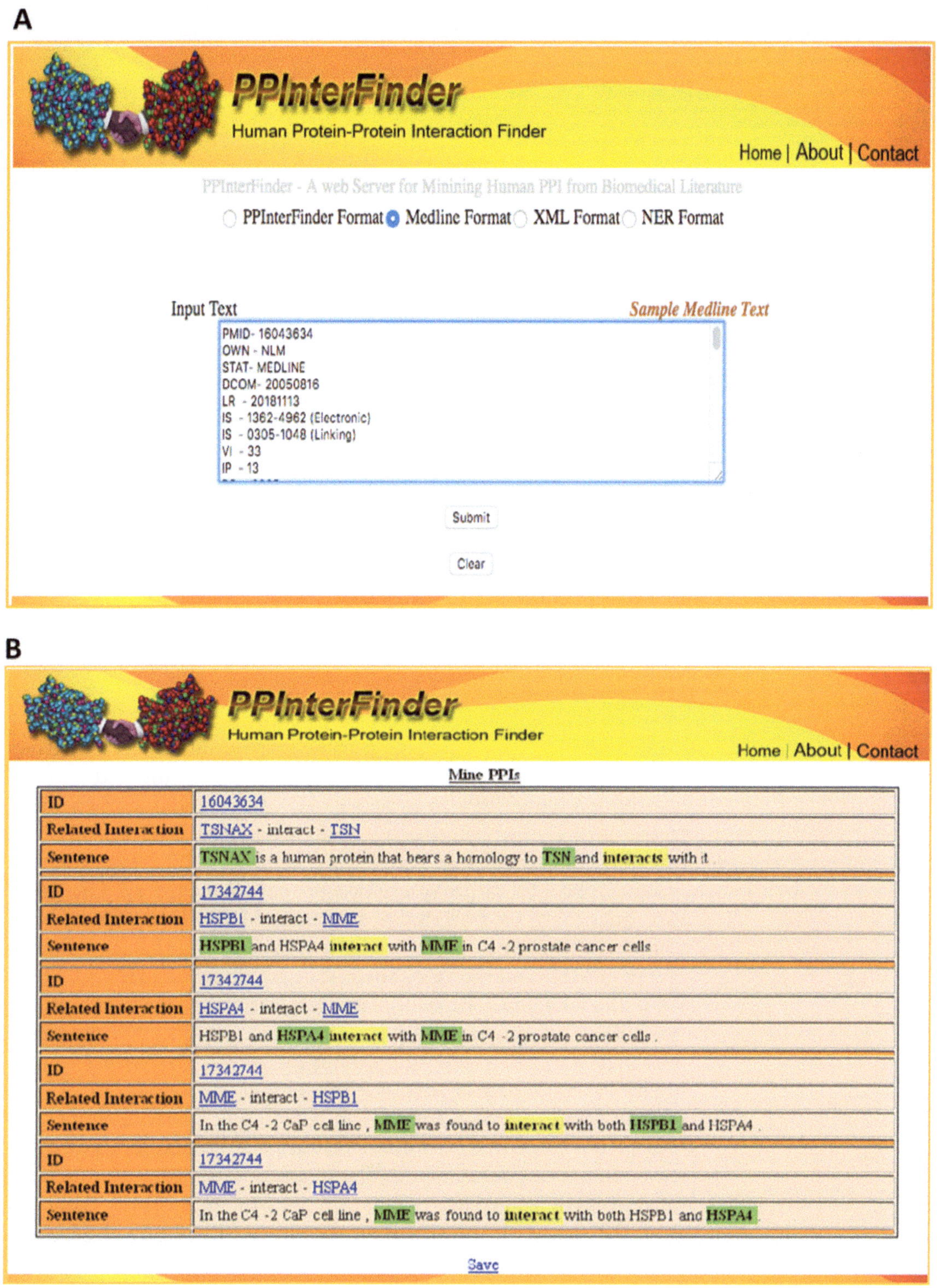

Fig. 4 PPInterFinder—a tool for extracting PPI

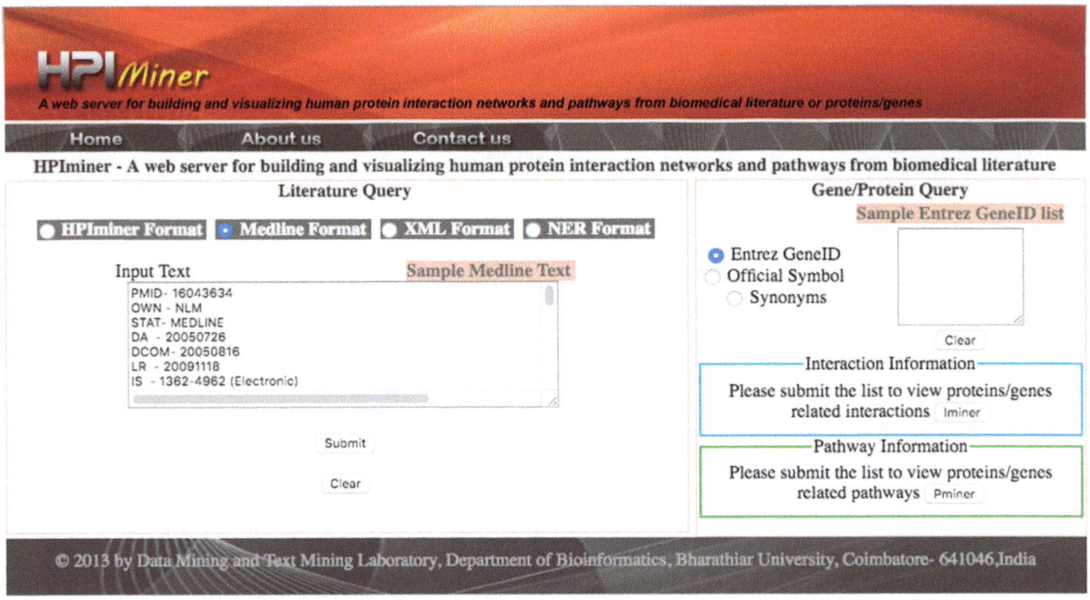

Fig. 5 HPIminer—a tool for visualizing PPI network

specific to protein A, pathways specific to protein B, and pathways involving both proteins.

We implemented the methodology as a web-based tool called HPIminer (http://www.biominingbu.org/HPIminer/). Like the other tools described in this pipeline, HPIminer is free to obtain and use. We included Cytoscape Web for constructing and visualizing the extracted PPI as network. HPIminer includes two interfaces: iMiner for visualizing protein related interactions and pMineR for visualizing pathways related to PPIs (Fig. 5). The input can be a literature query or a protein/gene query. The literature query can be PubMed abstracts in MEDLINE, XML, NER text, or HPIminer format. Similar to other tools, HPIminer format accepts PMID and the title/abstract tagged between ID and TEXT tags. The protein/gene query takes a list of Entrez Gene ID, official symbol, or synonyms as input. The literature query is processed with NAGGNER to identify protein mentions, ProNormz to normalize protein names, and PPInterFinder to extract PPIs. The protein/gene query retrieves the interacting proteins/genes for every input gene from the curated human PPI database. The pathways related to proteins in PPIs extracted with literature query or protein/gene query are retrieved from human pathway database. Both PPI network and pathways related to PPI can be visualized.

2.6 Beyond PPI Extraction

Text mining extends beyond extracting PPIs from published literature. Many sophisticated text-mining systems rely upon predefined templates for extracting PPIs within a sentence [2]. However, the information is available across the sentences in several PubMed articles (e.g., interaction between MIC26 and APOOL in PubMed ID, 25764979). Our simple and easy-to-implement tool called KinderMiner is capable of summarizing PPIs mentioned in one or multiple sentences [4]. Unlike the common text-mining pipeline, KinderMiner works on all PubMed articles, and the algorithm is independent of the standard corpora. A local version of the PubMed database is maintained within our institution and updated every month. The inputs are the key phrase (i.e., a protein of interest) and a list of target terms (i.e., a list of all the proteins). The system searches for significant relationships between the key phrase and target terms by counting the number of articles with key phrase alone, the number of articles with target term alone, and the number of articles with both. Simple statistics are performed to obtain target term: key phrase ratio and Fisher's exact test (FET) *p*-value. While the ratio is used to rank the pairs in descending order, FET *p*-value is used to filter the significant key phrase and target term pairs at a desired *p*-value, default being 1E-05.

We considered a list of 22 mitochondrial proteins and 68 manually curated PPIs [54] to illustrate the application of KinderMiner in predicting PPI. The mitochondrial proteins are considered as the key phrases, and a list of 19,181 HGNC protein-coding genes is considered as the target terms. Protein name, aliases, and gene symbol are retrieved from *gene_info*, a resource from NCBI (ftp://ftp.ncbi.nlm.nih.gov/gene/DATA/GENE_INFO) and used as the synonyms for key phrases and target terms. We evaluated the output at the default *p*-value and compared the performance with the standard databases such as STRING and PolySearch. KinderMiner identified many significant PPIs when compared to STRING and PolySearch (Table 1). Among the curated PPIs, KinderMiner predicted 26 PPIs, STRING retrieved 14 PPIs at high confidence (i.e., 0.700), and PolySearch retrieved 2 PPIs (Table 2). KinderMiner exhibits better recall than STRING and PolySearch. However, the number of known PPIs extracted by any system is low. The major reasons are: (1) PPIs were curated from full articles [54], and the information is not always present in the abstract, and (2) protein names and aliases in PubMed abstracts do not follow Entrez Gene nomenclature. Figure 6 shows the predicted PPIs for 22 mitochondrial proteins.

Among 22 mitochondrial proteins, APOOL is predicted to interact with 13 proteins (Table 1). KinderMiner identified five out of seven known PPIs and predicted eight new interacting proteins for APOOL (Table 2): CD46, PIEZO1, MIB1, MINOS1, APOO, C19orf70, MTX3, and MKI67. We manually queried PubMed to find at least one literature evidence that confirms the predictions (Fig. 6). Interaction of APOOL with MINOS1 is evident from [55]. Interaction of APOOL with CD46, APOO, and C19orf70 is available across multiple sentences

Table 1
PPIs for mitochondrial proteins

Mitochondrial proteins	Curated PPIs [54]	STRING Medium confidence (0.400)	STRING High confidence (0.700)	PolySearch 2.0	KinderMiner
APOOL	7	5	0	1	13
ATP5F1A	2	89	16	26	110
BOLA1	2	12	1	1	5
BOLA3	2	17	3	2	24
C6orf203	1	1	1	No entry	8
CCDC58	1	0	0	3	3
COQ8B	1	No entry	No entry	No entry	17
COQ9	1	13	11	7	43
ERGIC3	1	6	0	15	10
FMC1	2	1	0	3	6
IVD	2	15	3	188	235
LYRM4	3	32	6	2	25
LYRM7	2	3	0	2	18
MCCC2	1	14	4	13	65
NDUFA4	1	15	3	195	12
NDUFAF4	5	24	1	4	11
NDUFS3	25	67	11	63	118
OXCT1	1	98	2	38	56
OXCT2	1	43	1	20	93
PPM1K	3	99	1	7	104
PRELID1	1	7	2	2	6
SDHAF3	3	9	0	119	11

[56], and KinderMiner successfully predicted them. MICOS proteins form complex with MTX3, and APOOL is a member of MICO proteins [57]. The complex formation is possible through PPI [2]. The literature fails to support the interaction of APOOL with PIEZO1, MIB1, and MKI67. The articles related to these predictions are false positives due to abbreviation ambiguity between abbreviations related to protein names in PubMed articles and gene symbols.

Table 2
Evaluation using curated PPIs

Mitochondrial proteins	Curated PPIs [54]	STRING Medium confidence (0.400)	STRING High confidence (0.700)	PolySearch 2.0	KinderMiner
APOOL	7	1	0	0	5
ATP5F1A	2	0	0	0	0
BOLA1	2	0	0	0	0
BOLA3	2	2	0	0	1
C6orf203	1	0	0	No entry	0
CCDC58	1	0	0	0	0
COQ8B	1	No entry	No entry	No entry	1
COQ9	1	1	1	0	1
ERGIC3	1	0	0	0	0
FMC1	2	0	0	0	0
IVD	2	2	0	0	0
LYRM4	3	2	2	0	3
LYRM7	2	0	0	0	1
MCCC2	1	1	1	0	1
NDUFA4	1	0	0	0	0
NDUFAF4	5	2	1	1	1
NDUFS3	25	18	9	1	10
OXCT1	1	0	0	0	0
OXCT2	1	0	0	0	0
PPM1K	3	0	0	0	0
PRELID1	1	1	0	0	1
SDHAF3	3	0	0	0	1
All	68	30	14	2	26

KinderMiner is generalizable to summarize the relationship between biomedical concepts other than proteins. Identifying relevant transcription factors for cell reprogramming and retrieving potential drugs for lowering blood glucose gave promising and interesting results [4]. Although powerful, the current version of KinderMiner does have some limitations. These will be addressed in the near future (*see* **Note 1**).

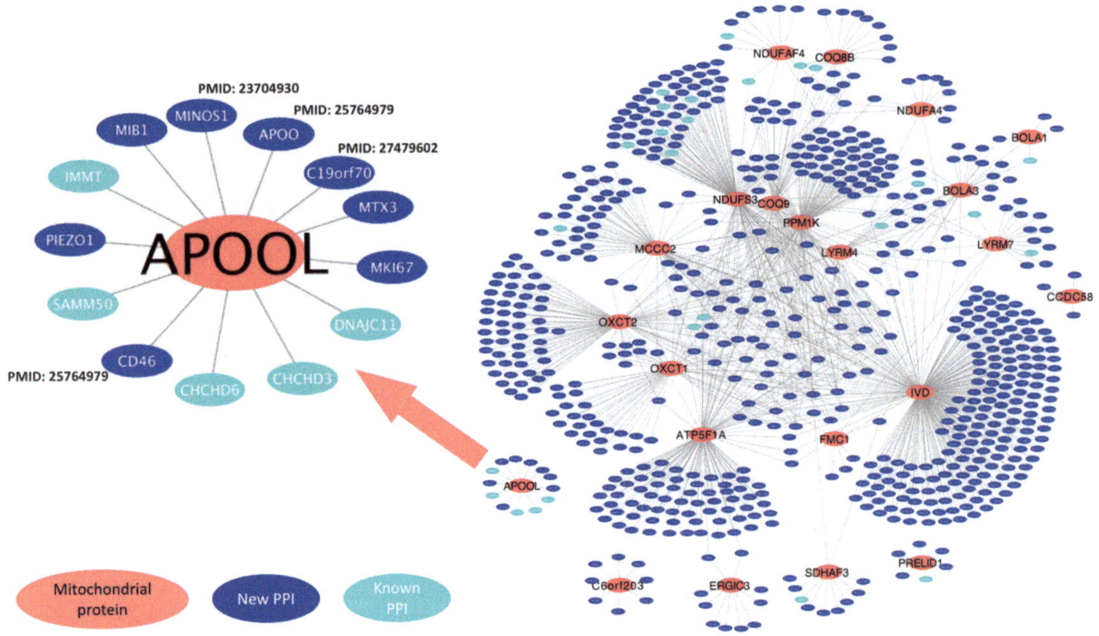

Fig. 6 PPI network for mitochondrial proteins

2.7 Generalizing to Organism of Interest

Our text-mining pipeline and KinderMiner are applicable to any organism of interest. Both require a dictionary of proteins specific to the organism of interest. The text-mining pipeline requires a new set of rules specific to organism of interest (*see* **Note 2**). We differentiate human proteins/genes from the rest by using HGNC in the normalization task. For other organisms, we recommend *gene_info* that includes 21,067,274 proteins/genes for 22,216 organisms (on May 22nd, 2018). While the proteins/genes are annotated with gene ID and gene symbol, the organisms are annotated with taxon ID. NCBI provides a resource on taxonomy (ftp://ftp.ncbi.nih.gov/pub/taxonomy/new_taxdump/new_taxdump.tar.gz) with 1,757,316 organisms. The resource includes taxon ID and organism name. Both the *gene_info* and taxonomy resources are updated daily. The organism-specific proteins are distinguished in the normalization module to facilitate the retrieval of organism-specific PPIs using PPInterFinder. For visualizing the PPI network, HPRD needs to be replaced with a curated PPI database specific to the organism of interest (if available) or removed.

2.8 Dictionary Update

We use various resources and databases in our text-mining pipeline and KinderMiner. We recommend updating them at regular intervals to keep up with the rapid growth of these databases. Most of the bioinformatics resources are updated on a daily, weekly, or monthly basis. UniProt is updated every month, gene2pubmed is updated every day, and HGNC is updated almost daily. Databases such as HPRD, KinBase, and KinWeb have not been updated since their first release. Periodic update of various dictionaries is mandatory because the gene symbols and aliases are often updated (e.g., gene symbol ADCK4 to COQ8B).

2.9 Evaluation Metrics and Methods

The performance of a text-mining system is evaluated on standard corpora using precision, recall, and *F*-score. A corpus is a selected subset of PubMed articles that are manually annotated by the domain experts. The process is expensive and time-consuming. Many standard corpora are available for named entity recognition, normalization, and PPI extraction [1]. Precision is defined as the portion of true PPIs identified by the system (i.e., relevance accuracy), recall is defined as the portion of true PPIs retrieved by the system (i.e., retrieval accuracy), and *F*-score is defined as the harmonic mean of precision and recall.

$$\text{Precision} = \frac{\text{TP}}{\text{TP} + \text{FP}}$$

$$\text{Recall} = \frac{\text{TP}}{\text{TP} + \text{FN}}$$

$$F - \text{score} = \frac{2 \times \text{Precision} \times \text{Recall}}{\text{Precision} + \text{Recall}}$$

TP (true positive) refers to the number of PPIs correctly extracted by the system, FP (false positive) refers to the number of PPIs incorrectly extracted by the system, and FN (false negative) refers to the number of PPIs that the system failed to extract. It is assumed that the performance remains the same when the text-mining pipeline is applied on all PubMed articles [33]. In the current world of big data, wet-lab scientists are motivated to use text mining to retrieve PPIs from all the articles at regular time intervals. The major challenges are to process millions of published articles and to evaluate the output. NCBI facilitates the downloading of PubMed articles from its ftp server (*see* **Note 3**). Evaluating extracted PPIs requires statistical methods other than common metrics of precision, recall, and *F*-score (*see* **Note 4**).

3 Notes

1. Currently KinderMiner is accessible only within the Morgridge Institute for Research. A web interface is under development and will be released in the near future. The system does not differentiate the positive and negative relationships between the protein pairs and, thus, can result in high recall with an increase in false positives. It performs exact string matching to identify protein/gene names, aliases, and symbols in PubMed articles. This is not an efficient approach because of the

orthographic variations of protein/gene names and aliases. The following improvements will be accomplished in the near future:

1.1 Partially differentiate positive and negative relationships: A simple approach to identify the negative relationship is to see whether target term or key phrase co-occurs with a negation keyword (i.e., "no," "not") in the same sentence or phrase.

1.2. Normalize protein/gene names and aliases: Normalization of protein/gene names and aliases within the dictionary and PubMed articles is mandatory to achieve their mapping. In the case of KinderMiner, normalization is possible only on the dictionary and requires retrieval of relevant articles to normalize their mentions. gene2-pubmed Is a good resource for retrieving relevant articles.

1.3. Introduce a relation term: The interaction between two proteins is confirmed with a relation term (e.g., interact, activate, bind) [2, 3]. A new flavor of KinderMiner to handle key phrase, target term, and relation term is under development. It is applicable to PPI prediction and other biological challenges such as disease comorbidity.

2. We identified the optimal feature weights efficiently from CRF model. We used two corpora for training and testing: JNLPBA2004 corpus that includes PubMed articles specific to human and BioCreative2004 task 1A corpus that includes PubMed articles with any organism. CRF model achieved 72.6% *F*-score on JNLPBA2004 corpus test data and 78.70% *F*-score BioCreative2004 task 1A corpus test data [12]. Thus, CRF model is applicable to any organism of interest. However, our rules are specific to human literature and may not be applicable to other organisms. The rules increased the performance of CRF models by 2.34% *F*-score on JNLPBA2004 corpus test data [12]. Their generation is time-consuming and requires experts' opinion. Many proteins overlap among organisms, and many PubMed articles include multiple organisms. For example, PMID:30404076 includes both humans and mice in medical subject heading (MeSH) annotation. Our approach can not differentiate whether a protein from the abstract is related to either one or both organisms. This is mandatory when our approach is applied on all PubMed articles. We suggest using gene2pubmed (ftp://ftp.ncbi.nlm.nih.gov/gene/DATA/gene2pubmed.gz), a resource from NCBI that provided genes and organisms annotations for PubMed articles and updated daily. It contains 1,137,216 PMIDs with

5,927,837 genes from 14,035 organisms (downloaded on May 11th, 2018).

3. PubMed database releases an annual baseline (ftp://ftp.ncbi.nlm.nih.gov/pubmed/baseline) in November and a daily update (ftp://ftp.ncbi.nlm.nih.gov/pubmed/updatefiles/). Both are XML files, and PubMed articles are extracted with a simple script. The articles are saved locally using Apache Lucene (http://lucene.apache.org/) or a similar application, which compresses the articles in binary format and indexes with PMID. 2017 PubMed baseline is ~32.3 GB and portable even to a laptop. Periodic update from the daily update is highly recommended to compete with the exponential volume of PubMed articles being published every day.
4. Most of the text-mining systems including ours are trained and tested on the experts' curated corpora that includes only a few hundred of PubMed articles. The systems are evaluated using precision, recall, and *F*-score. When applied on all PubMed articles, various statistical analysis are used [33]. A simple approach is to assign a score for each PPI and rank them in descending order to get the reliable ones at the top. KinderMiner ranks PPI based on target/keyphrase ratio and obtains top n PPIs on different FET *p*-value, default being 1E-05.

Acknowledgments

K.R., F.K., J.S., J.T., and R.S. acknowledge funding from the Morgridge Institute for Research and a grant from Marv Conney. I.R. acknowledges the GeoDeepDive Infrastructure, funded by NSF ICER 1343760.

References

1. Raja K, Patrick M, Gao Y, Madu D, Yang Y, Tsoi LC (2017) A review of recent advancement in integrating Omics data with literature mining towards biomedical discoveries. Int J Genomics 2017:10. https://doi.org/10.1155/2017/6213474
2. Raja K, Subramani S, Natarajan J (2013) PPInterFinder—a mining tool for extracting causal relations on human proteins from literature. Database (Oxford) 2013:bas052. https://doi.org/10.1093/database/bas052
3. Subramani S, Kalpana R, Monickaraj PM, Natarajan J (2015) HPIminer: a text mining system for building and visualizing human protein interaction networks and pathways. J Biomed Inform 54:121–131. https://doi.org/10.1016/j.jbi.2015.01.006
4. Kuusisto F, Steill J, Kuang Z, Thomson J, Page D, Stewart R (2017) A simple text mining approach for ranking pairwise associations in biomedical applications. AMIA Jt Summits Transl Sci Proc 2017:166–174
5. Liu Y, Liang Y, Wishart D (2015) PolySearch2: a significantly improved text-mining system for discovering associations between human diseases, genes, drugs, metabolites, toxins and more. Nucleic Acids Res 43(W1): W535–W542. https://doi.org/10.1093/nar/gkv383
6. Huang M, Zhu X, Hao Y, Payan DG, Qu K, Li M (2004) Discovering patterns to extract protein-protein interactions from full texts. Bioinformatics 20(18):3604–3612. https://doi.org/10.1093/bioinformatics/bth451

7. Chowdhary R, Zhang J, Liu JS (2009) Bayesian inference of protein-protein interactions from biological literature. Bioinformatics 25 (12):1536–1542. https://doi.org/10.1093/bioinformatics/btp245
8. Bui QC, Katrenko S, Sloot PM (2011) A hybrid approach to extract protein-protein interactions. Bioinformatics 27(2):259–265. https://doi.org/10.1093/bioinformatics/btq620
9. Jenssen TK, Laegreid A, Komorowski J, Hovig E (2001) A literature network of human genes for high-throughput analysis of gene expression. Nat Genet 28(1):21–28. https://doi.org/10.1038/88213
10. Ananiadou S, Mcnaught J (2005) Text mining for biology and biomedicine. Artech House, Inc., Boston
11. Kabiljo R, Clegg AB, Shepherd AJ (2009) A realistic assessment of methods for extracting gene/protein interactions from free text. BMC Bioinformatics 10:233. https://doi.org/10.1186/1471-2105-10-233
12. Raja K, Subramani S, Natarajan J (2014) A hybrid named entity tagger for tagging human proteins/genes. Int J Data Min Bioinform 10(3):315–328
13. Krallinger M, Valencia A (2005) Text-mining and information-retrieval services for molecular biology. Genome Biol 6(7):224. https://doi.org/10.1186/gb-2005-6-7-224
14. Blaschke C, Andrade MA, Ouzounis C, Valencia A (1999) Automatic extraction of biological information from scientific text: protein-protein interactions. In: Proceedings international conference on intelligent systems for molecular biology, pp 60–67
15. Tanabe L, Scherf U, Smith LH, Lee JK, Hunter L, Weinstein JN (1999) MedMiner: an internet text-mining tool for biomedical information, with application to gene expression profiling. BioTechniques 27 (6):1210–1214. 1216-1217
16. Aloy P, Russell RB (2004) Ten thousand interactions for the molecular biologist. Nat Biotechnol 22(10):1317–1321. https://doi.org/10.1038/nbt1018
17. Gao M, Skolnick J (2010) Structural space of protein-protein interfaces is degenerate, close to complete, and highly connected. Proc Natl Acad Sci U S A 107(52):22517–22522. https://doi.org/10.1073/pnas.1012820107
18. Zhou D, He Y (2008) Extracting interactions between proteins from the literature. J Biomed Inform 41(2):393–407. https://doi.org/10.1016/j.jbi.2007.11.008
19. Orchard S, Ammari M, Aranda B, Breuza L, Briganti L, Broackes-Carter F, Campbell NH, Chavali G, Chen C, del-Toro N, Duesbury M, Dumousseau M, Galeota E, Hinz U, Iannuccelli M, Jagannathan S, Jimenez R, Khadake J, Lagreid A, Licata L, Lovering RC, Meldal B, Melidoni AN, Milagros M, Peluso D, Perfetto L, Porras P, Raghunath A, Ricard-Blum S, Roechert B, Stutz A, Tognolli M, van Roey K, Cesareni G, Hermjakob H (2014) The MIntAct project--IntAct as a common curation platform for 11 molecular interaction databases. Nucleic Acids Res 42 (Database issue):D358–D363. https://doi.org/10.1093/nar/gkt1115
20. Ceol A, Chatr Aryamontri A, Licata L, Peluso D, Briganti L, Perfetto L, Castagnoli L, Cesareni G (2010) MINT, the molecular interaction database: 2009 update. Nucleic Acids Res 38(Database issue): D532–D539. https://doi.org/10.1093/nar/gkp983
21. Bader GD, Betel D, Hogue CW (2003) BIND: the biomolecular interaction network database. Nucleic Acids Res 31(1):248–250
22. Salwinski L, Miller CS, Smith AJ, Pettit FK, Bowie JU, Eisenberg D (2004) The database of interacting proteins: 2004 update. Nucleic Acids Res 32(Database issue):D449–D451. https://doi.org/10.1093/nar/gkh086
23. Szklarczyk D, Morris JH, Cook H, Kuhn M, Wyder S, Simonovic M, Santos A, Doncheva NT, Roth A, Bork P, Jensen LJ, von Mering C (2017) The STRING database in 2017: quality-controlled protein-protein association networks, made broadly accessible. Nucleic Acids Res 45(D1):D362–d368. https://doi.org/10.1093/nar/gkw937
24. Subramani S, Raja K, Natarajan J (2014) ProNormz--an integrated approach for human proteins and protein kinases normalization. J Biomed Inform 47:131–138. https://doi.org/10.1016/j.jbi.2013.10.003
25. Levy R, Andrew G (2006) Tregex and Tsurgeon: tools for querying and manipulating tree data structures. In: Proceedings of the fifth international conference on language resources and evaluation. doi:citeulike-article-id:3441831
26. Raja K, Natarajan J (2018) Mining protein phosphorylation information from biomedical literature using NLP parsing and support vector machines. Comput Methods Prog Biomed 160:57–64. https://doi.org/10.1016/j.cmpb.2018.03.022
27. Mukherjea S, Subramaniam LV, Chanda G, Sankararaman S, Kothari R, Batra VS, Bhardwaj DN, Srivastava B (2004) Enhancing a

biomedical information extraction system with dictionary mining and context disambiguation. IBM J Res Dev 48:693–702

28. Erhardt RA, Schneider R, Blaschke C (2006) Status of text-mining techniques applied to biomedical text. Drug Discov Today 11 (7–8):315–325. https://doi.org/10.1016/j.drudis.2006.02.011
29. Xia JR, Liu NF, Zhu NX (2008) Specific siRNA targeting the receptor for advanced glycation end products inhibits experimental hepatic fibrosis in rats. Int J Mol Sci 9(4):638–661
30. Hasegawa S, Harada K, Morokoshi Y, Tsukamoto S, Furukawa T, Saga T (2013) Growth retardation and hair loss in transgenic mice overexpressing human H-ferritin gene. Transgenic Res 22(3):651–658. https://doi.org/10.1007/s11248-012-9669-0
31. Park Y, Byrd RJ (2001) Hybrid text mining for finding abbreviations and their definitions. Paper presented at the 6th conference on empirical methods in natural language processing, Pittsburgh, USA
32. Schwartz AS, Hearst MA (2003) A simple algorithm for identifying abbreviation definitions in biomedical text. Pac Symp Biocomput:451–462
33. Raja K, Patrick M, Elder JT, Tsoi LC (2017) Machine learning workflow to enhance predictions of adverse drug reactions (ADRs) through drug-gene interactions: application to drugs for cutaneous diseases. Sci Rep 7 (1):3690. https://doi.org/10.1038/s41598-017-03914-3
34. Settles B (2005) ABNER: an open source tool for automatically tagging genes, proteins and other entity names in text. Bioinformatics 21 (14):3191–3192. https://doi.org/10.1093/bioinformatics/bti475
35. Tsuruoka Y, Tateishi Y, Kim J-D, Ohta T, McNaught J, Ananiadou S, Tsujii J (2005) Developing a robust part-of-speech tagger for biomedical text. In: Bozanis P, Houstis EN (eds) Advances in informatics. Springer, Berlin, Heidelberg, pp 382–392
36. Mika S, Rost B (2004) NLProt: extracting protein names and sequences from papers. Nucleic Acids Res 32(Web Server issue):W634–W637. https://doi.org/10.1093/nar/gkh427
37. Leaman R, Gonzalez G (2008) BANNER: an executable survey of advances in biomedical named entity recognition. Pac Symp Biocomput:652–663
38. Morgan AA, Lu Z, Wang X, Cohen AM, Fluck J, Ruch P, Divoli A, Fundel K, Leaman R, Hakenberg J, Sun C, Liu HH, Torres R, Krauthammer M, Lau WW, Liu H, Hsu CN, Schuemie M, Cohen KB, Hirschman L (2008) Overview of BioCreative II gene normalization. Genome Biol 9(Suppl 2):S3. https://doi.org/10.1186/gb-2008-9-s2-s3
39. The human protein/gene name dictionary from NCBI. http://www.ncbi.nlm.nih.gov/gene
40. The universal protein resource (UniProt) (2008) Nucleic acids research. 36(Database issue):D190–D195. https://doi.org/10.1093/nar/gkm895
41. Yates B, Braschi B, Gray KA, Seal RL, Tweedie S, Bruford EA (2017) Genenames.org: the HGNC and VGNC resources in 2017. Nucleic Acids Res 45(D1):D619–d625. https://doi.org/10.1093/nar/gkw1033
42. Manning G, Whyte DB, Martinez R, Hunter T, Sudarsanam S (2002) The protein kinase complement of the human genome. Science (New York, NY) 298(5600):1912–1934. https://doi.org/10.1126/science.1075762
43. Milanesi L, Petrillo M, Sepe L, Boccia A, D'Agostino N, Passamano M, Di Nardo S, Tasco G, Casadio R, Paolella G (2005) Systematic analysis of human kinase genes: a large number of genes and alternative splicing events result in functional and structural diversity. BMC Bioinformatics 6(Suppl 4):S20. https://doi.org/10.1186/1471-2105-6-s4-s20
44. Morgan AA, Lu Z, Wang X, Cohen AM, Fluck J, Ruch P, Divoli A, Fundel K, Leaman R, Hakenberg J, Sun C, H-h L, Torres R, Krauthammer M, Lau WW, Liu H, Hsu C-N, Schuemie M, Cohen KB, Hirschman L (2008) Overview of BioCreative II gene normalization. Genome Biol 9(Suppl 2):S3–S3. https://doi.org/10.1186/gb-2008-9-s2-s3
45. Koike A, Takagi T (2004) Gene/protein/family name recognition in biomedical literature. Paper presented at the HLT-NAACL 2004 workshop: biolink 2004, linking biological literature, ontologies and databases (BioLink 2004)
46. Henry VJ, Bandrowski AE, Pepin A-S, Gonzalez BJ, Desfeux A (2014) OMICtools: an informative directory for multi-omic data analysis. Database (Oxford) 2014:bau069. https://doi.org/10.1093/database/bau069
47. Temkin JM, Gilder MR (2003) Extraction of protein interaction information from unstructured text using a context-free grammar. Bioinformatics 19(16):2046–2053
48. Ono T, Hishigaki H, Tanigami A, Takagi T (2001) Automated extraction of information on protein-protein interactions from the biological literature. Bioinformatics 17 (2):155–161

49. Oughtred R, Stark C, Breitkreutz BJ, Rust J, Boucher L, Chang C, Kolas N, O'Donnell L, Leung G, McAdam R, Zhang F, Dolma S, Willems A, Coulombe-Huntington J, Chatr-Aryamontri A, Dolinski K, Tyers M (2018) The BioGRID interaction database: 2019 update. Nucleic Acids Res. https://doi.org/10.1093/nar/gky1079

50. Kanehisa M, Goto S (2000) KEGG: Kyoto encyclopedia of genes and genomes. Nucleic Acids Res 28(1):27–30

51. Croft D, O'Kelly G, Wu G, Haw R, Gillespie M, Matthews L, Caudy M, Garapati P, Gopinath G, Jassal B, Jupe S, Kalatskaya I, Mahajan S, May B, Ndegwa N, Schmidt E, Shamovsky V, Yung C, Birney E, Hermjakob H, D'Eustachio P, Stein L (2011) Reactome: a database of reactions, pathways and biological processes. Nucleic Acids Res 39 (Database issue):D691–D697. https://doi.org/10.1093/nar/gkq1018

52. Caspi R, Altman T, Dreher K, Fulcher CA, Subhraveti P, Keseler IM, Kothari A, Krummenacker M, Latendresse M, Mueller LA, Ong Q, Paley S, Pujar A, Shearer AG, Travers M, Weerasinghe D, Zhang P, Karp PD (2012) The MetaCyc database of metabolic pathways and enzymes and the BioCyc collection of pathway/genome databases. Nucleic Acids Res 40(Database issue):D742–D753. https://doi.org/10.1093/nar/gkr1014

53. Goel R, Harsha HC, Pandey A, Prasad TS (2012) Human protein reference database and human proteinpedia as resources for phosphoproteome analysis. Mol BioSyst 8 (2):453–463. https://doi.org/10.1039/c1mb05340j

54. Floyd BJ, Wilkerson EM, Veling MT, Minogue CE, Xia C, Beebe ET, Wrobel RL, Cho H, Kremer LS, Alston CL, Gromek KA, Dolan BK, Ulbrich A, Stefely JA, Bohl SL, Werner KM, Jochem A, Westphall MS, Rensvold JW, Taylor RW, Prokisch H, Kim JP, Coon JJ, Pagliarini DJ (2016) Mitochondrial protein interaction mapping identifies regulators of respiratory chain function. Mol Cell 63 (4):621–632. https://doi.org/10.1016/j.molcel.2016.06.033

55. Weber TA, Koob S, Heide H, Wittig I, Head B, van der Bliek A, Brandt U, Mittelbronn M, Reichert AS (2013) APOOL is a cardiolipin-binding constituent of the Mitofilin/MINOS protein complex determining cristae morphology in mammalian mitochondria. PLoS One 8 (5):e63683. https://doi.org/10.1371/journal.pone.0063683

56. Anand R, Strecker V, Urbach J, Wittig I, Reichert AS (2016) Mic13 is essential for formation of crista junctions in mammalian cells. PLoS One 11(8):e0160258. https://doi.org/10.1371/journal.pone.0160258

57. Huynen MA, Muhlmeister M, Gotthardt K, Guerrero-Castillo S, Brandt U (2016) Evolution and structural organization of the mitochondrial contact site (MICOS) complex and the mitochondrial intermembrane space bridging (MIB) complex. Biochim Biophys Acta 1863(1):91–101. https://doi.org/10.1016/j.bbamcr.2015.10.009

Chapter 3

Construction of Functional Protein Networks Using Domain Profile Associations

Jung Eun Shim and Insuk Lee

Abstract

Proteins are major functional molecules that physically and functionally interact to carry out cellular processes. The physical interactions are generally mediated by domain-level interactions. Thus, novel protein-protein interactions can be predicted using various computational methods based on domain-domain interactions, using resolved structures of protein complexes. Functional protein interactions can be inferred based on shared domains between proteins, since proteins involved in the same biological processes tend to harbor common domains. We recently developed a method of inferring functional interactions between proteins using associations between their domain compositions, which can be represented as domain profiles. Since the method requires only protein domain annotations, it can be easily applied to any species with a sequenced genome. Here, we describe in detail the method of generating domain profiles for proteins and measuring the association between them to infer functional interactions between proteins. We also demonstrate that domain profile association can be used to successfully construct a large-scale functional network of human proteins.

Key words Protein network, Protein domain, Domain profile, Weighted mutual information

1 Introduction

Protein domains are a useful sequence feature for the study of protein functions and biological processes, because a domain is a structural and functional unit that mediates many basic biochemical processes and protein-protein interactions (PPIs). It is relatively easy to model protein domains simply based on recurring protein sequences. A major database of protein domains, InterPro v.70, contains (as of September, 2018) more than 35,000 unique entries representing recurring protein sequences [1]. This large public repository of domain information has been utilized for studies of the functions of individual proteins and also the interactions between proteins.

Several computational algorithms have been developed to infer PPIs using domain information [2]. Many of the previous

Stefan Canzar and Francisca Rojas Ringeling (eds.), *Protein-Protein Interaction Networks: Methods and Protocols*, Methods in Molecular Biology, vol. 2074, https://doi.org/10.1007/978-1-4939-9873-9_3,

domain-based approaches, however, aimed to identify novel PPIs using domain-domain interactions (DDIs) extracted from PPIs with resolved molecular structures for the interfacing domains [3]. These DDIs have been deposited in integrated databases [4, 5]. Therefore, one might expect that PPIs can be inferred simply based on the known domain-level interactions between proteins [6, 7]. However, we previously found that PPI networks based on naïve score summation of all DDIs between proteins cannot construct accurate protein networks [8].

Another approach to identify protein interactions using domain information is domain profile association [9, 10]. Many proteins having similar biological processes share identical domains, because the domains are the functional units. Therefore, we hypothesized that the tendency of domain sharing between two proteins can support their functional association. It is relatively easy to identify domains from each protein due to the well-developed domain sequence databases and scanning software [11]. Using the identified domains, we can represent each protein as a domain profile, which is a vector of presence or absence information of domains in the protein. There could be many different ways to measure associations between domain profiles, and different measurements will infer different protein networks.

We recently developed weighted mutual information (WMI), which measures the information association between two domain profiles based on mutual information (MI) with more weight given for rarer domains, assuming that rarer domains contain more specific functional information [8]. Here, we present a detailed description of the method for constructing functional protein networks based on domain profile association with practical guidelines. Many of the computational procedures introduced here have been implemented and deposited at https://netbiolab.org/w/WMI. We also demonstrate the applicability of this method by constructing a large-scale network of human proteins.

2 Materials

2.1 Domain Information of Proteins

Many proteins are pre-annotated for domains. A domain annotation file, protein2ipr.dat, is freely downloadable from the InterPro ftp server (ftp://ftp.ebi.ac.uk/pub/databases/interpro/current) [1]. This file contains domain annotations for all reported proteins. Therefore, domain information for each protein of the target species can be extracted from the protein2ipr.dat file (*see* **Note 1**). However, there is a simpler way to obtain domain information for the proteins. The BioMart web tool (http://asia.ensembl.org/biomart/martview/) makes it easier to obtain domain annotation data for the major species such as human and mouse using the following steps:

1. Choose the Ensembl Gene database and a species dataset. For example, to generate a human domain profile, select the "Human gene (GRCh38.p12)" from the "Ensembl Genes 93 database."
2. In the "Attributes" part, choose the gene or protein ID that you want in the "External" tab and "InterPro ID" in the "PROTEIN DOMAINS AND FAMILIES" tab.
3. View the results via the "results" button at the top, and download it in the *.tsv format. Check "Unique results only" because duplicate records may be included. Additionally, records with null value in either gene (or protein) or domain need to be removed.

2.2 Domain Profiles

Using the mapping file between proteins and domains, we can generate the domain profile. A stack of domain profiles represents a protein-by-domain matrix $\boldsymbol{M}$, in which $\boldsymbol{M}[i, j]$ is 1 if protein p_i contains domain d_j and 0 otherwise (*see* **Note 2**).

3 Methods

To measure similarity or information association between profiles of binary values, we may use MI, dice coefficient, Jaccard coefficient, cosine coefficient, overlap coefficient, and kappa score [12]. Among these measures, MI does not require a priori choosing of any particular model; thus, it is considered a more general model. Also, it is very popular because of the robustness and accuracy for sparse data. Domain profiles are very sparse because only a small fraction of the known domains occur in each protein. Therefore, we chose MI as a strategy to measure domain profile association.

3.1 Measuring Domain Profile Associations

The overall procedure of measuring domain profile associations is summarized in Fig. 1. For a generated domain profile matrix, the associations between individual domain profiles can be measured by the following procedures:

1. Compute the "domain-specific weight (ω_j)."
   To give more weight to rarer domains, which tend to be associated with more specific biological processes, we developed the WMI method. For a given $n \times m$ protein domain matrix $\boldsymbol{M}$, the domain-specific weight ω_j for each domain d_j is defined by

$$\omega_j = \frac{\sum_{k=1}^{n} \sum_{l=1}^{m} M[k, l]}{\sum_{k=1}^{n} M[k, j]}$$

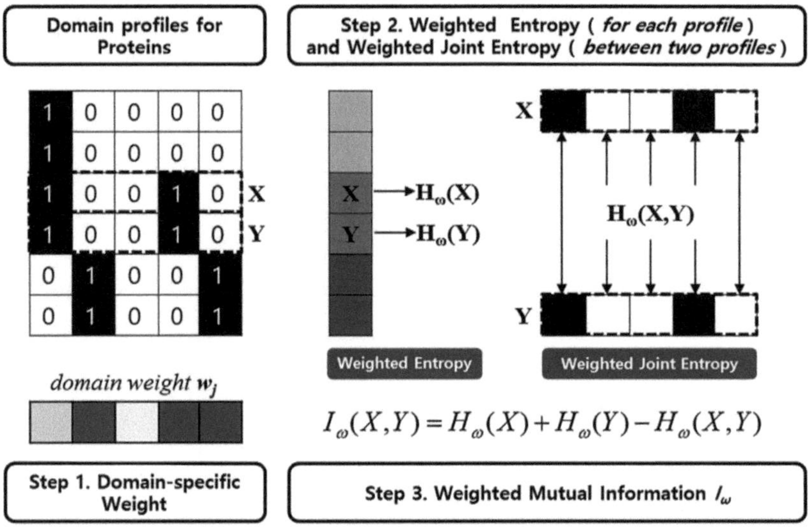

Fig. 1 Overview of the measurement of the association between domain profiles. Using domain profiles for proteins, we can calculate domain-specific weight of each domain by taking the inverse function of the frequency of each domain among proteins (**Step 1**). The heavier weight is indicated as a darker shade. Similar to the mutual information (MI) calculation, entropies for each of two domain profiles and the joint entropy between them need to be calculated first to measure weighted mutual information (WMI) (**Step 2**). However, WMI differs from MI in taking into account the domain-specific weight (weighted entropy and weighted joint entropy). Finally, WMI for protein X and Y is calculated by combining the entropies of the two domain profiles and the joint entropy (**Step 3**)

2. Compute the "weighted entropy (H_ω)" indicating the complexity of domain composition of each protein with different weights across domains and the "weighted joint entropy ($H_\omega(X, \Upsilon)$)" representing domain sharing between proteins. For two proteins X and Υ, the weighted entropy of each protein, $H_\omega(X)$ and $H_\omega(\Upsilon)$, and weighted joint entropy between two proteins, $H_\omega(X, \Upsilon)$, are calculated by the following equation:

$$H_\omega(X) = -\sum_{i=1}^{k_X} p(X_i) \cdot \log p(X_i), \quad H_\omega(\Upsilon) = -\sum_{i=1}^{k_\Upsilon} p(\Upsilon_i) \cdot \log p(\Upsilon_i)$$

$$H_\omega(X, \Upsilon) = -\sum_{i=1}^{k_X} \sum_{j=1}^{k_\Upsilon} p(X_i, \Upsilon_j) \cdot \log p(X_i, \Upsilon_j)$$

where k_X and k_Υ are the cardinalities of the outcomes of X and Υ, respectively.

3. Compute the "weighted mutual information (I_ω)" between protein X and Υ by the following equation:

$$I_\omega(X, \Upsilon) = H_\omega(X) + H_\omega(\Upsilon) - H_\omega(X, \Upsilon)$$

4. A protein pair with a larger I_{ω} value is more likely to be functionally associated.

The source code for calculating weighted mutual information based on the domain profile matrix is freely available from www.netbiolab.org/w/WMI (*see* **Note 3**).

3.2 Removal of Paralogs

Paralogs are homologous protein pairs that originate from gene duplication events within a species to create new functions or diversify existing functions. Different paralogs tend to have different functions. However, paralogs have high scores for domain profile associations because they have similar domain profiles. For this reason, it is necessary to remove the paralogs from the inferred protein functional associations by the following steps:

1. Measure sequence similarity between proteins based on *E*-values using the following BLAST command:

```
> blastp -db [protein reference-sequence file] -query
[protein reference-sequence file] -out [output filename]
```

2. If the *E*-value is <1e-03, remove the protein pairs from the inferred functional protein interactions to consider them as paralogs.
3. If gene ID, rather than protein ID, was used for domain profiles, convert gene ID into protein ID, and then compare the blast *E*-value of each protein pair.

3.3 Benchmarking Functional Protein Networks

To evaluate the quality of the inferred functional protein interactions, we used gold-standard data that comprised protein pairs with shared biological process or pathway annotations, such as the Gene Ontology (GO) biological process [13] and Kyoto Encyclopedia of Genes and Genomes (KEGG) [14] (*see* **Note 4**). The gold-standard data were used to calculate the likelihood of being functionally associated conditioned on the data as log likelihood score (LLS) [15] by the following equation:

$$\mathrm{LLS} = \ln\left(\frac{P(L|E)/P(\sim L|E)}{P(L)/P(\sim L)}\right)$$

where $P(L|E)$ and $P(\sim L|E)$ represent the frequencies of positive (L) and negative ($\sim L$) gold-standard protein pairs for the given supporting evidence (e.g., high WMI score) and $P(L)$ and $P(\sim L)$ represent the frequencies of all positive and negative gold-standard protein pairs, respectively. In practice, inferred functional PPIs sorted by descending order of WMI score are binned by 1000 pairs, and then the likelihood score is computed for each bin. If the likelihood of functional association is equal to that of random protein pairs, the LLS value will be zero. Protein interaction networks with high quality will have high LLS. One caveat is that the overall likelihood of a protein being part of a network by domain profile association is often very high. This could be attributable to the existence of protein families, which have very similar domain profiles. If there are too many protein interactions with high LLS values for a particular species, the domain profile association network should be interpreted with caution (*see* **Note 5**).

4 Case Study: Construction of a Functional Network of Human Proteins Using Domain Profiles

Here we present a functional network of human (*Homo sapiens*) proteins to illustrate how to construct a functional protein network using domain profile associations.

1. Download the human domain occurrence profile from BioMart website by the method described in Subheading 2.1. Next, create a domain profile matrix with values of 0 or 1, depending on whether a particular protein harbors a particular domain, and format it as a matrix delimited by tab delimiter with InterPro ID as column heading and protein ID as row heading. In this study, we employed Entrez gene identifiers instead of any protein ID to construct gene-centric protein network. As a result, we can generate a human domain profile matrix comprising 17,361 coding genes and 8386 domains.
2. Calculate the WMI scores between all the protein pairs using the source code available from www.netbiolab.org/w/WMI, and sort them in descending order of the score (Fig. 2a).
3. Run self-BLAST for human proteins. If the *E*-value between two proteins is <1e-3, the two proteins are regarded as paralogs, and the corresponding links are removed from the network.
4. Calculate the LLS for each bin of 1000 inferred functional protein interactions. Find a nonlinear regression between WMI and LLS; then assign LLS to individual protein interactions using the regression model. The optimal regression is often found with a sigmoid function. The threshold of LLS is determined as the point at which LLS values of individual bins of 1000 protein interactions encounter a negative LLS value (Fig. 2b). Subsequently, take only protein interactions with LLS values that pass the threshold to obtain a network of high quality.
5. Benchmark the inferred functional protein network with the gold-standard protein pairs (Fig. 2c). To avoid overtraining, benchmark the network with an independent set of protein pairs from that used to train the network.

6. The inferred functional protein network can be analyzed with various network algorithms and visualized with network layout applications, such as SAFE [16], with Cytoscape viewing software [17] (Fig. 2d).

A

Gene A	Gene B	*WMI*
1373	790	0.0068040486
7156	8940	0.0067930207
5431	55703	0.0066525735
3549	6469	0.0066525735
113451	4953	0.0064795367
5313	5315	0.0060264457
2584	2585	0.0059812960
2597	26330	0.0059563180
50846	6469	0.0052396129
55288	89941	0.0051581402
10914	64895	0.0051580195
4953	51582	0.0050665739
113451	51582	0.0050665739
4836	9397	0.0049384299

⋮

B

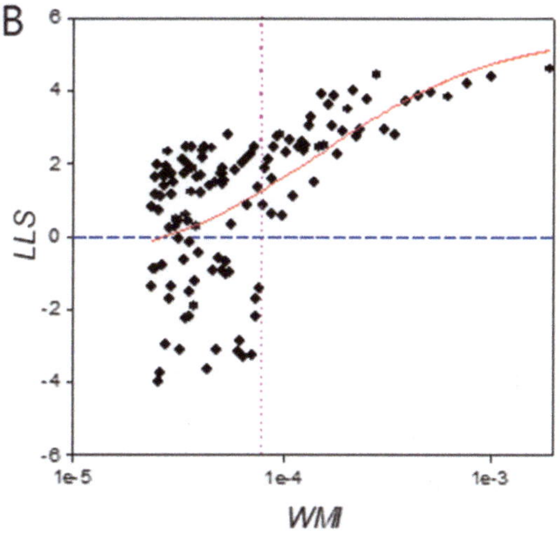

C

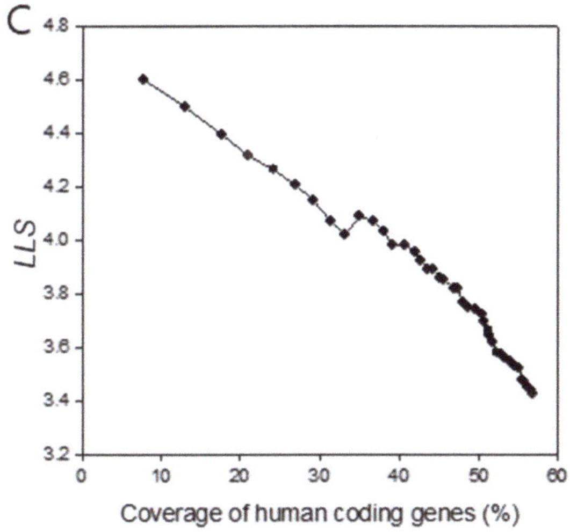

D

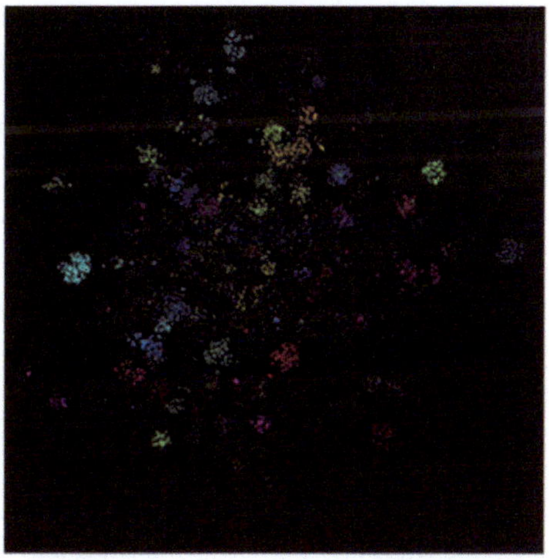

Fig. 2 Construction of a functional network of human proteins using domain profile associations. (**a**) Protein pairs for inferred functional interactions are listed in descending order of WMI scores. (**b**) A scatter plot of mean WMI and log likelihood score (LLS) for each bin of 1000 protein pairs. A regression curve was fit between the mean WMI and LLS. Threshold of LLS was selected where the bin of negative LLS was first encountered. Based on the threshold, 42,000 protein pairs were selected that were likely to be associated with each other, forming a network of proteins. Using the regression model, LLS were assigned to individual protein pairs with WMI score. (**c**) Cumulative LLS (*y*-axis) for every 1000 of the inferred functional protein interactions were used to represent the benchmark curve for the given genome coverage (*x*-axis). In this case study, the final network of 42,000 protein pairs covered more than 57% of the entire human coding gene set, with a cumulative LLS of 3.4, which represents an approximately 30-fold higher probability of association than association by random chance. (**d**) A network of the human proteins visualized by the SAFE layout algorithm and the Cytoscape network viewer

5 Notes

1. We can extract domain profiles for proteins of interest from the protein2ipr.dat file by a two-step procedure. The first step is to download protein2ipr.dat.gz from the InterPro download site ftp://ftp.ebi.ac.uk/pub/databases/interpro/current, and then uncompress the .gz file. The protein2ipt.dat file contains domain information for all known proteins. The second step is to obtain UniProt protein IDs of the target species and to extract them from the file.
2. The domain profile is a binary matrix which has 0 s and 1 s as values. The rows are for genes (gene ID), and the columns for InterPro entries (domain IDs). If a protein has a domain, the corresponding cell of the matrix is assigned as 1, otherwise as 0. The matrix should be completely filled up with no null value, and cells should be separated by a tab delimiter. The gene ID of the domain profile can be transformed so that it can be mapped to other gene IDs to be mainly used in the downstream analysis.
3. The source code provided from www.netbiolab.org/w/WMI can be compiled by the g ++ compiler to generate an executable file. Alternatively, you can convert an already created compiled file into an executable file using the '`chmod 744 [filename]`' command. This program should be run on a Unix-affiliated operating system, such as Linux or MAC OS. For the input file created in the Windows operating system, '^M' at the end of the line must be removed via the '`se ff = unix`' command.
4. To compile gold-standard functional protein interactions, we generally use databases of pathway or biological process annotations such as the GO [13] and KEGG pathway [14]. These annotation databases are often hierarchically organized with different granularity of definitions. If we choose pathway terms of a broad definition at high levels of the hierarchy, their coverage might be extensive but is at the cost of specificity. On the other hand, if we choose pathway terms of a narrower definition at lower levels of hierarchy, the information specificity can be increased but at the cost of coverage. Therefore, it is necessary to carefully determine the level of annotation hierarchy by comparing the number of genes of the selected pathway.
5. A protein family is a set of proteins with similar sequence features, such as domains, formed by gene duplications and generally share similar biochemical functions. Protein networks with high WMI scores are often generated for organisms with a high percentage of gene families with similar domain composition, such as plants.

Acknowledgments

This work was supported by the National Research Foundation of Korea (NRF) grant funded by the Korean Government (MSIT) (NRF-2018M3C9A5064709, NRF-2018R1A5A2025079) to I.L.

References

1. Finn RD, Attwood TK, Babbitt PC, Bateman A, Bork P, Bridge AJ, Chang HY, Dosztanyi Z, El-Gebali S, Fraser M, Gough J, Haft D, Holliday GL, Huang H, Huang X, Letunic I, Lopez R, Lu S, Marchler-Bauer A, Mi H, Mistry J, Natale DA, Necci M, Nuka G, Orengo CA, Park Y, Pesseat S, Piovesan D, Potter SC, Rawlings ND, Redaschi N, Richardson L, Rivoire C, Sangrador-Vegas A, Sigrist C, Sillitoe I, Smithers B, Squizzato S, Sutton G, Thanki N, Thomas PD, Tosatto SC, Wu CH, Xenarios I, Yeh LS, Young SY, Mitchell AL (2017) InterPro in 2017-beyond protein family and domain annotations. Nucleic Acids Res 45(D1):D190–D199. https://doi.org/10.1093/nar/gkw1107
2. Reimand J, Hui S, Jain S, Law B, Bader GD (2012) Domain-mediated protein interaction prediction: From genome to network. FEBS Lett 586(17):2751–2763. https://doi.org/10.1016/j.febslet.2012.04.027
3. Deng M, Mehta S, Sun F, Chen T (2002) Inferring domain-domain interactions from protein-protein interactions. Genome Res 12 (10):1540–1548. https://doi.org/10.1101/gr.153002
4. Yellaboina S, Tasneem A, Zaykin DV, Raghavachari B, Jothi R (2011) DOMINE: a comprehensive collection of known and predicted domain-domain interactions. Nucleic Acids Res 39(Database issue):D730–D735. https://doi.org/10.1093/nar/gkq1229
5. Kim Y, Min B, Yi GS (2012) IDDI: integrated domain-domain interaction and protein interaction analysis system. Proteome Sci 10(Suppl 1):S9. https://doi.org/10.1186/1477-5956-10-S1-S9
6. Sprinzak E, Margalit H (2001) Correlated sequence-signatures as markers of protein-protein interaction. J Mol Biol 311 (4):681–692. https://doi.org/10.1006/jmbi.2001.4920
7. Wojcik J, Schachter V (2001) Protein-protein interaction map inference using interacting domain profile pairs. Bioinformatics 17(Suppl 1):S296–S305
8. Shim JE, Lee I (2016) Weighted mutual information analysis substantially improves domain-based functional network models. Bioinformatics 32(18):2824–2830. https://doi.org/10.1093/bioinformatics/btw320
9. Chothia C, Gough J, Vogel C, Teichmann SA (2003) Evolution of the protein repertoire. Science 300(5626):1701–1703. https://doi.org/10.1126/science.1085371
10. Moore AD, Bjorklund AK, Ekman D, Bornberg-Bauer E, Elofsson A (2008) Arrangements in the modular evolution of proteins. Trends Biochem Sci 33(9):444–451. https://doi.org/10.1016/j.tibs.2008.05.008
11. Jones P, Binns D, Chang HY, Fraser M, Li W, McAnulla C, McWilliam H, Maslen J, Mitchell A, Nuka G, Pesseat S, Quinn AF, Sangrador-Vegas A, Scheremetjew M, Yong SY, Lopez R, Hunter S (2014) InterProScan 5: genome-scale protein function classification. Bioinformatics 30(9):1236–1240. https://doi.org/10.1093/bioinformatics/btu031
12. Byrt T, Bishop J, Carlin JB (1993) Bias, prevalence and kappa. J Clin Epidemiol 46 (5):423–429
13. The Gene Ontology C (2017) Expansion of the Gene Ontology knowledgebase and resources. Nucleic Acids Res 45(D1): D331–D338. https://doi.org/10.1093/nar/gkw1108
14. Kanehisa M, Furumichi M, Tanabe M, Sato Y, Morishima K (2017) KEGG: new perspectives on genomes, pathways, diseases and drugs. Nucleic Acids Res 45(D1):D353–D361. https://doi.org/10.1093/nar/gkw1092
15. Lee I, Date SV, Adai AT, Marcotte EM (2004) A probabilistic functional network of yeast genes. Science 306(5701):1555–1558. https://doi.org/10.1126/science.1099511
16. Baryshnikova A (2016) Systematic functional annotation and visualization of biological networks. Cell Syst 2(6):412–421. https://doi.org/10.1016/j.cels.2016.04.014
17. Cline MS, Smoot M, Cerami E, Kuchinsky A, Landys N, Workman C, Christmas R, Avila-

Campilo I, Creech M, Gross B, Hanspers K, Isserlin R, Kelley R, Killcoyne S, Lotia S, Maere S, Morris J, Ono K, Pavlovic V, Pico AR, Vailaya A, Wang PL, Adler A, Conklin BR, Hood L, Kuiper M, Sander C, Schmulevich I, Schwikowski B, Warner GJ, Ideker T, Bader GD (2007) Integration of biological networks and gene expression data using cytoscape. Nat Protoc 2 (10):2366–2382. https://doi.org/10.1038/nprot.2007.324

Chapter 4

Reconstruction of Protein–Protein Interaction Networks Using Homology-Based Search: Application to the Autophagy Pathway of Aging in *Podospora anserina*

Ina Koch, Oliver Philipp, Andrea Hamann, and Heinz Osiewacz

Abstract

The integration of the available experimental data represents a main problem in systems biology. In particular, in medical sciences, many new data became available, but often data are incomplete and of different quality and quantity. Here, we describe a method for the automatic derivation of protein–protein interaction networks based on homology search, which is applicable to arbitrary pathways and species. We implemented the method as a freely available open-source R package. To demonstrate the application of the method, we consider the autophagy pathway in the filamentous fungus *Podospora anserina*, which represents an established model organism to unravel the mechanisms of biological aging. Further, we apply network analysis methods to prove the reliability of the network.

Key words Network reconstruction, Protein–protein interactions, Homology search, Autophagy, Aging, *Podospora anserina*, R package, Network analysis, Reliability of networks, Data integration

1 Introduction

Biological function is often not performed by a single protein only but by interacting proteins or protein complexes. Thus, there is a great interest in experimental detection of protein–protein interactions (PPI) and in their exploration. In the past, many PPI databases for various organisms have been developed, e.g., [1, 2]. For some well-established model organisms, species-specific databases are available, e.g., [3, 4], which contain also PPI data. For the majority of organisms, there is no or only little information on PPI data available. Therefore, different theoretical approaches have been developed to predict PPI. These are often based on the homology of protein sequences. There are methods that apply

Ina Koch and Oliver Philipp shared first-authorship.

Stefan Canzar and Francisca Rojas Ringeling (eds.), *Protein-Protein Interaction Networks: Methods and Protocols*, Methods in Molecular Biology, vol. 2074, https://doi.org/10.1007/978-1-4939-9873-9_4,

supervised learning to filter and score the predicted interactions, using additional biological data, such as functional annotation, co-expression, and/or text-mining data [5]; for a recent review, *see* [6]. Approaches that predict PPI based on sequence data or network topology information often provide only precomputed data [2, 7] work for a set of predefined organisms [8, 9] or require pairs of proteins in the target organism [10, 11]. Furthermore, most of the methods do not supply information about the underlying reference PPI. But often it becomes necessary to easily access the entire underlying information about the predicted PPI to interpret and experimentally validate the interaction.

The development of our method was motivated by our interest in modeling of aging processes in the filamentous fungus *Podospora anserina* [12–15]. Initially, we found only little information on PPI data for this organism. The KEGG database [16] provides small subnetworks of some specific, mainly metabolic, pathways. Also, the STRING database [17] contains only some predicted PPI for *P. anserina*.

Here, we describe an approach to predict PPI for specific pathways or biological processes in an organism for which the complete DNA sequence is available without the need for predefinition of putative proteins and their interactions. The approach is supported by the R package Path2PPI [18]. Path2PPI does not require any presumptions about the pathway in the target species and can be applied to every fully sequenced organism. Furthermore, each prediction can be attributed to the reference PPI.

In this chapter, we start with the description of PPI databases, before we explain Path2PPI. Then, we discuss the verification of the network, using network analysis techniques, and explain possible extensions of a PPI network to identify functional modules. To illustrate an example, we provide the predicted PPI network for autophagy processes in *P. anserina*, which recently has been identified to play a key role in aging of this model organism [19–23]. This network was extended by experimental data, comprising the transcriptome data as well as newly generated PPI data from a yeast two-hybrid analysis. Based on this topological and functional analysis, the most important proteins were determined, and functional modules were identified, which correspond to the different sub-pathways of autophagy. Based on the transcriptome data analysis, different genes were identified, which are continuously up- or downregulated during aging, showing that autophagy-associated genes are significantly often co-expressed during aging [24]. This predicted network provides a systems-biological view on autophagy and enables the analysis of the relationship of autophagy and aging.

2 Materials

2.1 Protein–Protein Interaction Databases

The knowledge on PPI serves as basis to understand biological function. There exist many experimental methods to detect and investigate PPI [25]. These methods differ in quality, sensitivity, effort, and costs. Thus, a broad range of different databases exists, which comprise PPI data of different sources deduced from various organisms. Species-specific PPI databases provide PPI data of only one particular organism. Generally, these databases are only available for some well-studied organisms, e.g., *Saccharomyces cerevisiae* [26], *Drosophila melanogaster* [5], or *Homo sapiens* [4]. In contrast, species-independent PPI databases provide PPI data for various species and organisms. For example, the KEGG database provides information on biological pathways, including similar or evolutionary conserved pathways of different organisms [16]. The BioGRID database also contains a vast amount of manually curated PPI data for many different organisms [1]. Some PPI databases provide predicted PPI data, resulting from data mining as well as statistical approaches. These databases enable to obtain PPI data even for organisms, where only little experimental information is available. In contrast, the major drawbacks of these databases are their limitation to a predefined set of organisms and that they represent only a snapshot of the latest available data. The quality and reliability strongly depend on the specific assumptions and settings made by the database provider. The STRING database [2], one of the most established databases, provides, additionally to experimental data, predicted PPI data based on different sources and methods, e.g., text mining, correlation studies, and biological experiments. To compile the knowledge of the PPI databases, *meta* databases have been developed, which integrate the available data of other PPI databases into a new PPI database. Typically, these databases use rather experimental than predicted data and combine the advantages of the different species-specific and species-independent databases. For example, iRefIndex [27] is such a *meta* database, which provides PPI data of the major PPI databases, e.g., BIND [28], BioGRID [29], and HPRD [4].

2.2 Verification of the Predicted Network Using Topological Network Features

Experimental data are often incomplete and exhibit different qualities and quantities dependent on the method used. The transfer and compilation of interactions from different sources to one PPI network must rely on well-defined rules and assumptions. Before starting any analysis with the predicted model, it has to be sure that the model cannot have been randomly formed but exhibits "real-world" properties. The computation of topological network features for the model and for a set of randomly generated models represents a common way to do this [30].

We considered the following topological features: (1) the diameter, (2) the modules and modularity, (3) the clustering coefficient, and (4) the vertex degree and vertex betweenness.

1. The diameter of a network is defined as the longest path of all shortest paths between each two vertices. "Real-world" networks, such as metabolic or PPI networks, have a small diameter, exhibiting the "small-world" property [31].
2. Modules are subgraphs, in which the number of edges is much higher than between them. We assume that in biological networks, a module implements a pathway or sub-pathway with a certain biological function. To measure the quality of modularization, we compute the modularity according to [32].
3. The clustering coefficient (or transitivity) measures the probability that the adjacent vertices of a vertex are connected. The global transitivity is defined as the ratio of the triangles and the connected triples in the network. If n is the number of triangles and t the number of connected triplets, then the global clustering coefficient c is computed by $c = 3n/t$ [33].
4. The degree of a vertex or node is the number of adjacent vertices. A high-vertex degree indicates a hub in the network. Generally, PPI networks exhibit the scale-free property, i.e., the degree distribution follows a power law such that these networks consist of some vertices with a high degree and many vertices with a low degree.

The betweenness centrality of a vertex is the number of shortest paths that pass through the vertex, considering the shortest paths of every vertex pair. The higher the betweenness of a vertex the more important it is, for example, for information flows.

2.3 Prediction of Protein–Protein Interaction Networks Using Path2PPI

Path2PPI [18] computes new interactions in the target species and scores based on the degree of homology and the number of reference species with that interaction. It provides the choice of up to seven of the most established model organisms (*Homo sapiens*, *Mus musculus*, *Rattus norvegicus*, *Saccharomyces cerevisiae*, *Escherichia coli*, *Caenorhabditis elegans*, and *Drosophila melanogaster*). The software uses the interaction repository iRefIndex [27] to find the corresponding relevant interactions for sets of proteins. It also requires results of NCBI BLAST+ searches [34] of all reference species against the target species.

3 Methods

3.1 Protein–Protein Interaction Networks

PPI networks are undirected, labeled graphs, where the vertices represent the proteins and the edges molecular interactions between the proteins; *see* Fig. 1. Often, a PPI network describes a

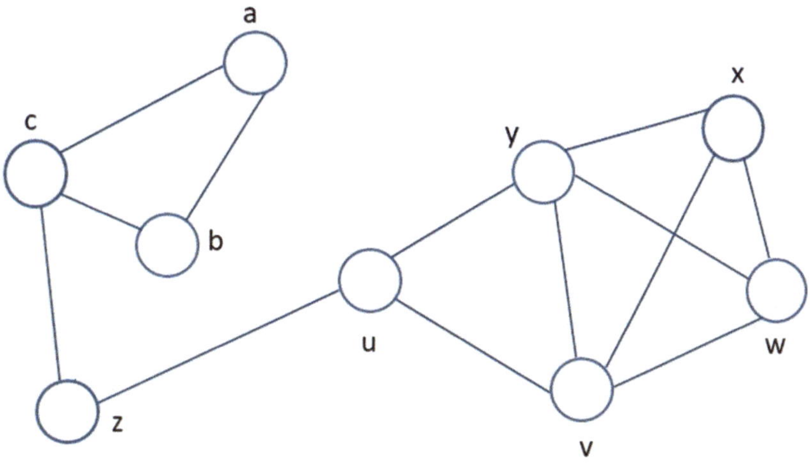

Fig. 1 Representation of a PPI network. Each vertex represents a protein, and the edges indicate interacting proteins. The vertex sets {*a*, *b*, *c*}, {*v*, *w*, *x*, *y*}, and {*u*, *v*, *y*} form complete subgraphs, i.e., all possible edges between the vertices of each set exist. Vertex *z* does not belong to a complete subgraph

specific biological pathway. In most cases, PPI networks serve as the first mathematical model of a biological pathway, where no detailed data, for example, kinetic data, are available. The aim is the prediction on the dynamics of the pathways based on the topology, on simulations, and on in silico knockout or perturbation analyses.

A PPI network is an undirected, labeled graph $G = (V, E)$, consisting of a finite set of vertices, V, where each vertex, $v \in V$, corresponds to a protein, and a finite set of edges, E, where each edge, $e = (v, u) \in E$ with $u, v \in V$, describes an interaction of protein v with protein u. For example, if a protein v forms a complex with the proteins u and $t \in V$, then all three vertices are connected by the edges (v, u), (v, t), and (u, t).

Further properties are possible, depending on specific requirements:

- We can weight an edge according to a specific scoring function, $w(e)$, which, for example, gives a probability score of a predicted interaction.
- To distinguish different types of interactions, we define different edge types as $G = (V, E, f)$, where each edge is colored according to the coloring function, f, i.e., we assign a discrete value to each edge.
- If interactions are directed and the required information is available, then the PPI network consists of directed edges with ordered pairs of vertices $(v; u)$, e.g., if protein v activates protein u.
- In contrast to simple graphs, PPI networks can contain loops, $e = (v, v)$, which represent self-interacting proteins.

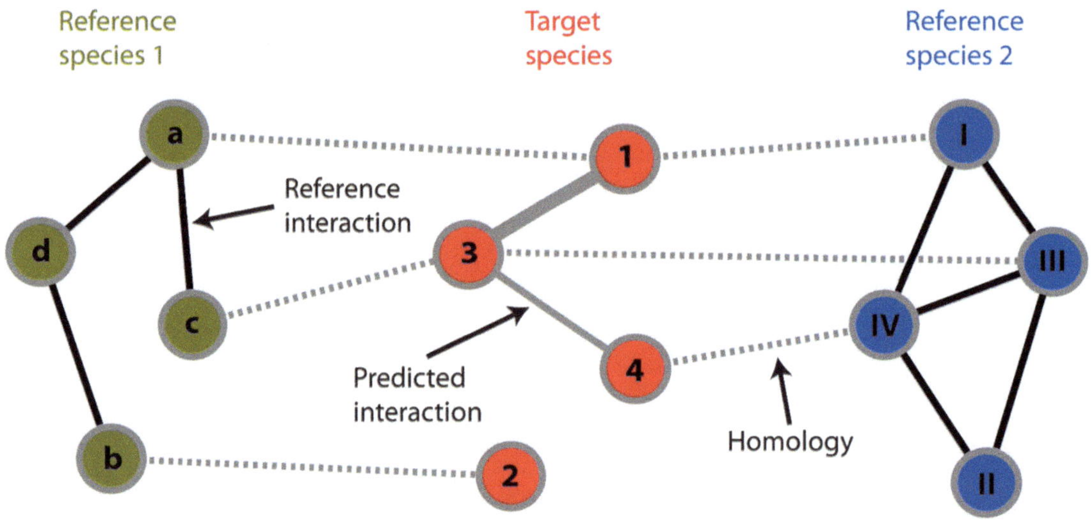

Fig. 2 Homology-based prediction of PPI networks. We assume that, if two proteins are interacting in a reference species, two homologous proteins (gray dotted edges) in the target species may interact as well (gray solid edges). Path2PPI also considers that the probability of an interaction in the target species increases if the corresponding interaction was found in more than one reference species indicated by a thick gray edge, for example, between the vertices 1 and 3 of the target species PPI network

3.2 Prediction of Protein–Protein Interaction Networks Using Path2PPI

Path2PPI [18] predicts protein–protein interactions based on homology search with fully sequenced organisms. Figure 2 schematically illustrates the principle.

A major advantage of Path2PPI is the easy access to the underlying reference interactions, i.e., all information provided by iRefIndex [27], e.g., source database, interaction type, and reference publication. Furthermore, Path2PPI provides a direct visualization in R of the predicted PPI network based on the Igraph package [35]. It contains a comprehensive tutorial and explains a case study to predict interactions of the induction step of autophagy in *P. anserina* based on homology search in human and *S. cerevisiae*. There are three types of visualization methods available, the normal, detailed, and hybrid (Fig. 3a–c).

Path2PPI provides detailed information about each interaction. The results are given as data frame or as Igraph objects, which can be used for subsequent analyses in R or in advanced analysis tools like Cytoscape [36]. Path2PPI is available via the Bioconductor project [37].

3.3 Verification of the Predicted Network Using Topological Network Features

To decide whether the predicted network significantly differs from randomly generated networks, we compared the topological network features, diameters, modules and modularity, cluster coefficient, degree distribution, and betweenness, with the median value of thousand networks of randomly chosen interaction partners of the same proteins. We used two sets according to the feature

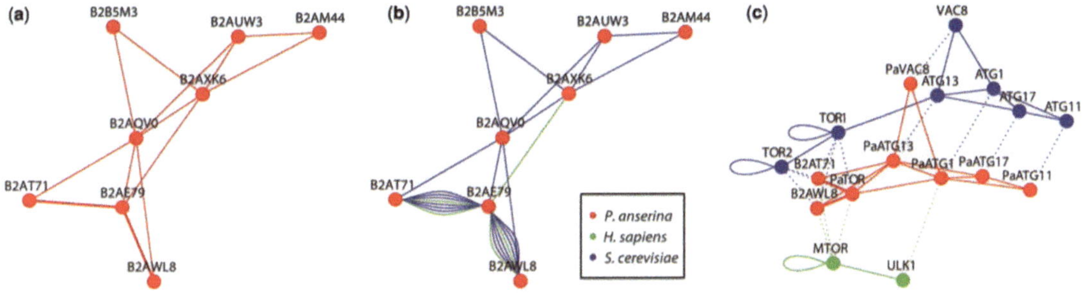

Fig. 3 The three visualization modes for the predicted PPI network of autophagy induction in *P. anserina* based on homology search in human and *S. cerevisiae*. (**a**) Normal view. The edge thickness corresponds to the scores, and the number of reference species shows the PPI. (**b**) Detailed view. The edges are colored according to the corresponding reference species. Multiple edges represent multiple findings of a PPI. (**c**) Hybrid network representation. This representation includes the PPI networks of the reference species together with the predicted PPI network of the target species. Solid, colored lines stand for a PPI. Dotted, colored edges show homology relation of the reference proteins to the target species

considered. In the first set (*set 1*), the vertex degrees remain unchanged, i.e., the number of adjacent edges of each vertex is the same. We only randomly changed the corresponding interacting partners. For the second set (*set 2*), we changed also the number of adjacent edges and randomly choose the interaction partners, only preserving the total number of edges. We computed the topological features for the predicted network and for the random networks, using the Igraph package [35] with the default settings. To compare the diameter of the predicted network with the median diameter of the randomly rewired networks (*set 1*), and since the set of diameters of the random networks represents a discrete distribution, we computed the relative frequency of each diameter, which can be interpreted as an empirical *p*-value. For module detection and modularity computation, we applied a random walk approach [38] and a hierarchical agglomeration algorithm [32], respectively. For the statistical significance, we applied the Shapiro-Wilk test on the modularities of the random networks to check whether they are normally distributed. We computed the *p*-value of the modularity of the predicted network, using the normal distribution of the random modularities (*set 1*). For the computation of empirical *p*-values, we applied appropriate statistical tests for the corresponding topological feature. The detailed results are described in [31]. We applied the same test procedure as for the modularity to compute the *p*-value for the cluster coefficient of the predicted network. To compare the cumulative distributions of the vertex degrees and the betweennesses of the random networks (*set 2*) with those of the predicted network, we used the Kolmogorov-Smirnov test. This well-known statistical test computes a *p*-value for the probability that two samples were drawn from the same distribution. A

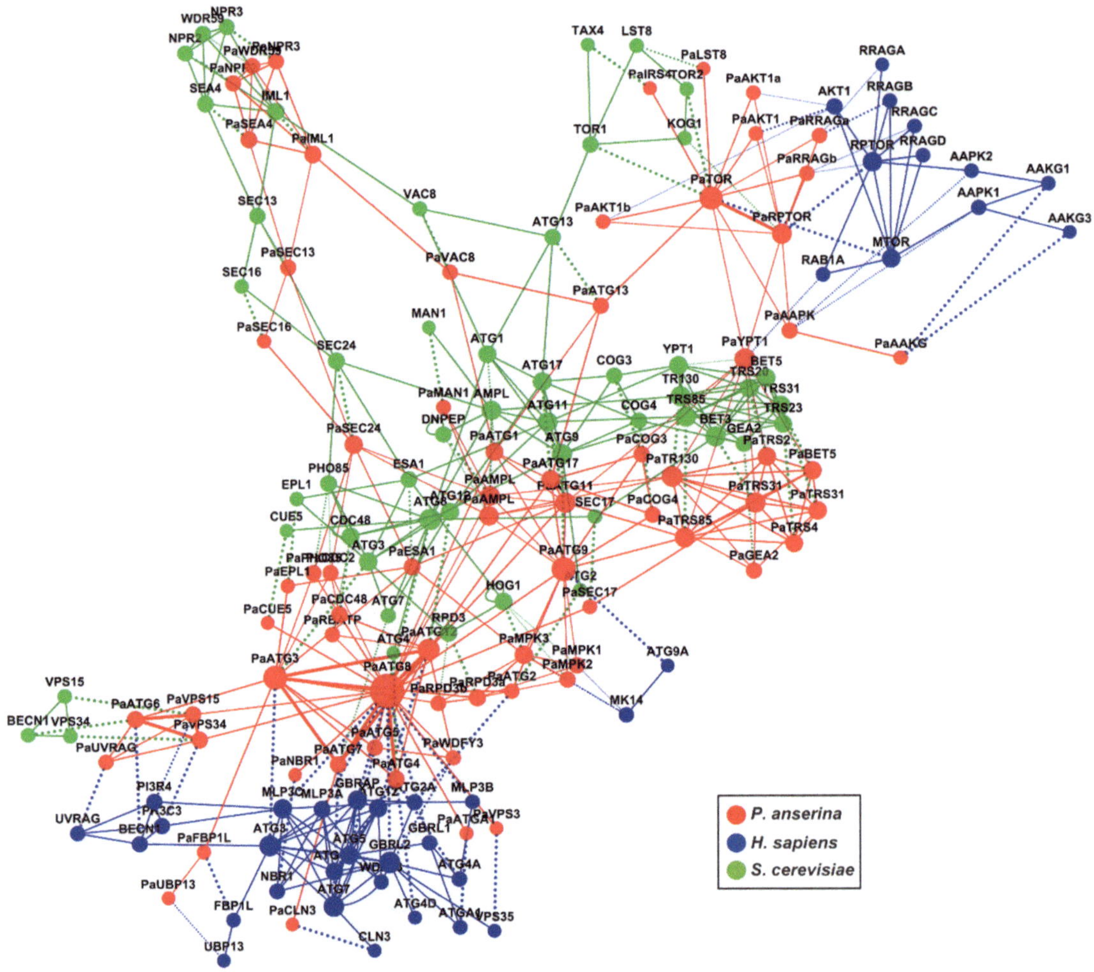

Fig. 4 The hybrid network, combining both reference PPI networks from human (blue) and *S. cerevisiae* (green) with the predicted PPI network of *P. anserina* (red). The homologous relationships between two proteins are depicted as dotted lines

significantly low *p*-value indicates that the betweennesses and degree distributions of the random networks are significantly different from those of the predicted network.

3.4 Application to P. anserina: The Autophagy Network

The predicted PPI network consists of 89 proteins, which participate in 188 interactions; *see* Fig. 4.

Additionally, we integrated the age-dependent expression profiles of the corresponding transcripts from the former transcriptome study to indicate which of the autophagy-related genes are downregulated or upregulated during aging [14]; *see* Fig. 5. Module computation detected seven modules. According to the presence of important marker proteins, we could assign functions to each module. For example, one module (red), containing the TOR

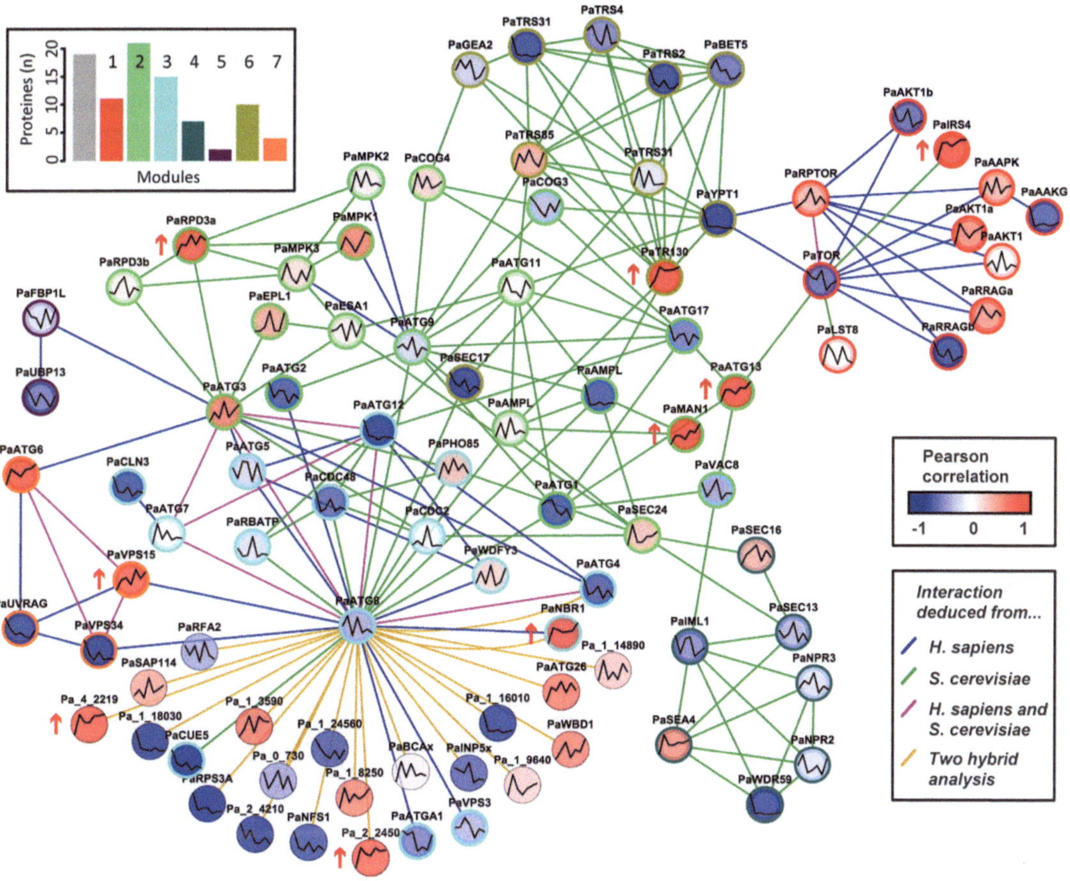

Fig. 5 Predicted autophagy PPI network of *P. anserina* with age-dependent expression profiles. Solid lines are color-coded and indicate the species that exhibits the interaction (green edges for yeast and blue for human). If a PPI was found in both species, the edge is colored purple. If a PPI was found in the yeast two-hybrid analysis, the edge is colored *yellow*. The vertices are color-coded according to the Pearson correlation coefficient of the corresponding age-dependent expression profiles. The more the expression of a transcript was increased (decreased) during aging, the more red (blue) is the vertex. Red arrows indicate transcripts, which were continuously upregulated during aging according to [14]. Hub vertices are those with many edges, here, for example, PaATG8 and PaSEC24. The number of proteins per module—from module 1 (red) to module 7 (orange)—are depicted in the bar plot in the upper left corner. The gray bar corresponds to proteins, which were not assigned to a module, but subsequently added. Vertices of the same module are indicated by the same color of the border

kinase, corresponds to the initiation processes. Another module (green) describes the induction process. A third module (turquois) covers the autophagosome nucleation and/or elongation step, comprising the proteins PaATG5, PaATG7, PaATG8, and PaATG12. Considering the degree distribution and the betweenness, we identified many vertices with a small number of neighbors and vertices with a high vertex degree. This was in accordance with the scale-free property of the "real world" of PPI networks.

References

1. Chatr-aryamontri A, Breitkreutz B-J, Oughtred R et al (2015) The BioGRID interaction database: 2015 update. Nucleic Acids Res 43(D1):470–478
2. Franceschini A, Szklarczyk D, Frankild S et al (2013) STRING v9.1: protein-protein interaction networks, with increased coverage and integration. Nucleic Acids Res 41 (D1):808–815
3. Güldener U, Münsterkötter M, Oesterheld M et al (2006) MPact: the MIPS protein interaction resource on yeast. Nucleic Acids Res 34 (D):436–441
4. Prasad TK, Goel R, Kandasamy K et al (2009) Human protein reference database - 2009 update. Nucleic Acids Res 37(suppl 1):67–772
5. Yu J, Pacifico S, Liu G et al (2008) DroID: the Drosophila Interactions Database, a comprehensive resource for annotated gene and protein interactions. BMC Genomics 9:461. https://doi.org/10.1186/1471-2164-9-46
6. Rao VS, Srinivas K, Sujini GN et al (2014) Protein-protein interaction detection: methods and analysis. Int J Proteomics 2014:147648. https://doi.org/10.1155/2014/147648
7. Pesch R, Zimmer R (2013) Complementing the eukaryotic protein interactome. PLoS One 8:e66635. https://doi.org/10.1371/journal.pone.0066635
8. Deng Y, Gao L, Wang B (2013) ppiPre: predicting protein-protein interactions by combining heterogeneous features. BMC Systems Biol 7(Suppl 2):S8. https://doi.org/10.1186/1752-0509-7-S2-S8
9. Wiles AM, Doderer M, Ruan J et al (2010) Building and analyzing protein interactome networks by cross-species comparisons. BMC Systems Biol 4:36. https://doi.org/10.1186/1752-0509-4-36
10. Chen CC, Lin CY, Lo YS et al (2009) PPISearch: a web server for searching homologous protein–protein interactions across multiple species. Nucleic Acids Res 37:W369–W375
11. Murakami Y, Mizuguchi K (2014) Homology-based prediction of interactions between proteins using averaged one-dependence estimators. BMC Bioinf 15:213. https://doi.org/10.1186/1471-2105-15-213
12. Figge MT, Reichert A, Meyer-Hermann M et al (2012) Deceleration of fusion-fission cycles improves mitochondrial quality control during aging. PLoS Comput Biol 8:e1002576. https://doi.org/10.1371/journal.pcbi.1002576
13. Osiewacz HD, Hamann A, Zintel S (2013) Assessing organismal aging in the filamentous fungus *Podospora anserina*. Methods Mol Biol 965:439–462. https://doi.org/10.1007/978-1-62703-239-1_29
14. Philipp O, Hamann A, Servos J et al (2013) A genome-wide longitudinal transcriptome analysis of the aging model *Podospora anserina*. PLoS One 8:e83109. https://doi.org/10.1371/journal.pone.0083109
15. Hamann A, Osiewacz HD (2018) *Podospora anserina*: a filamentous fungus with a strong mitochondrial etiology of aging. In: Conn M (ed) Handbook of models on human aging, 2nd edn. Elsevier, Amsterdam, pp 431–444
16. Kanehisa M, Goto S, Sato Y et al (2014) Data, information, knowledge and principle: back to metabolism in KEGG. Nucleic Acids Res 42 (D1):199–205
17. Szklarczyk D, Franceschini A, Wyder S et al (2015) STRING v10: protein-protein interaction networks, integrated over the tree of life. Nucleic Acids Res 43(D1):447–452
18. Philipp O, Osiewacz HD, Koch I (2016) Path2PPI: an R package to predict protein-protein interaction networks for a set of proteins. Bioinformatics 32(9):1427–1429
19. Knuppertz L, Hamann A, Pampaloni F et al (2014) Identification of autophagy as a longevity-assurance mechanism in the aging model *Podospora anserina*. Autophagy 10:822–834
20. Kramer P, Jung A, Hamann A et al (2016) Cyclophilin D is involved in the regulation of autophagy and affects the lifespan of *P. anserina* in response to mitochondrial oxidative stress. Front Genet 7:165. https://doi.org/10.3389/fgene.2016.00165
21. Warnsmann V, Osiewacz HD (2016) Curcumin treatment affects mitochondrial respiration, induces autophagy and increases lifespan of the fungal aging model *Podospora anserina*. Endocytobiosis and Cell Res 27:20–26
22. Knuppertz L, Warnsmann V, Hamann A et al (2017) Stress-dependent opposing roles for mitophagy in aging of the ascomycete *Podospora anserina*. Autophagy 13:1037–1052. https://doi.org/10.1080/15548627.2017.1303021
23. Knuppertz L, Osiewacz HD (2017) Autophagy compensates impaired energy metabolism in CLPXP deficient *Podospora anserina* strains and extends healthspan. Aging Cell 16:704–715. https://doi.org/10.1111/acel.1260

24. Philipp O, Hamann A, Osiewacz HD et al (2017) The autophagy interaction network of the aging model *Podospora anserina*. BMC Bioinformatics 18:196. https://doi.org/10.1186/s12859-017-1603-2
25. Xing S, Wallmeroth N, Berendzen KW et al (2016) Techniques for the analysis of protein-protein interactions in vivo. Plant Physiol 171(2):727–758
26. Cherry JM, Hong EL, Amundsen C et al (2012) *Saccharomyces* genome database: the genomics resource of budding yeast. Nucleic Acids Res 40(D1):D700–D705
27. Razick S, Magklaras G, Donaldson IM (2008) iRefIndex: a consolidated protein interaction database with provenance. BMC Bioinformatics 9:405. https://doi.org/10.1186/1471-2105-9-405
28. Bader GD, Betel D, Hogue CWV (2003) BIND: the biomolecular interaction network database. Nucleic Acids Res 31(1):248–250
29. Stark C, Breitkreutz B-J, Reguly T et al (2006) BioGRID: a general repository for interaction datasets. Nucleic Acids Res 34(1):535–539
30. Newman M (2010) Networks: an introduction. Oxford University Press, Oxford
31. Zhu X, Gerstein M, Snyder M (2017) Getting connected: analysis and principles of biological networks. Genes Dev 21(9):1010–1024
32. Clauset A, Newman MEJ, Moore C (2004) Finding community structure in very large networks. Phys Rev E 70(6):066111. https://doi.org/10.1103/PhysRevE.70.066111
33. Barrat A, Barthelemy M, Pastor-Satorras R et al (2004) The architecture of complex weighted networks. Proc Natl Acad Sci U S A 101(11):3747–3752
34. Camacho C, Coulouris G, Avagyan V et al (2009) BLAST+: architecture and applications. BMC Bioinformatics 10:421. https://doi.org/10.1186/1471-2105-10-421
35. Csárdi G, Nepusz T (2006) The Igraph software package for complex network research. InterJournal Compl Syst 1695(5):1–9
36. Cline MS, Smoot M, Cerami E et al (2007) Integration of biological networks and gene expression data using Cytoscape. Nat Protoc 2:2366–2382. https://doi.org/10.1038/nprot.2007.324
37. Huber W, Carey VJ, Robert Gentleman R et al (2015) Orchestrating high-throughput genomic analysis with bioconductor. Nat Methods 12(2):115–121
38. Pons P, Latapy M (2006) Computing communities in large networks using random walks. J Graph Algorithms Appl 10(2):191–218

Chapter 5

Predicting Interacting Protein Pairs by Coevolutionary Paralog Matching

Thomas Gueudré, Carlo Baldassi, Andrea Pagnani, and Martin Weigt

Abstract

Even if we know that two families of homologous proteins interact, we do not necessarily know, which specific proteins interact inside each species. The reason is that most families contain paralogs, i.e., more than one homologous sequence per species. We have developed a tool to predict interacting paralogs between the two protein families, which is based on the idea of inter-protein coevolution: our algorithm matches those members of the two protein families, which belong to the same species and collectively maximize the detectable coevolutionary signal. It is applicable even in cases, where simpler methods based, e.g., on genomic co-localization of genes coding for interacting proteins or orthology-based methods fail. In this method paper, we present an efficient implementation of this idea based on freely available software.

Key words Protein–protein interaction, Coevolution, Predicting interacting paralogs, Direct coupling analysis, Paralog matching

1 Introduction

The experimental identification of protein–protein interactions is an arduous task [1, 2]. Large-scale yeast two-hybrid assays and protein affinity mass spectrometry assays are well-documented techniques. However high false-positive and false-negative error rates hamper a clear picture of the interactome at large. The outburst of sequence -data availability in genomic databases triggered a parallel effort in the development of in silico techniques such as colocalization of genes on the genome (e.g., operons in bacteria) [3, 4], the Rosetta stone method (domains fused to a single protein in some genome are expected to interact in other genomes) [5, 6], and also coevolutionary techniques like phylogenetic profiling (correlated presence or absence of interacting proteins in genomes) [7] or similarities between phylogenetic trees of groups of orthologous proteins (cf. the mirrortree method) [8, 9]. Here we present an approach [10], based on a simultaneous construction of the joint

Stefan Canzar and Francisca Rojas Ringeling (eds.), *Protein-Protein Interaction Networks: Methods and Protocols*, Methods in Molecular Biology, vol. 2074, https://doi.org/10.1007/978-1-4939-9873-9_5,

multiple sequence alignment (MSA) and detection of interprotein coevolution using direct-coupling analysis [11, 12], to solve the problem of matching paralogs (cf. also [13] for a related method). The method is based on the idea that the correct matching of interacting paralogs maximizes the interprotein coevolutionary signal; cf. Figure 1 for an illustration of interprotein residue

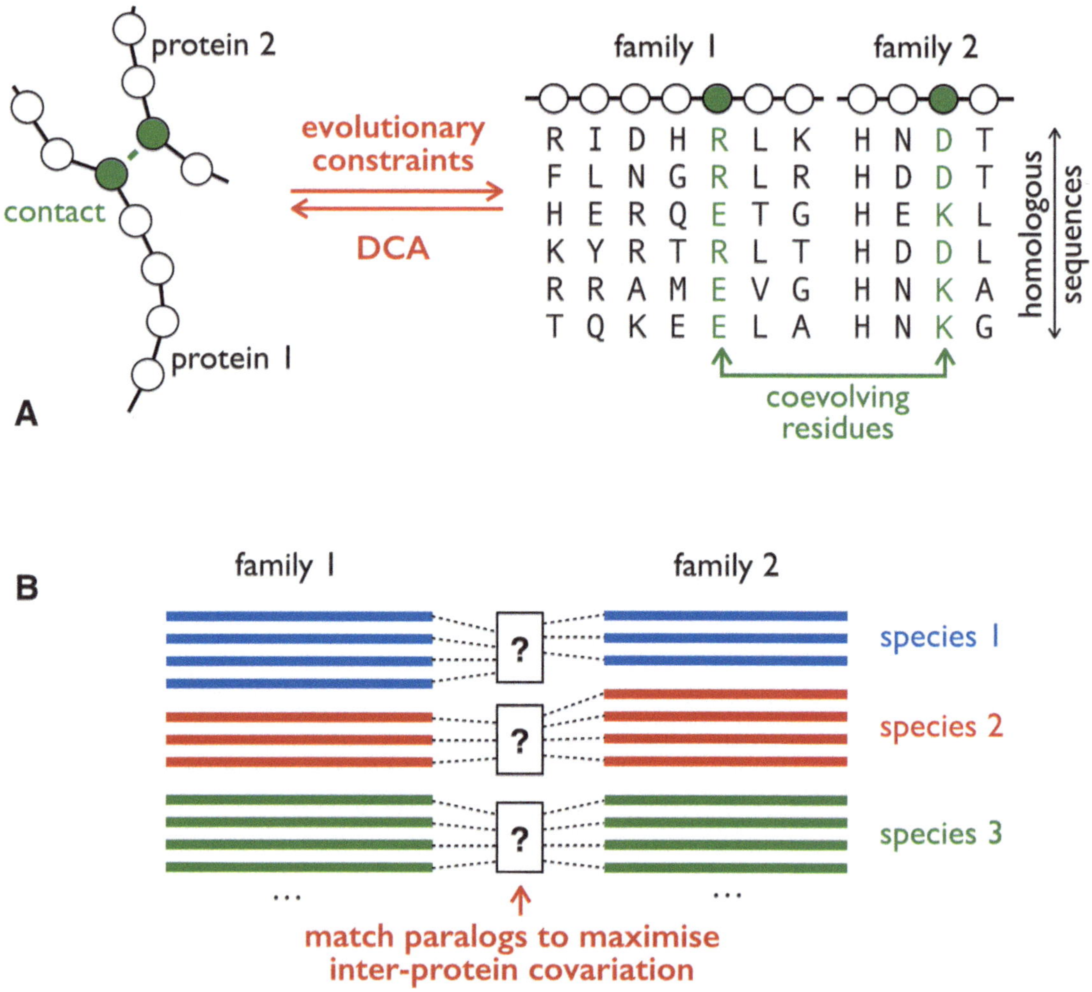

Fig. 1 Illustration of inter-protein coevolution and paralog matching. (**a**) Evolutionary conserved contacts between interacting proteins lead to coevolution, i.e., to a correlated amino-acid usage in the MSA columns corresponding to the contacting residue positions. Direct-coupling analysis (DCA) and other coevolutionary modeling approaches use this idea to detect correlations in amino-acid occurrences between MSA columns and to relate them to interprotein residue–residue contacts. (**b**) The application of DCA requires a joint MSA of the two protein families. Frequently only two separate MSAs are available, which contain several paralogous sequences per species (represented by colored lines in the figure). Our paralog-matching algorithm aims at matching paralogs (and thus identifying specifically interacting protein pairs) by maximizing the detectable interprotein covariation in the matched joint MSA

coevolution and the paralog matching problem. The method has been implemented as a Julia package available at https://github.com/Mirmu/ParalogMatching.jl.

When developing this algorithm, a particular attention was devoted to memory and computational efficiency, in order for the algorithm to be well-adapted to large-scale studies. The computational bottleneck is matrix inversion (whose complexity grows cubically with size), for which well-optimized libraries (BLAS and LAPACK) exist and are called by Julia. The size of the matrix to be inverted is 20 L × 20 L, with L being the length of the two concatenated alignment sequences. This inversion has to be redone for each batch of species processed by the algorithm. Compared to this operation, the cost of the matching per se is almost negligible. Therefore, very large domains can be challenging to process, although we successfully matched sequences of up to 1200 amino acids. On the other hand, the number of sequences in the FASTA files is not a limitation, and (usually) the larger the alignment, the better. This method is also well-adapted for systems- biology studies, where thousands of domain matchings have to be investigated.

2 Materials

1. Two multiple sequence alignments in (possibly gzip-compressed) FASTA format. As detailed below, suitable information about the species should be provided in the header of each entry of the FASTA file. The main idea of the method is to perform a statistical analysis of two given multiple sequence alignments, each containing one protein family. Each family should comprise several species, and each species may have several sequences belonging to the family (paralogs). The algorithm tries to associate (match) interacting partners from the two families within each species.
2. Computing software: A fully functional version of Julia language (version 0.6 or higher): https://github.com/JuliaLang/julia.
3. Computing hardware: Julia can be installed on a quite extensive number of operating systems (Linux 2.6.18+, macOS 10.8+, Windows 7+, FreeBSD 11.0+) and platforms (*see* https://github.com/JuliaLang/julia#currently-supported-platforms for the most updated list of supported platforms). The package is routinely tested against Julia 0.6 and higher on Linux, OS X, and Windows.

3 Methods

The paralog-matching analysis protocol consists of several steps:

1. Julia can be either installed from source download and compilation (https://github.com/JuliaLang/julia#source-download-and-compilation) or by binary installation (https://github.com/JuliaLang/julia#binary-installation).
2. ParalogMatching.jl is a Julia package, which can be installed from within Julia REPL (the interactive shell of Julia accessible via the Julia command) as

   ```
   julia> Pkg.clone("https://github.com/Mirmu/ParalogMatching.
   jl")
   ```

   Dependencies will be installed automatically. However, at least one linear programming solver supported by MathProgBase (http://mathprogbasejl.readthedocs.io/en/latest/) should be installed. The list of available solvers is accessible at the JuliaOpt page (http://www.juliaopt.org/#packages). If unsure, use.

   ```
   julia> Pkg.add("Clp")
   ```

   or

   ```
   julia> Pkg.add("GLPKMathProgInterface")
   ```

   which are free and widely tested open-source solvers.
3. To verify that everything is properly installed, run the command

   ```
   julia> Pkg.test("ParalogMatching")
   ```

   After some screen printout, which will be explained below, the program should output the message.

   ```
   INFO: ParalogMatching tests passed
   ```

4. Usage: The ParalogMatching.jl module can be loaded as

   ```
   julia> using ParalogMatching
   ```

5. In the following we will assume to be in a directory containing the two FASTA files of the multiple alignments to be matched: "A.fasta.gz," and "B.fasta.gz." In the GitHub repository, we already provided two MSAs for testing purposes (*cf.* https://github.com/Mirmu/ParalogMatching.jl/tree/master/test). Note that in general the files need not to be compressed: the

program will automatically detect the correct format. The FASTA format is not particularly restrictive, and therefore files may come in many forms (*cf.* https://en.wikipedia.org/wiki/FASTA_format). It is, however, important for the header to hold enough metadata to perform the matching. The Julia package loads the FASTA and parses sequences and headers, assuming the header to be preceded by a ">" symbol (and not a semicolon, as it is sometimes but rarely seen).

A typical (single-line) example of header is provided by the European database, UniProt:

```
>db|UniqueIdentifier|EntryName ProteinName OS=Organism-
Name OX=OrganismIdentifier [GN=GeneName ]PE=ProteinExis-
tence SV=SequenceVersion
```

The various fields are separated by the "|" and come in no particular order. As a comparison, the header of the FASTA provided as a test in the Julia Package is structured as follows:

```
>tr|J1Z5B4|J1Z5B4_VIBCL/258-450
```

where "tr" stands for UniProtKB/TrEMBL database (irrelevant for our purposes), while the block after the first "|" is mnemonic protein identification code, and the "VIBCL" after the underscore is a mnemonic species identification code, with at most five alphanumeric characters (for a precise reference, *cf.* https://www.uniprot.org/help/entry_name). Finally, the block after "/" relates to the position of the fragment on the protein over which it has been identified.

The functions provided in the package automatically attempt to parse the species names and the UniProt ID, using a standard format specification. Nonetheless, due to the variety of those headers, it also provides a flexible way to parse custom headers, through *regular expressions* (regex's). In Julia, regex's follow the Perl dialect (*cf.* https://perldoc.perl.org/perlre.html) and are represented as a string preceded by the character "r." One of the most useful constructs for header parsing in our experience follows the pattern "([^/]+)/," which reads as "capture all characters except the slash, provided there are any, up to the first slash." The capture group is denoted by the parentheses; the [^/] is a class that matches any character except (^) the slash (/); the + symbol means "one or more."

In its simplest form, the header needs to have at least two capture groups, the first one returning the UniProt ID and the second the species name (additional capture groups are ignored). Any additional group is ignored. To make sure the identifier and the species fall into the proper capture groups, the better practice is to pass "id" and "species" as labels of

Table 1
Exemplary header metadata and corresponding regular expression (regex) to parse it

<table>
<tr><th>Header</th><th>Regex</th></tr>
<tr><td>>SPECIES/OTHERINFO/LOCATION</td><td>r"^(?<species>[^/]+)/[^/]+/(?<id>.+)"</td></tr>
<tr><td>>LOCATION:OTHERINFO/SPECIES/OTHERINFO</td><td>r"^(?<id>[^:]+):[^/]+/(?<species>[^/]+)/.+"</td></tr>
<tr><td>>SPECIES_LOCATION|OTHERINFO</td><td>r"^(?<species>[^_]+)_(?<id>[^|]+)\|.+"</td></tr>
</table>

those groups, which is done by inserting the directives "?<id>" or "?<species>" inside the capture group. In Table 1 we display some examples of common headers with their corresponding regex. (The initial carets "^" are used to anchor the match at the beginning of the header, and the syntax ".+" is used to match any non-empty sequence of characters.)

6. The module provides a high-level interface, with one function which performs all operations and writes the output to a file

```
paralog_matching(infile1::AbstractString,
                 infile2::AbstractString,
                 outfile::AbstractString;
                 cutoff = 500,
                 batch = 1,
                 strategy = "covariation",
                 lpsolver = nothing)
```

The *positional* arguments infile1, infile2, and outfile are strings containing, respectively, the input file name of the first, second, and output alignments (e.g., "A.fasta.gz," "B.fasta.gz," "matchedAB.fasta.gz"). There are optional *keyword* arguments:

(a) Cutoff: Used to discard all species for which there are more than a certain number of paralogous sequences in either alignment. Use 0 to disable this filter entirely. Default value = 500.

(b) Batch: specifies the number of species that should be matched before updating the underlying model; smaller values give better results but increase the computational time. Default value = 1.

(c) Strategy: the strategy to use when computing the matching. Available strategies:

- "Covariation": uses the Gaussian DCA model and performs a maximum matching on the scores (default).

- "Greedy": same as "covariation" but performs a greedy matching.
- "Random": produces a random matching. Useful for null models.
- "Genetic": tries to use the UniProt ID information to determine which sequences belong to the same operon (only used for testing, not a valid general strategy).

(d) pseudo_count: gives the amount of regularization used for the inversion of the correlation matrix. A default of 0.5 gives sensible results; its value must be strictly between 0.0 and 1.0.

(e) lpsolver: Linear programming solver used when performing the matching with the "covariation" strategy. The default uses a solver provided by the GLPK library (https://github.com/JuliaOpt/GLPK.jl). This can be overridden by passing, e.g., lpsolver = GurobiSolver(OutputFlag = false) or similar (see the documentation for MathProgBase https://github.com/JuliaOpt/MathProgBase.jl).

The following information is printed on the standard output, at each step of the matching process between alignments A and B:

```
Recomputing the model
batch of species


YERFR
([1])
YERIN
([1, 4])
PICSI
([1, 2], [2, 3])
RIEAD
([1, 2], [2, 3])
```

For efficiency reasons, the matching is built in batches: the smaller the batch size, the more accurate the final result. Yet, in our experience, batches of around five species do not degrade much the performance of the algorithm, while significantly reducing its computational time. In the above example, the matching is built in batches of four species (this specific batch contains the species "YERFR," "YERIN," "PICSI," and "RIEAD"). The tuples printed below each name are the matchings obtained between alignments A and B, for each species. Sequences are labeled with numbers (1-based). For example, for the species RIEAD, the algorithm decided to associate the first and second sequences of alignment A with the third and second sequences of alignment B, respectively.

Besides writing the output to a file, the paralog_matching function returns two values: the first one is a specialized object which contains the "harmonized" alignments (i.e., two filtered version of the original alignments, in which only the sequences belonging to species which exist in both alignments are kept); the second one is the matching between those two harmonized alignments, expressed as a list in which the i-th element denotes which sequence of the second alignment should be matched to the i-th sequence of the first out (with the sentinel value 0 denoting a no-match).

Therefore, the precise mapping at the end can be found either by inspecting the objects returned by the method paralog_matching or simply by reading the sequence names in the concatenated FASTA written as output file.

7. Once the concatenated FASTA is obtained, it can be treated as any other FASTA for further analysis. Especially, one can infer contact points between the two proteins through coevolution signals. As explained above, the paralog matching itself relies on those signals to associate interacting partners. But because it needs to perform such inference fast, we relied on a Gaussian direct coupling analysis, very similar to the one found in the GaussDCA Julia package (https://github.com/carlobaldassi/GaussDCA.jl). While this method has an excellent trade-off between accuracy and speed, other approaches, slower but potentially more accurate, have been developed. State-of-the-art results can be reached with pseudo-likelihood methods, such as the one found in the Julia package (https://github.com/pagnani/PlmDCA).

Acknowledgments

A.P. and M.W. acknowledge funding by the EU H2020 research and innovation program MSCA-RISE-2016 under grant agreement No. 734439 INFERNET.

References

1. Shoemaker BA, Panchenko AR (2007) Deciphering protein–protein interactions. Part I. Experimental techniques and databases. PLoS Comput Biol 3(3):e42
2. Rao VS, Srinivas K, Sujini GN, Kumar GN (2014) Protein-protein interaction detection: methods and analysis. Int J Proteomics 2014:147648
3. Dandekar T, Snel B, Huynen M, Bork P (1998) Conservation of gene order: a fingerprint of proteins that physically interact. Trends Biochem Sci 23(9):324–328
4. Galperin MY, Koonin EV (2000) Who's your neighbor? New computational approaches for functional genomics. Nat Biotechnol 18 (6):609–613

5. Marcotte CJV, Marcotte EM (2002) Predicting functional linkages from gene fusions with confidence. Appl Bioinforma 1(2):93–100
6. Marcotte EM et al (1999) Detecting protein function and protein-protein interactions from genome sequences. Science 285(5428):751–753
7. Pellegrini M, Marcotte EM, Thompson MJ, Eisenberg D, Yeates TO (1999) Assigning protein functions by comparative genome analysis: protein phylogenetic profiles. Proc Natl Acad Sci U S A 96(8):4285–4288
8. Pazos F, Valencia A (2001) Similarity of phylogenetic trees as indicator of protein–protein interaction. Protein Eng 14(9):609–614
9. Juan D, Pazos F, Valencia A (2008) High-confidence prediction of global interactomes based on genome-wide coevolutionary networks. Proc Natl Acad Sci U S A 105(3):934–939
10. Gueudré T, Baldassi C, Zamparo M, Weigt M, Pagnani A (2016) Simultaneous identification of specifically interacting paralogs and interprotein contacts by direct coupling analysis. Proc Natl Acad Sci U S A 113 (43):12186–12191
11. Szurmant H, Weigt M (2018) Inter-residue, inter-protein and inter-family coevolution: bridging the scales. Curr Opin Struct Biol 50:26–32
12. Cocco S, Feinauer C, Figliuzzi M, Monasson R, Weigt M (2018) Inverse statistical physics of protein sequences: a key issues review. Rep Prog Phys 81(3):032601
13. Bitbol AF, Dwyer RS, Colwell LJ, Wingreen NS (2016) Inferring interaction partners from protein sequences. Proc Natl Acad Sci U S A 113(43):12180–12185

Chapter 6

A Web-Based Protocol for Interprotein Contact Prediction by Deep Learning

Xiaoyang Jing, Hong Zeng, Sheng Wang, and Jinbo Xu

Abstract

Identifying residue–residue contacts in protein–protein interactions or complex is crucial for understanding protein and cell functions. DCA (direct-coupling analysis) methods shed some light on this, but they need many sequence homologs to yield accurate prediction. Inspired by the success of our deep-learning method for intraprotein contact prediction, we have developed RaptorX-ComplexContact, a web server for interprotein residue–residue contact prediction. Given a pair of interacting protein sequences, RaptorX-ComplexContact first searches for their sequence homologs and builds two paired multiple sequence alignments (MSA) based on genomic distance and phylogeny information, respectively. Then, RaptorX-ComplexContact uses two deep convolutional residual neural networks (ResNet) to predict interprotein contacts from sequential features and coevolution information of paired MSAs. RaptorX-ComplexContact shall be useful for protein docking, protein–protein interaction prediction, and protein interaction network construction.

Key words Interprotein contact prediction, Protein–protein interaction (PPI) prediction, Protein interaction network, Protein complex, Deep learning (DL), Direct-coupling analysis (DCA), Multiple sequence alignment (MSA), Protein docking

1 Introduction

Proteins play various roles in cellular and biochemical processes by physically interacting with other proteins or forming protein complexes [1, 2]. Studying protein–protein interactions (PPIs) at residue level is crucial for understanding protein functions in organisms. Experimental techniques have been greatly improved to determine protein complex structure, but they are still low throughput and costly [3, 4]. Therefore, developing effective computational methods to elucidate the 3D structure of a PPI or complex from its sequence is demanded. Interprotein

Xiaoyang Jing, Hong Zeng and Sheng Wang contributed equally to this work.

Stefan Canzar and Francisca Rojas Ringeling (eds.), *Protein-Protein Interaction Networks: Methods and Protocols*, Methods in Molecular Biology, vol. 2074, https://doi.org/10.1007/978-1-4939-9873-9_6,

residue–residue contact prediction serves as one of the important intermediate steps for this.

Recent progress indicates that long-range intraprotein contacts may allow accurate 3D structure modeling for a single protein chain [5, 6]. Several recent studies have also reported that interprotein contacts are useful for structure modeling of a PPI or protein docking [7–9]. To maintain the stability of a protein structure, the substitutions of spatially proximal residues would occur in pairs [10]. Based on this observation, coevolution analysis, especially the direct-coupling analysis (DCA), makes use of statistical methods to predict intraprotein contacts [11–13] or interprotein contacts [7, 8, 14, 15] by identifying coevolved residues in multiple sequence alignments (MSAs). Nevertheless, DCA needs a lot of homologous sequences to yield accurate prediction. It is very challenging to find many sequence homologs for an interacting protein pair, especially for eukaryotic species [16], so DCA methods usually have low accuracy for interprotein contact prediction. To deal with the situation when not many sequence homologs are available, we have developed a deep convolutional residual neural network for contact prediction. Our deep-learning (DL) method for intraprotein contact prediction [6, 17] greatly outperformed the DCA methods and achieved the state-of-the-art performances in both CASP12 [18] and CASP13 [19]. In addition to coevolution information, our DL method makes use of contact occurrence patterns and even global context information learned from native contact maps to greatly improve prediction accuracy so that it is still effective even with only dozens of sequence homologs [19].

Here we present RaptorX-ComplexContact, a web server for de novo interprotein residue–residue contact prediction. The underlying algorithm of this server is the deep-learning model composed of two deep residual neural networks (ResNet) [6, 17]. Our DL method makes use of both sequential features and coevolution information to reduce the number of required sequence homologs and to greatly improve performances over pure DCA methods [6]. We trained our DL model by individual protein chains, so there is no overlap between our training data and test data. RaptorX-ComplexContact uses both genomic and phylogeny information to identify interlogs and build better MSAs for a pair of interacting proteins from eukaryotes, which further enhances its performance.

2 Materials

The following are required and optional materials for the use of RaptorX-ComplexContact server:

1. A personal computer with Internet connection and a web browser with JavaScript enabled. RaptorX-ComplexContact server is compatible with three popular web browsers: Google Chrome, Firefox, and Internet Explorer. Nevertheless, the former two browsers may be slightly better than the third one in visualizing the prediction results.
2. The amino-acid sequences or multiple sequence alignments (MSAs) of the query protein pair in FASTA format. Only the MSAs generated by HHblits are systematically tested although in principle any MSAs shall work.
3. The amino-acid sequences or multiple sequence alignments (MSAs) could also be uploaded to the server as text files.
4. The job name and email address are optional, but a valid email address is strongly recommended since it can facilitate job management and result retrieval.

3 Methods

3.1 Job Submission

1. Open the hyperlink http://raptorx.uchicago.edu/ComplexContact/ in the web browser.
2. From the menu at the top of the page, select "New job."
3. In the "Job Identification" field of the form, input a job name (default is "my job") and an email address that will be used for notification upon job completion. The job name and email address are optional, but they are useful for job retrieval. A user account uniquely identified by the email address is automatically created when a user submits the first job.
4. In the "First Sequence" and "Second Sequence" fields of the form, input no more than 20 sequence pairs or a pair of multiple sequence alignments (MSAs) of the query interacting protein pair.
5. If your input is a pair of MSAs, the check box at the bottom of "Job Identification" field shall be checked.
6. Confirm that the required and optional inputs are correct; then click on the "Submit" button to submit a new job to the server (for information on running times, *see* **Note 1**).

The required inputs for RaptorX-ComplexContact server are (1) a batch of no more than 20 sequence pairs or (2) a pair of MSAs for the query interacting protein pair. Either sequence pairs or an MSA pair can be submitted as long as they are in FASTA format. The pair of MSAs should be formatted as follows:

1. Each protein sequence in the MSAs occupies two lines (in FASTA format). The first contains name and annotations,

and the second is the primary sequence. All the sequences should have the same length when gaps are counted.

2. The first sequences in the MSAs are assumed to be the query proteins for which interprotein contacts will be predicted. These two sequences should not contain any gaps. The remaining protein sequences may contain gaps represented by "-."
3. At most 20,000 sequences are allowed in a single MSA. If a file is used to submit an MSA, the file size should be smaller than 5 Mb.

Figure 1 shows an input example for a pair of putatively interacting proteins (PDB IDs 2A1TS and 2A1TR). In addition to using the web interface, users can also submit a job using the publicly available program Curl, as shown in Fig. 2. In this way, the job

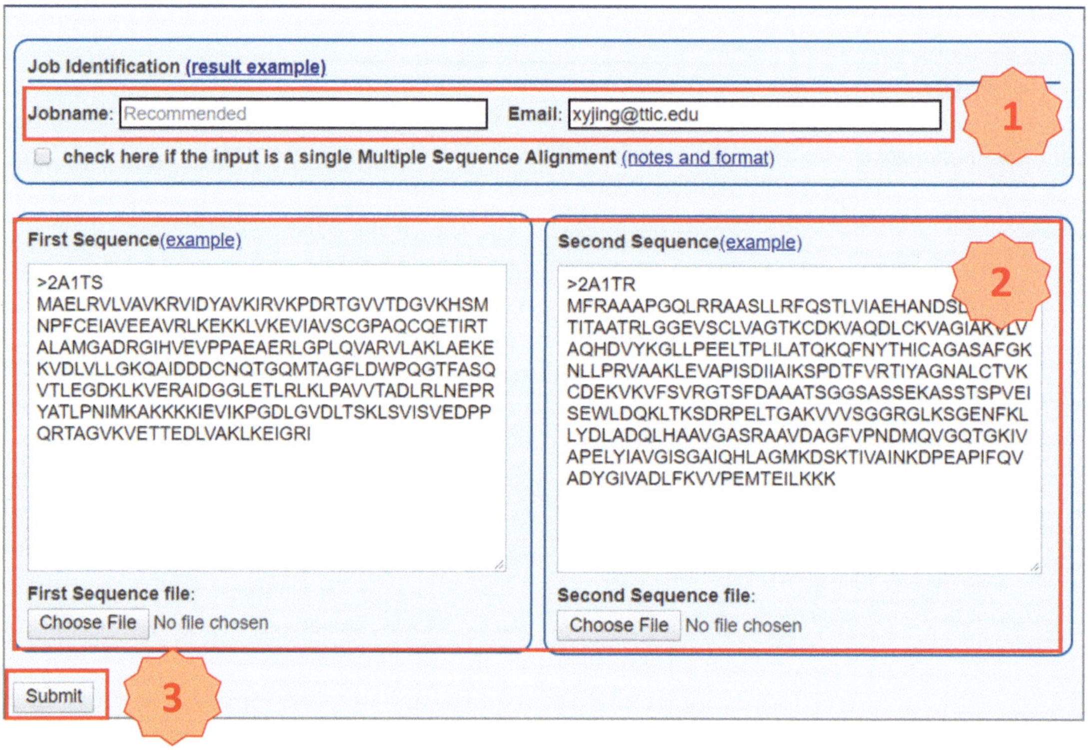

Fig. 1 The RaptorX-ComplexContact job submission page. (1) Job name and user email address (optional), (2) a pair of sequences in FASTA format (sequences can also be submitted in a file), and (3) Submit button

Fig. 2 Using Curl to submit a job to RaptorX-ComplexContact

name and email address are also optional. The job retrieval URL will be returned to the screen after submission. When Curl is used to submit jobs, only one sequence pair is allowed at a time. However, users can run Curl multiple times to submit many jobs. At any time, each user is allowed to have up to 500 unfinished jobs in the RaptorX server.

3.2 RaptorX-ComplexContact Implementation

3.2.1 Overview

As shown in Fig. 3, given a pair of interacting protein sequences A and B submitted by a user, RaptorX-ComplexContact first searches for homologous sequences and builds MSAs for A (MSA_A) and B (MSA_B) using HHblits [20]. Then RaptorX-ComplexContact uses two strategies (genome- and phylogeny based) to concatenate MSA_A and MSA_B into two paired MSAs consisting of only interlogs, denoted as MSA_g and MSA_p. Based on the MSAs of interlogs, RaptorX-ComplexContact predicts two interprotein contact maps using our deep-learning model (trained by single protein chains) and calculates their average as the final prediction.

3.2.2 Concatenate Multiple-Sequence Alignments for a Protein Pair

For each protein sequence, RaptorX-ComplexContact generates an MSA by running HHblits with eight iterations and *E*-value = 1E-20 to search through the UniClust30 HHM library created in October 2017. The other parameters for running HHblits are set to "-maxfilt 100000000 -diff inf -all -neffmax 20." If the user submits a pair of MSAs as input, the server will use the submitted MSAs to predict contacts instead of generating new MSAs. Let MSA_A and MSA_B denote the MSAs for proteins A and B, respectively. RaptorX-ComplexContact employs two strategies to concatenate MSA_A and MSA_B into a paired MSA consisting of only interlogs. One is based on genomic distance, and the other makes use of phylogeny information. Since not every protein in MSA_A and MSA_B can be paired, the unpaired proteins are removed after concatenation.

Concatenating MSAs by genomic distance. In prokaryotes and some eukaryotes, interacting genes are often co-located on the chromosome into operons [21]. If the intergenic distance of two proteins is less than a given threshold (20 is used here), RaptorX-ComplexContact assumes that they form an interacting pair. This strategy is also employed by EVcomplex [8] and Gremlin-Complex [7], while some details in their methods are different.

Figure 4 shows an example of genome-based MSA concatenation. The left (orange) and right (blue) MSAs contain five and six protein chains, respectively, and the intergenic distance threshold used here is three. There are six contigs containing six copies of proteins (orange blocks) in the left MSA and eight copies of proteins (blue blocks) in the right MSA. Gray blocks are coding DNA sequences not belonging to these two MSAs. Based upon intergenic distance, four protein pairs can be formed, as shown by the red double arrows.

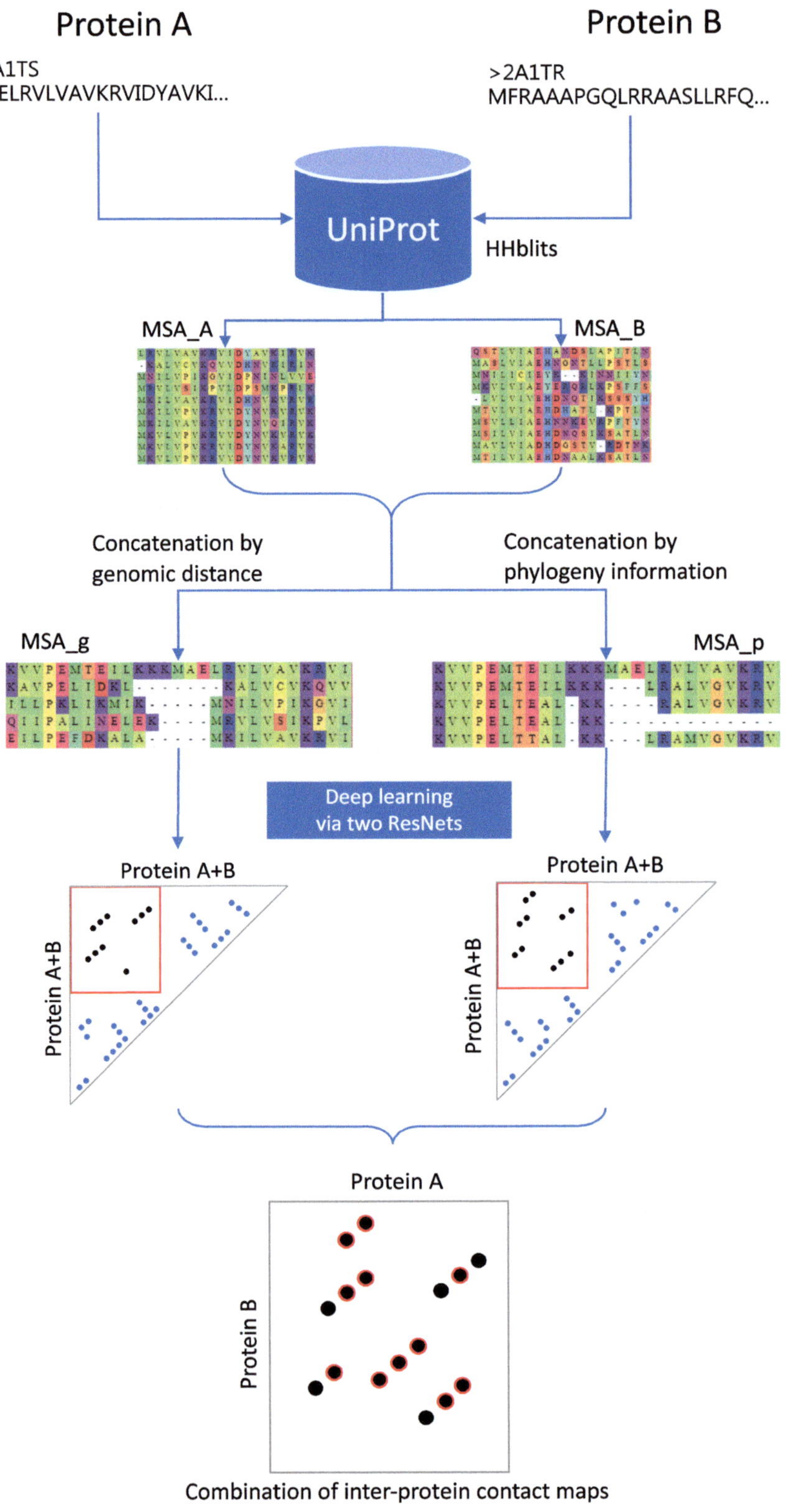

Fig. 3 The workflow of RaptorX-ComplexContact

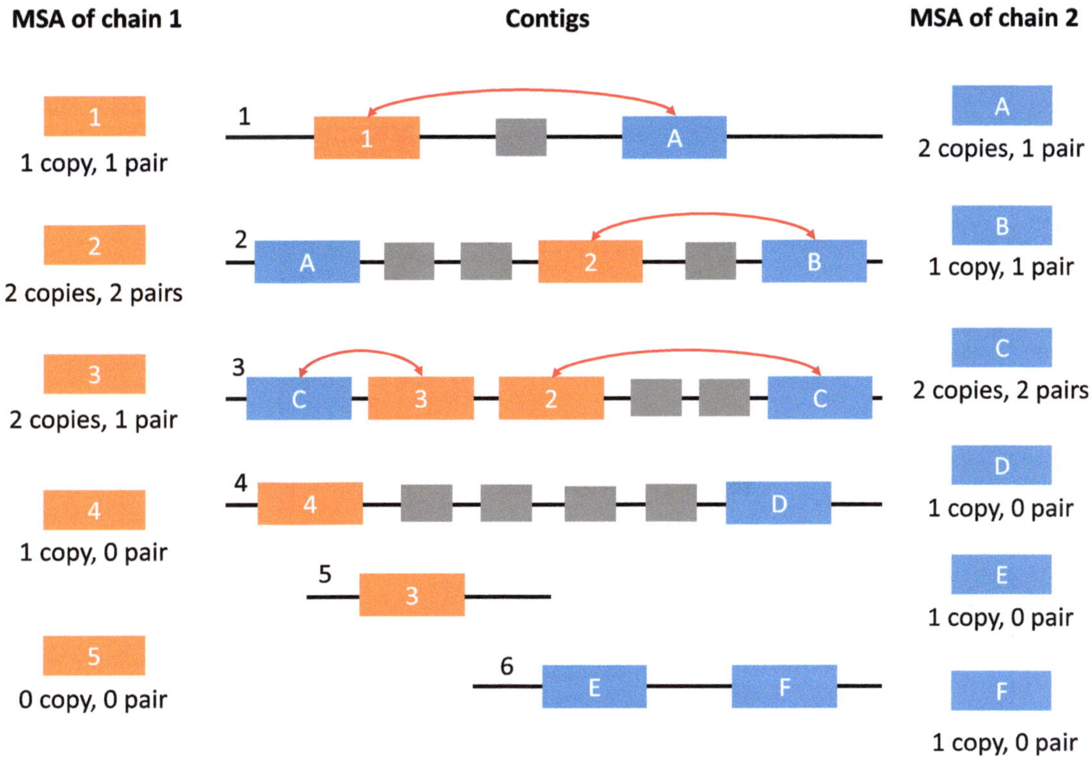

Fig. 4 An example of genome-based MSA concatenation

Concatenating MSAs by phylogeny information. For a protein pair in eukaryotes, their individual MSAs may contain abundant paralogs, and their genes may interact even if they are not close by genomic distance. This makes it very challenging to concatenate two individual MSAs. To address this problem, RaptorX-ComplexContact concatenates two individual MSAs using phylogeny information. According to the phylogeny tree in Taxonomy Database [22], RaptorX-ComplexContact first groups proteins in each MSA by their species (or sub-species if possible). Then it sorts proteins of a specific species/subspecies in each MSA by their sequence similarity (from high to low) to their respective query proteins. Let $p_1, p_2, \ldots, p_m$ and $q_1, q_2, \ldots, q_n$ be the sorted proteins of a specific species in two MSAs, respectively. p_i and q_i will be paired together where i ranges from 1 to the minimum of m and n.

Figure 5 shows an example of phylogeny-based MSA concatenation. The left MSA contains eight proteins, and the right one has nine proteins (including the query protein). The proteins are divided into four groups. Proteins in the same group are sorted by their sequence similarities with the query, that is, proteins with higher sequence similarities will be paired and the others will be removed. Finally, we obtain a paired MSA consisting of seven interlogs (including the query itself), as shown by the red double arrows.

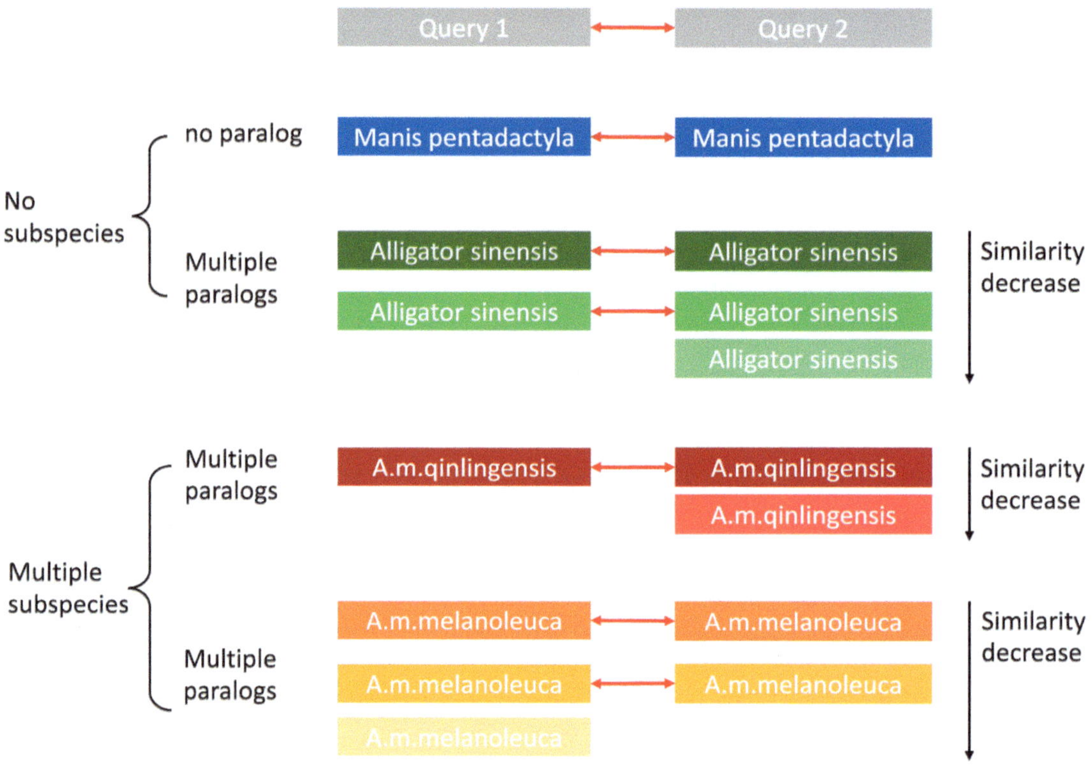

Fig. 5 An example of phylogeny-based MSA concatenation

Our experimental results [23] show that the phylogeny-based and genome-based methods for MSA construction are complementary to each other. The phylogeny-based strategy works better for eukaryotes, while the genomic-based strategy works better for prokaryotes. Their combination can improve the performance on eukaryotes.

3.2.3 Deep Learning for Interprotein Contact Prediction

The deep-learning (DL) model used in RaptorX-ComplexContact is formed mainly by two deep residual neural networks (ResNet) [24]. The first ResNet handles sequential features by a series of 1D convolutional transformation to capture long-range sequential context of each residue in the target sequence. The second ResNet handles the pairwise features including the output of the first network, coevolution information, and pairwise potential by a series of 2D convolutional transformation to capture long-range 2D context of a residue pair. Finally, based on the output of the second ResNet, logistic regression is used to predict the probability of any two residues forming a contact. Please see our previous work [6, 18] for a detailed description of this DL model.

The sequential features include protein sequence profile (i.e., position-specific scoring matrix), three-state secondary structure, and three-state solvent accessibility predicted by our in-house tool

RaptorX-Property [25]. For each residue in the paired sequence, the dimension of sequence profile is 20; both the dimension of predicted secondary structure and solvent accessibility is 3, thus we have a 2D matrix with dimension $L \times 26$ ($20 + 3 + 3 = 26$) to represent the paired sequence where L is the total sequence length. The pairwise features include the direct coevolution information produced by CCMpred [13] and mutual information derived from paired MSAs. Each paired MSA concatenated by phylogeny-based or genome-based methods is fed into CCMpred to calculate direct coevolution strength between any two residues in the MSA. Mutual information and pairwise potential between any two columns in a paired MSA are also calculated as features. Please see our previous work [6] for details. The DL model used in RaptorX-ComplexContact was originally developed for intraprotein contact prediction and trained using only individual protein chains, while it is also very effective for interprotein contact prediction, as shown by our experimental results [23, 26].

3.3 Job Retrieval

A user may retrieve jobs: (1) by JobID, (2) by sequence, and (3) by email address. Each job is assigned one unique job ID and one URL for job retrieval. Users can use the job ID or sequence to retrieve their jobs by clicking the "Job Status" link at the top of the web page, through which users may find a job by one of its submitted sequences. If an email is provided in submission, users will be notified by email once their jobs are done. Users can also retrieve all of their jobs by the "My Jobs" link at the top of the web page using the provided email (*see* **Notes 2–4**).

3.4 Outputs and Result Download

As shown in Fig. 6, the results page includes three sections: (I) input sequence pair, (II) predicted complex contact map, and (III) multiple sequence alignment. Section I shows the original input sequence pair with index information. Section II displays the predicted contact map, the job status information, a panel for contact image zooming and dragging, a panel for downloading the predicted complex contact map, and a panel for downloading the detailed prediction results. Figure 7 shows an example result file downloaded from RaptorX-ComplexContact. This file includes a list of residue pairs with top-predicted contact probability (arranged in descending order) Section III shows the number of homologous sequences in each MSA and two paired MSAs generated by genome- and phylogeny-based methods visualized by MSAViewer [27] (*see* **Note 5**).

3.5 Quality Assessment of the Predicted Probability

RaptorX-ComplexContact predicts the probability of any two interprotein residues forming a contact, as shown in Fig. 7. Since there is no clear threshold of probability to distinguish between contact and noncontact residues, to provide some valuable references to users, here we show our assessment results [26] of the top

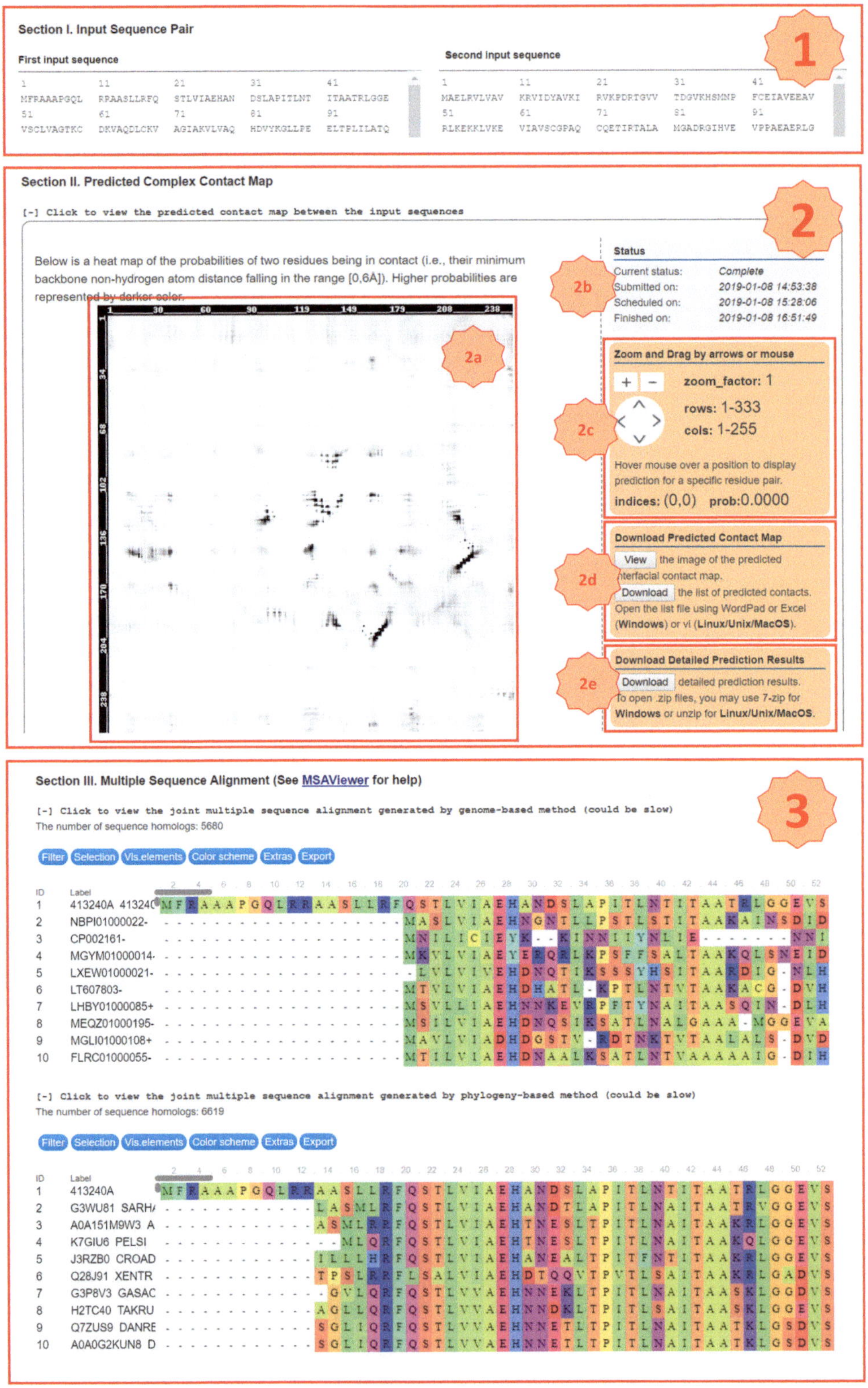

Fig. 6 The result page of RaptorX-ComplexContact server. (1) Section I, input sequence pair; (2) Section III, predicted complex contact map, (2a) the predicted complex contact map image, (2b) the job status information, (2c) buttons to zoom and drag the contact image, (2d) the view and download buttons of contact image, (2e) the download button of detailed prediction results; and (3) Section III, two paired MSAs generated by genome-based and phylogeny-based methods, respectively

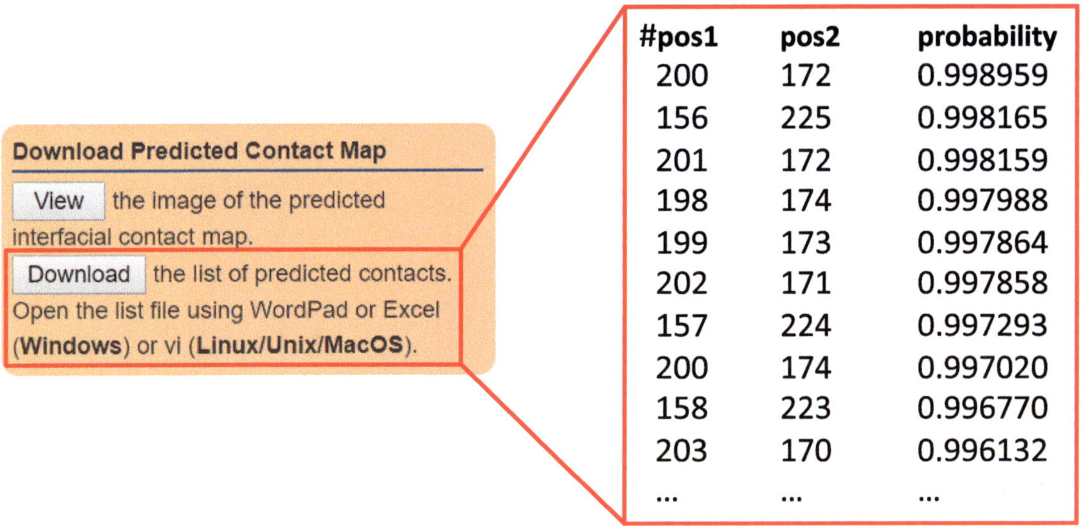

Fig. 7 The detailed prediction results downloaded from RaptorX-ComplexContact server. Columns "pos1" and "pos2" show the indices of two residues (of proteins A and B) predicted to form a contact, respectively. Column "probability" shows the predicted probability of two residues being in contact

Table 1
Precision and recall of the top 50 predicted interfacial contacts by ComplexContact calculated from 3Dcomplex data

Probability	Precision	Recall
0.95	0.70	0.03
0.90	0.57	0.06
0.85	0.45	0.09
0.80	0.35	0.12
0.75	0.27	0.14
0.70	0.22	0.17
0.65	0.19	0.21
0.60	0.16	0.27
0.55	0.14	0.35
0.50	0.13	0.44

50 predicted probability values on 3D complex data [28]. Table 1 shows the precision and recall for a list of probability values produced by RaptorX-ComplexContact. For example, when the predicted probability is >0.90, the precision is 0.57. For more assessment results, please refer to [23, 26].

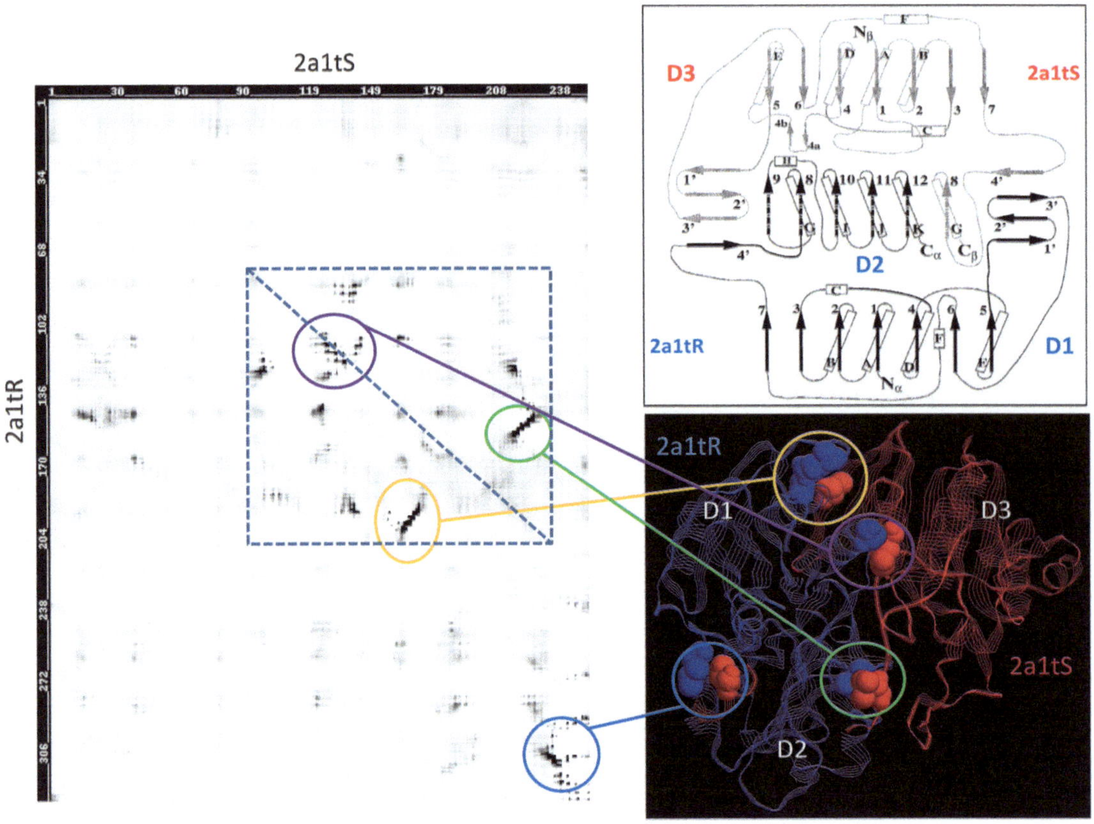

Fig. 8 Predicted intermolecular residue–residue contact map of human ETF protein: 2a1tR (alpha subunit consisting domain D1 and D2) and 2a1tS (beta subunit consisting domain D3). Four top-predicted regions and their corresponding positions on the crystal structure are shown in blue, green, purple, and blue circles, respectively. Dotted lines show the interface between D1 and D3, which forms a pseudo twofold symmetry architecture, as illustrated in topology view in the upper right part. (figure taken from [30])

3.6 Case Study

Human electron-transferring flavoprotein (ETF) is a ubiquitous electron carrier, via which the electrons are shuttled to the respiratory chain [29]. Human ETF is a 63-kDa heterodimer (PDB, 2a1t) and folds into three distinct domains: two domains (D1 and D2) belong to the alpha subunit (2a1tR), and the third domain (D3) is contributed by the beta subunit (2a1tS) [29]. The key intermolecular contacts between 2a1tR and 2a1tS reside in the interface between D1 and D3, as well as the interface between D2 and D3, respectively. Although there is no sequence similarity between D1 and D3, they share an extremely similar fold by a pseudo twofold axis [30], as shown in the upper right part in Fig. 8. Specifically, the two domains are tightly interacted by a three-stranded antiparallel beta-sheet (marked as 1 to 3) with a fourth strand (marked as 4) coming from the pseudo-symmetry partner, which strongly stabilizes the quaternary structure of the ETF protein [30]. The interface between D2 and D3 (marked as helix K and G) folds in a manner reminiscent of bacterial flavodoxins [30].

We submit the amino-acid sequences of 2a1tR and 2a1tS to RaptorX-ComplexContact. Strikingly, the top predicted intermolecular residue–residue contacts with high confidence are consistent with the native interactions between D1 and D3, as shown in the purple, green, and yellow circles in Fig. 8. Moreover, our predicted intermolecular contact map also reveals the pseudo twofold symmetry (*see* dotted lines in Fig. 8). Finally, our prediction successfully recovers the interface between D2 and D3, as shown in the blue circles in Fig. 8.

4 Notes

1. The running time of a job depends on the length of its two sequences and the number of homologous sequences detected by the server. For a protein pair of 250 residues, about 1 h is needed to finish the job after it is scheduled to run. When there are many waiting jobs (or jobs of long sequences) in the queue, it may take a few hours. Overall, our server can finish 100–200 jobs per day.
2. If your email box cannot receive a large attachment file, please submit a SMALL number of sequences in a batch. Otherwise, the result package may be too big to be received.
3. A single job may contain at most 20 sequence pairs, and each user can have no more than 500 unfinished jobs at any time. Further, the results of a job are guaranteed to be stored for only 14 days after the job is completed, although empirically most jobs are stored for 6 months to 1 year. To store your jobs for a much longer time for publications, please contact the RaptorX team through the "Inquiry & Bug Report" link.
4. Please save the assigned JobID for job retrieval, especially when an email address is not provided in submission.
5. To provide a better overview of the results, Section III is not automatically opened when the result page is loaded. To show the MSAs, please click on the corresponding subsection.

Acknowledgments

This work was supported by National Institutes of Health grant R01GM089753 to JX and National Science Foundation grant DBI-1564955 to JX.

References

1. Jones S, Thornton JM (1996) Principles of protein-protein interactions. Proc Natl Acad Sci 93:13–20
2. Alberts B (1998) The cell as a collection of protein machines: preparing the next generation of molecular biologists. Cell 92:291–294

3. Lensink MF, Velankar S, Kryshtafovych A et al (2016) Prediction of homoprotein and heteroprotein complexes by protein docking and template-based modeling: a CASP-CAPRI experiment. Proteins 84:323–348
4. Lensink MF, Velankar S, Wodak SJ (2017) Modeling protein–protein and protein–peptide complexes: CAPRI 6th edition. Proteins 85:359–377
5. Kim DE, DiMaio F, Yu-Ruei Wang R et al (2014) One contact for every twelve residues allows robust and accurate topology-level protein structure modeling. Proteins 82:208–218
6. Wang S, Sun S, Li Z et al (2017) Accurate de novo prediction of protein contact map by ultra-deep learning model. PLoS Comput Biol 13:e1005324
7. Ovchinnikov S, Kamisetty H, Baker D (2014) Robust and accurate prediction of residue–residue interactions across protein interfaces using evolutionary information. elife 3:e02030
8. Hopf TA, Schärfe CP, Rodrigues JP et al (2014) Sequence co-evolution gives 3D contacts and structures of protein complexes. elife 3:e03430
9. Yu J, Andreani J, Ochsenbein F, Guerois R (2017) Lessons from (co-) evolution in the docking of proteins and peptides for CAPRI rounds 28–35. Proteins 85:378–390
10. Gromiha MM, Selvaraj S (2004) Inter-residue interactions in protein folding and stability. Prog Biophys Mol Biol 86:235–277
11. Jones DT, Buchan DW, Cozzetto D, Pontil M (2011) PSICOV: precise structural contact prediction using sparse inverse covariance estimation on large multiple sequence alignments. Bioinformatics 28:184–190
12. Marks DS, Hopf TA, Sander C (2012) Protein structure prediction from sequence variation. Nat Biotechnol 30:1072
13. Seemayer S, Gruber M, Söding J (2014) CCMpred—fast and precise prediction of protein residue–residue contacts from correlated mutations. Bioinformatics 30:3128–3130
14. Gueudré T, Baldassi C, Zamparo M et al (2016) Simultaneous identification of specifically interacting paralogs and interprotein contacts by direct coupling analysis. Proc Natl Acad Sci 113:12186–12191
15. Weigt M, White RA, Szurmant H et al (2009) Identification of direct residue contacts in protein–protein interaction by message passing. Proc Natl Acad Sci 106:67–72
16. Rodriguez-Rivas J, Marsili S, Juan D, Valencia A (2016) Conservation of coevolving protein interfaces bridges prokaryote–eukaryote homologies in the twilight zone. Proc Natl Acad Sci 113:15018–15023
17. Wang S, Li Z, Yu Y, Xu J (2017) Folding membrane proteins by deep transfer learning. Cell Syst 5:202–211.e3
18. Wang S, Sun S, Xu J (2018) Analysis of deep learning methods for blind protein contact prediction in CASP12. Proteins 86:67–77
19. Xu J (2018) Distance-based protein folding powered by deep learning. arXiv preprint arXiv:181103481
20. Remmert M, Biegert A, Hauser A, Söding J (2012) HHblits: lightning-fast iterative protein sequence searching by HMM-HMM alignment. Nat Methods 9:173
21. Feinauer C, Szurmant H, Weigt M, Pagnani A (2016) Inter-protein sequence co-evolution predicts known physical interactions in bacterial ribosomes and the Trp operon. PLoS One 11:e0149166
22. Federhen S (2011) The NCBI taxonomy database. Nucleic Acids Res 40:D136–D143
23. Zhou T, Wang S, Xu J (2017) Deep learning reveals many more inter-protein residue-residue contacts than direct coupling analysis. bioRxiv:240754
24. He K, Zhang X, Ren S, Sun J (2016) Deep residual learning for image recognition. In: Proceedings of the IEEE conference on computer vision and pattern recognition, pp 770–778
25. Wang S, Li W, Liu S, Xu J (2016) RaptorX-property: a web server for protein structure property prediction. Nucleic Acids Res 44: W430–W435
26. Zeng H, Wang S, Zhou T et al (2018) ComplexContact: a web server for inter-protein contact prediction using deep learning. Nucleic Acids Res 46(W1):W432–W437
27. Yachdav G, Wilzbach S, Rauscher B et al (2016) MSAViewer: interactive JavaScript visualization of multiple sequence alignments. Bioinformatics 32:3501–3503
28. Levy ED, Pereira-Leal JB, Chothia C, Teichmann SA (2006) 3D complex: a structural classification of protein complexes. PLoS Comput Biol 2:e155
29. Toogood HS, van Thiel A, Scrutton NS, Leys D (2005) Stabilisation of non-productive conformations underpins rapid electron transfer to ETF. J Biol Chem 280(34):30361–30366
30. Roberts DL, Frerman FE, Kim J-JP (1996) Three-dimensional structure of human electron transfer flavoprotein to 2.1-Å resolution. Proc Natl Acad Sci 93:14355–14360

Chapter 7

Visual Analysis of Protein–Protein Interaction Docking Models Using COZOID Tool

Jan Byska, Adam Jurcik, Katarina Furmanova, Barbora Kozlikova, and Jan J. Palecek

Abstract

Networks of protein–protein interactions (PPI) constitute either stable or transient complexes in every cell. Most of the cellular complexes keep their function, and therefore stay similar, during evolution. The evolutionary constraints preserve most cellular functions via preservation of protein structures and interactions. The evolutionary conservation information is utilized in template-based approaches, like protein structure modeling or docking. Here we use the combination of the template-free docking method with conservation-based selection of the best docking model using our newly developed COZOID tool.

We describe a step-by-step protocol for visual selection of docking models, based on their similarity to the original protein complex structure. Using the COZOID tool, we first analyze contact zones of the original complex structure and select contact amino acids for docking restraints. Then we model and dock the homologous proteins. Finally, we utilize different analytical modes of our COZOID tool to select the docking models most similar to the original complex structure.

Key words Protein–protein interactions, Contact zone, Contact residue, Multiple sequence alignment, Conservation rate, Protein docking, COZOID tool, Visual selection

1 Introduction

Protein–protein interactions (PPIs) lie at the heart of most cellular processes. Their networks constitute either stable or transient complexes in every cell. While computational predictions of multiprotein PPI networks remain a challenging task, methods for analysis of a pair of binding partners developed significantly [1]. These methods either predict protein regions involved in the particular PPI or, even more precisely, elucidate the configuration of the complex. Binding site prediction is usually based on many properties, for instance, amino acid composition, solvation, electrostatics,

Stefan Canzar and Francisca Rojas Ringeling (eds.), *Protein-Protein Interaction Networks: Methods and Protocols*, Methods in Molecular Biology, vol. 2074, https://doi.org/10.1007/978-1-4939-9873-9_7,

or hydrophobicity [2]. Prediction of the spatial configuration of the complex can be either template-based or template-free. Despite the limited number of protein–protein complexes in the Protein Data Bank, docking templates can be found for complexes representing almost all the known protein–protein interactions, provided the components themselves have a known structure or can be homology-built [3].

Although there is a contrast between stable assemblies and more versatile complexes (such as those involved in stress response or cellular signaling; [4]), evolutionary constraints preserve most cellular functions via preservation of protein structures and interactions. The protein–protein interactions place constraints on sequence divergence which result in interface positions being more conserved than other surface positions [5–7]. Cross-species interactome mapping demonstrates that coevolution of interacting proteins is remarkably prevalent [8, 9] and studies of interface geometry conservation for interdomain interfaces indicated that the interaction geometry is largely similar when the interface sequence identity is higher than 30% [10–12]. Therefore, it is possible to use evolutionary information to predict interactions, both at the global level of the interactome and at the detailed level of protein–protein interfaces [13]. With the use of protein sequence evolutionary coupling information derived from carefully generated multiple sequence alignments, interactions can be predicted with high accuracy and at the resolution of single residues [14]. Furthermore, with the combination of 3D docking methods and evolutionary information, it is possible to acquire protein complex models with atomic resolution.

Here, we describe the combination of docking method with conservation-based selection of the best docking model using our COZOID tool [15]. We exemplify the workflow using the Nse1–Nse3 dimer, as these proteins belong to highly conserved Kite family of proteins (which constitute the essential part of the SMC complexes; [16]). The crystal structure of the human Nse1–Nse3 dimer [17] is used to make a model of the yeast *S. pombe* Nse1–Nse3 complex (*see* **Note 1**; [18, 19]). The (multiple) sequence alignments and their evolutionary information are used in several steps in the protocol: (a) homology modeling (Subheading 3.3), (b) docking constraint setting (Subheadings 3.1 and 3.2), and (c) contact residues scoring (Subheading 3.4). Implementation of the multiple sequence alignments and conservation rates into the COZOID tool arms it for evolution-based selection of docking models. With this protocol, students and researchers can prepare their own protein complex models of interest if their homologous template structures exist.

2 Databases and Tools

Here we offer some useful databases and tools for our protocol. However, except for the COZOID tool, the other tools are optional. For example, there are other docking servers, like HADDOCK, available (instead of PyDock).

1. The RCSB PDB protein structure database: http://www.rcsb.org/.
2. The UniProt database: https://www.uniprot.org/.
3. The CLUSTAL server for protein sequence alignment: https://www.ebi.ac.uk/Tools/msa/clustalo/.
4. Link for the Rate4Site tool download: https://www.tau.ac.il/~itaymay/cp/rate4site.html.
5. Link for the COZOID tool download: http://decibel.fi.muni.cz/cozoid/.
6. The I-TASSER server for protein modeling: https://zhanglab.ccmb.med.umich.edu/I-TASSER/.
7. The PyDOCK server for protein docking: https://life.bsc.es/pid/pydockweb/.

3 Methods

In this section, we describe a step-by-step protocol for visual selection of docking models, based on their similarity to the original protein complex structure. We apply and show our workflow to prepare docking models of the yeast *S. pombe* Nse1–Nse3 dimer, based on the human NSE1–NSE3 crystal structure [17, 19]. Using the COZOID tool, we first analyze contact zones of the original human NSE1–NSE3 dimer structure and select contact amino acids for docking restraints (Subheadings 3.1 and 3.2.). Then we model and dock the yeast Nse1–Nse3 proteins (Subheading 3.3). Finally, we utilize different modes of our COZOID tool to select the yeast Nse1–Nse3 docking models (Subheading 3.4) most similar to the original human NSE1–NSE3 complex. The exemplary data can be downloaded at http://decibel.fi.muni.cz/cozoid/ (*see* **Note 2**). In the following text, we only briefly describe operations required to follow this Nse1–Nse3 complex example (*see* **Note 3**).

3.1 The Reference Protein Complex Analysis

First, we investigate the interaction between the human NSE1 and NSE3 proteins (present in the crystal structure 3NW0; [17]) using the COZOID tool. The goal of this part is to understand the spatial conformation of the reference complex, including detailed information about the amino acids in the contact zone. Creating the mental image of the interaction between the reference proteins is important as it helps in the consequent steps, when exploring the vast space of possible mutual configurations of the docked proteins (Subheading 3.4). In addition, the list of contact amino acids is instrumental in setting restraints for a docking run (Subheadings 3.2 and 3.3).

1. First, download the PDB file (ID: 3NW0) of the protein complex of interest to your computer (*see* **Note 4**).
2. Open the downloaded PDB file in the COZOID tool (*see* **Note 5**) using the menu *File >> Open Structure(s)* (*see* **Note 6**).
3. Explore the spatial orientation of the proteins by rotating the 3D view using the mouse while pressing the left button (*see* **Note 7**).
4. To analyze the contact between the protein partners within the complex, switch the representation to the molecular surface (*Visualization >> Surface*), and color it by the *Contact zone* information. The coloring setting is available in one of the tabs in the right panel (*Structure >> Coloring*). In this panel, select the option *Contact zone* under the *Color* property (Fig. 1).
5. To reduce the visual occlusion in the contact zone, activate the *Exploded* view (Fig. 2a) or *Open Book* view (Fig. 2b; *see* **Note 8**).

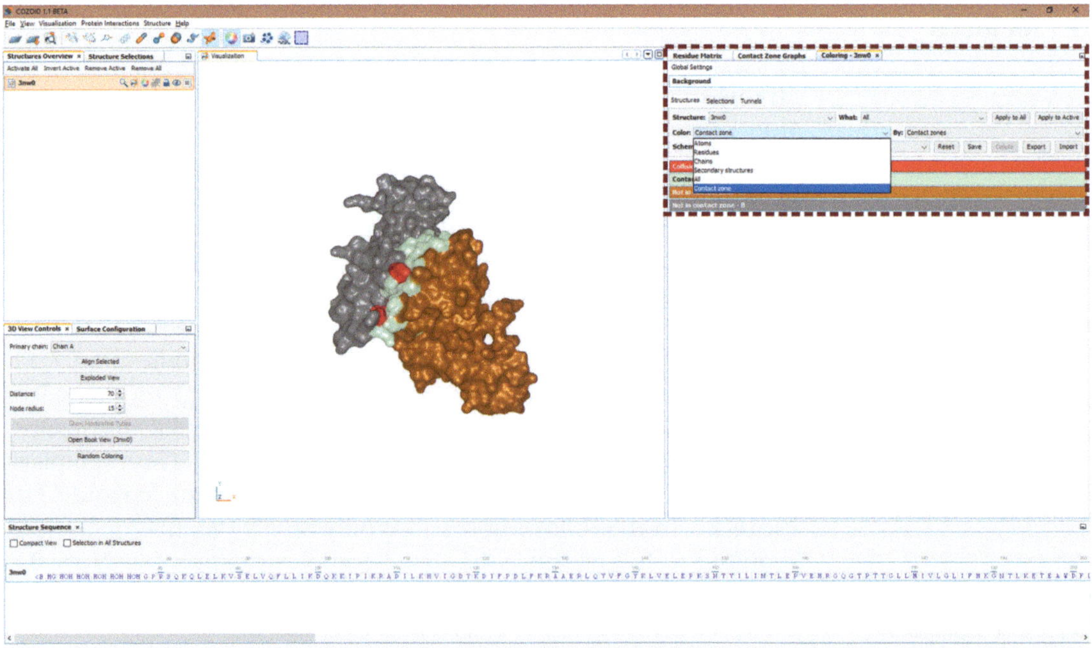

Fig. 1 Visualization of the protein complex interface. The contact zone is colored by light green. Red color depicts areas where the amino acids from both proteins are very close to each other. The parts that are not in the contact are depicted by brown (NSE1) and gray (NSE3). The red rectangle shows the coloring menu

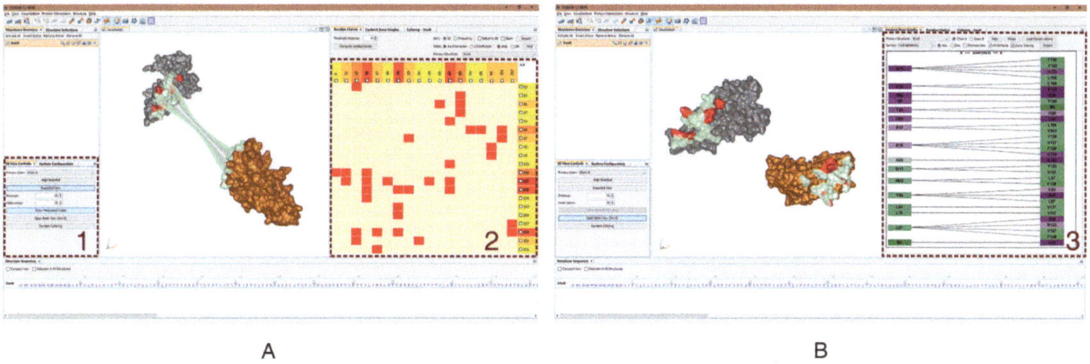

Fig. 2 Visualization of the contact zone. The Exploded View (**a**) enlarges the distance between the paired proteins. The Open Book View (**b**) rotates the contact zones toward the camera. The 3D View Controls (rectangle #1), Residue Matrix (rectangle #2), and Contact Zone Graphs (rectangle #3) panels are red framed

These views can be initiated from the *3D View Controls* panel (*Protein Interactions>> 3D View Controls*), situated on the left side (Fig. 2a). Explore the geometry of the contact zones in the 3D view. When hovering the mouse cursor over a part of the surface for a second, you will get a name–number of the amino acid occupying that area.

6. To obtain very detail information about the interacting amino acids, open the *Residue Matrix* panel (*Protein Interactions >> Residue Matrix*; Fig. 2a). In addition, use the *Contact Zone Graph* mode (*Protein Interactions >> Contact Zone Graphs*; Fig. 2b; *see* **Note 9**) to explore individual contacts and to sort the amino acids by a selected biochemical property. To assess the conservation properties of the individual contact residues, calculate conservation rates as described in Subheading 3.2.

3.2 Protein Homology Rating

Our workflow and selection process are based on the evolutionary conservation of the contact residues. Therefore, we next compute and explore the conservation rates of the contact amino acids identified in the previous step (Subheading 3.1). The conservation rates are calculated using external tools (like Clustal and Rate4Site) and further utilized to explore contact zone between protein partners using the COZOID tool. The goal of this part is to investigate the similarities across multiple species as the functional amino acids tend to be conserved throughout the evolution [7]. This information is later used to set docking restraints (Subheading 3.3) and score docking models (Subheading 3.4).

1. Download the sequence (FASTA) files of the interacting proteins (i.e., NSE1 and NSE3; *see* **Note 10**) of multiple species such that they represent distinct evolutionary steps between the

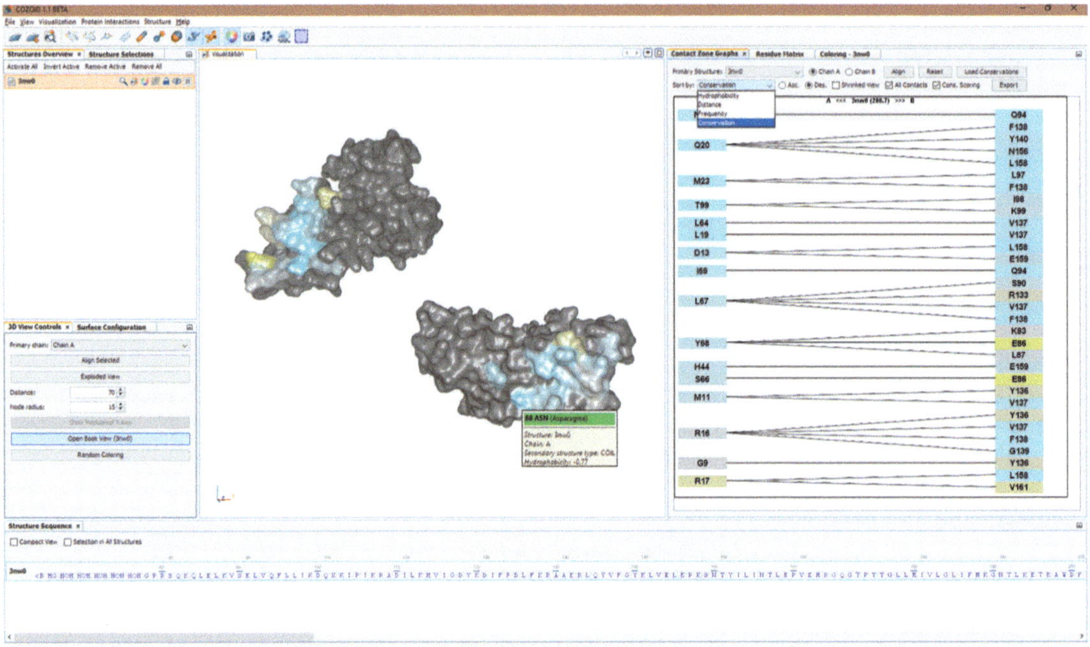

Fig. 3 Visualization of conservation rates of the contact residues. The visualization of the contact amino acids using Open Book view and Contact Zone graph sorted and colored by their conservation. When hovering with mouse cursor over the molecular surface, the information about corresponding amino acids is revealed

yeast *S. pombe* and the human (e.g., model organisms like human, mouse, chicken, fish, frog, yeast).

2. Prepare separate multiple sequence alignments (MSA) for each interacting partner sequence (*see* **Note 11**) either manually or using, e.g., the CLUSTAL server. Prepare the alignments in FASTA format.
3. Make sure that the similarity of the protein sequences is higher than 30% which is required for proper protein structure modeling and docking (*see* **Note 12**).
4. Calculate the conservation rates of the individual amino acids present in MSAs of each interacting partner using the Rate4Site tool [20]. Download and install the tool (*see* **Note 13**). Start a command line (or terminal), and process the alignments using the following command:

```
rate4site -s [alignment in FASTA] -a [name of your human sequence] -o [NSE1.res or NSE3.res]
```

5. Load the files with the conservation rate results to the COZOID tool. Press the *Load Conservations* button on the *Contact Zone Graphs* panel (*Protein Interactions* >> *Contact Zone Graphs* >> *Load Conservations*). In the popup window,

load the respective files for each protein by clicking on the buttons with three dots (use *NSE1.res* file for the chain A and *NSE3.res* file for the chain B).

6. List the contact amino acids according to their conservation rate using *Sort by* drop-down menu on the *Contact Zone Graph* panel (Fig. 3). Switch the primary chain between A (NSE1) and B (NSE3) to identify the most conserved contact amino acids in each chain (*see* **Note 14**).
7. Open the *Coloring* panel (*Structure* >> *Coloring*), and set the *Color* to *Contact zone* and *By* to *Conservation*. Activate the surface representation, and visually inspect the conservation of the partner contact zones in 3D using the *Exploded* and/or *Open Book* views (*see* **step 5** in Subheading 3.1).
8. Select the most conserved contact amino acids either directly by clicking on the surface in the 3D views or by clicking on their names on *Contact Zone Graph* panel (*see* **Note 15**). Use this information as the restraining (constraint) parameter for docking in Subheading 3.3.

3.3 Protein Structure Modeling of Homologs and Their Docking

In the previous step, we have identified the contact amino acids that are well-conserved across multiple species. In this step, we will model the structure of the yeast Nse1 and Nse3 proteins and dock these homologous proteins using most conserved contact residues for restraint. The docking run will provide tens or hundreds of complex configuration models which are then analyzed in Subheading 3.4.

1. Upload the sequence of the model organism of your choice (*see* **step 1** in Subheading 3.2) to homology modeling server, and run the modeling (e.g., using the I-TASSER server).
2. On the I-TASSER web page (https://zhanglab.ccmb.med.umich.edu/I-TASSER/; [21]), either copy and paste your protein sequence or upload it from your computer (*see* **Note 16**). Fill in your e-mail address, password (*see* **Note 17**), and the project name (*see* **Note 18**), and click on the *Run I-TASSER* button (*see* **Note 19**).
3. Wait for the I-TASSER results E-mail and open the web link provided in its body. From the web page, download the tarball file with the modeling results, unzip the file, and use the best model for docking (*see* **Note 20**).
4. Model both partner proteins in this way (repeat **steps 1–3**).
5. Upload homology models of both protein partners to the docking server (e.g., the PyDock server). On the PyDock website (https://life.bsc.es/pid/pydockweb/; [2]), fill in the project name and your e-mail address, upload your homology protein models (*see* **Note 21**), and click on the *Continue* button.

6. On the next page, enter the restraining parameters (*see* **Note 22**) either based on the original structure data (**step 8** in Subheading 3.2) or based on your experimental data (e.g., from mutagenesis; [19]).
7. Run the docking project by submitting the job, and wait for the docking results (*see* **Note 23**).
8. Download the zipped file with the docking results from the web page when finished (*see* **Note 24**). Unpack the zipped file, and analyze the docking models using the COZOID tool (Subheading 3.4).

3.4 Analysis of Docking Models Using COZOID Tool

In this section, we aim to identify the best docking models computed in Subheading 3.3. We will visually compare and validate the individual docks in order to score them in an informed way. For this purpose, we will again utilize the conservation rates computed in Subheading 3.2, as well as the comparative visualization modes provided by the COZOID tool. Using these visual and conservation-based selection modes, we will obtain the homologous protein complex configuration (yeast Nse1–Nse3 dimer) which has the best match to the original human NSE1–NSE3 dimer.

1. Upload the docking results along with the original protein complex structure to the COZOID tool (**step 2** in Subheading 3.1). First load the reference human *3nw0.pdb* (if not already loaded) and then all the docking models (Subheading 3.3; *see* **Note 25**).
2. Align all the individual configurations in 3D by selecting all loaded configurations in *Structures Overview* (*Structures Overview >> Activate All*) panel and clicking on the *Align Selected* button in *3D View Controls.*
3. Explore the (dis)similarities between the individual configurations in the 3D view (*see* **Note 26**).
4. To align sequences of the original protein complex structure (human NSE1–NSE3) and the sequences of the homologous models (yeast Nse1–Nse3), click on the *Align* button on *Contact Zone Graphs* panel (*Contact Zone Graphs >> Align*). In the popup dialog (*Sequence alignment*), load the FASTA or clustal files (e.g., *NSE1.fasta* for chain A and *NSE3.fasta* for Chain B). Once the files are loaded, select the proper sequence alignments for both involved proteins, based on the current species you are investigating (i.e., H.s./S.p. for chain A and H.s./S.p. for chain B) and click *Align* (Fig. 4).
5. Open the *Residue matrix* panel, and select those docking models fitting to (some of) the original restraining criteria, e.g., Q18. The filtering can be done by selecting "CZ Definition"

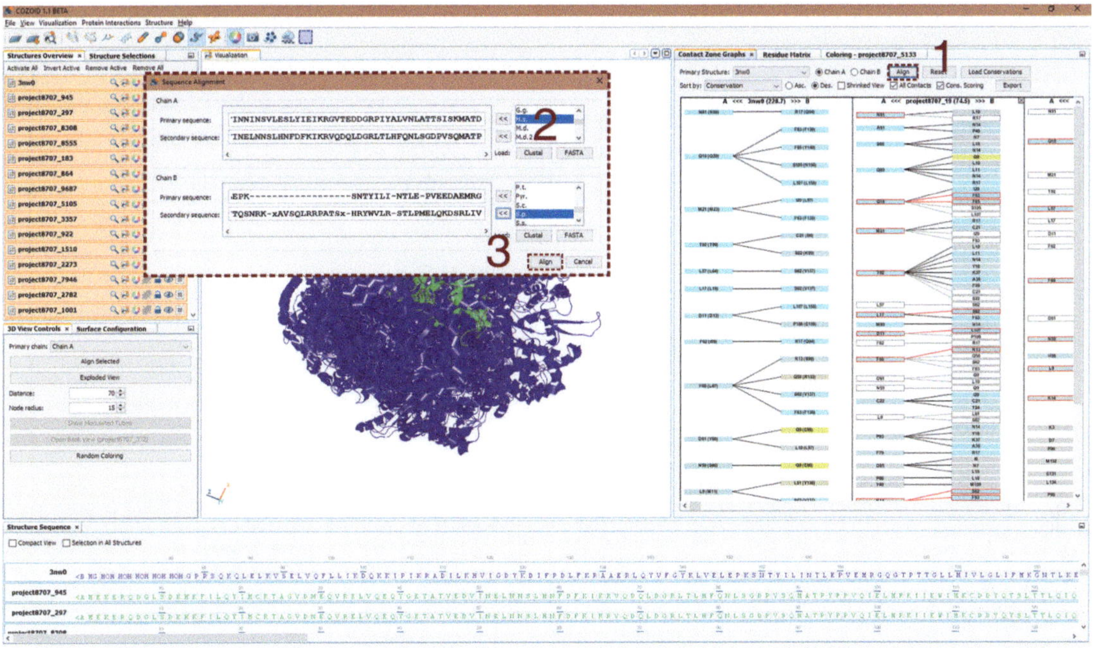

Fig. 4 Sequence alignment. The figure shows the individual steps to align the yeast models (S.p.; secondary sequence) to the primary 3nw0 human structure (H.s.; primary sequence) using sequences from FASTA files

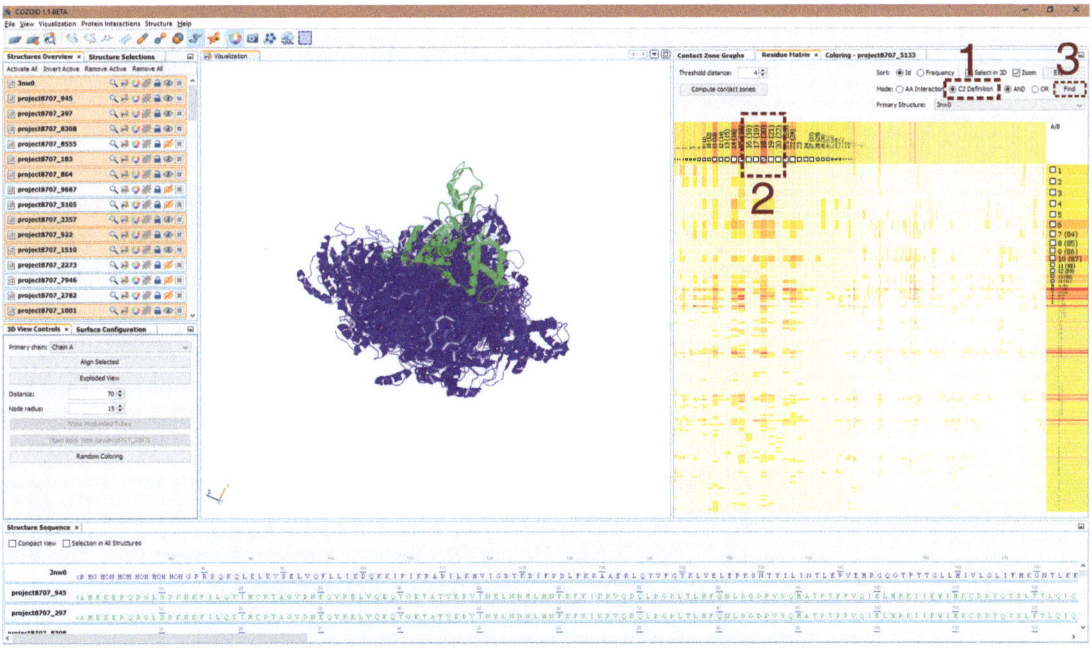

Fig. 5 Selection of docking models in the Residue Matrix panel. The docking models can be filtered based on the amino acids (e.g., those used as restraints) selected in the Residue Matrix

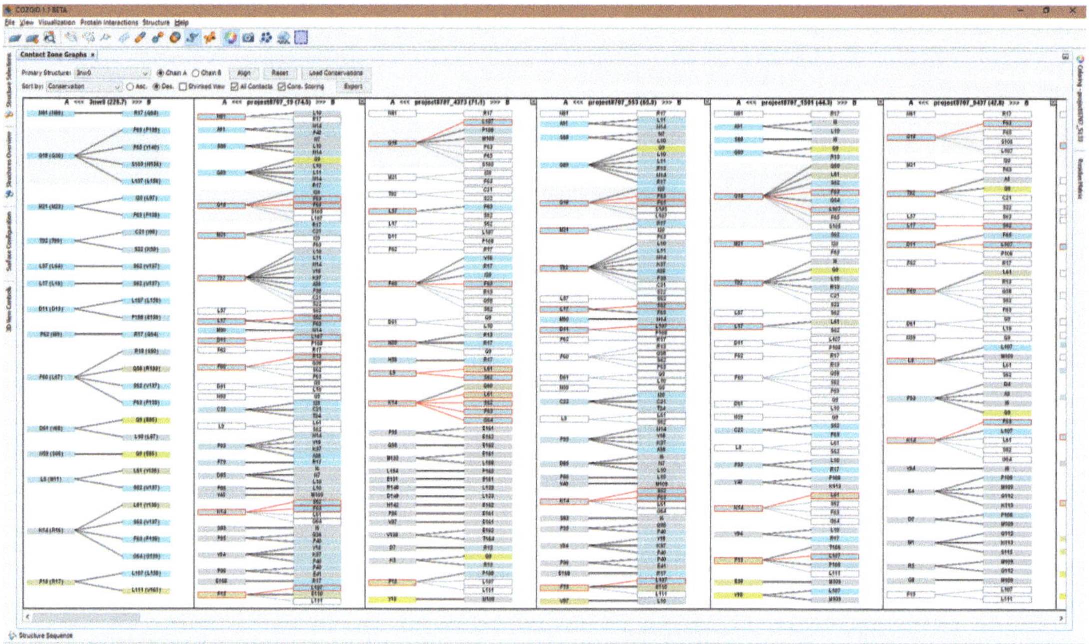

Fig. 6 Scoring of docking models according to their conserved contacts. The visualization showing the best docking models ranked according to the conservation of contact residues and individual contacts. In the Contact Zone Graphs, the amino acids present in both crystal and model are framed in red; the amino acid missing in the model is depicted in white; interaction present in both crystal and model is in red (two red frames connected by the red line)

mode and ticking the check box next to the contact residue name in the top row. Click on the *Find* button which will hide configurations that do not contain the specified contact residue (*see* **Note 27**; Fig. 5).

6. Switch to the *Contact Zone Graphs* panel. If necessary, load the conservation rates (*see* **step 5** in Subheading 3.2), and sort the amino acids in graphs by conservation under the *Sort by* drop-down menu (Fig. 6).
7. Observe the (dis)similarities between the interaction pairs in the reference crystal (on the left; *see* **Note 28**) and the individual docking models. If an amino acid is present in contact in both crystal and model, it is framed in red. The conserved contacts (between two red-framed contact residues) are highlighted by the red lines.
8. Use the best model(s) (ordered from left to right) for further analysis of the protein–protein interaction (e.g., to design mutagenesis of the contact zones; [19]), or use the *Open Book view* mode to prepare figures for your publication; (*see* **Note 29**).

4 Notes

1. *S. pombe* interactions are significantly better conserved in human as compared to *S. cerevisiae* interactions [8].
2. Download zipped Test Data file http://decibel.fi.muni.cz/cozoid/. The Test Data folder contains 3nw0.pdb original NSE1–NSE3 complex structure (directory *human*), yeast Nse1–Nse3 PyDock docking results (directory *yeast*), protein alignment, and conservation rate files.
3. For more information, please refer to the full user guide available at http://decibel.fi.muni.cz/cozoid/.
4. The PDB files can be downloaded from http://www.rcsb.org/.
5. Download the zipped COZOID 1.1 version file http://decibel.fi.muni.cz/cozoid/. Unpack COZOID files and double-click on the **cozoid.exe** or **cozoid64.exe** file (depending on your computer) to start the program (directory *Cozoid* >> *Bin*).
6. Alternatively, you can directly download the PDB file inside the COZOID tool using its PDB ID through the menu *File* >> *Download Structure*.
7. By default, the protein complex is represented as the cartoons with the chain A (NSE1) depicted in green and chain B (NSE3) in violet.
8. The Exploded View enlarges the distance between the paired proteins, while the Open Book View rotates the contact zones towards the camera. The Exploded and Open Book views cannot be active at the same time.
9. Both Residue Matrix and Contact Zone Graph show contact amino acids within a user-defined threshold distance (by default 4 Ångströms). While Residue Matrix provides a more general overview, Contact Zone Graphs allow sorting the amino acids by a selected biochemical property in each protein individually. The list of contact amino acids can be exported into csv file using the *Export* button on the Residue Matrix. In addition, the list of all contacts can be exported into csv file using the *Export* button on the Contact Zone Graph.
10. The FASTA files with sequences of the protein parts used for the protein complex crystallization (i.e., human proteins in 3NW0) can be downloaded from http://www.rcsb.org/. To retrieve orthologous sequences of representative model organisms, you can use the UniProt database https://www.uniprot.org/.
11. It is advisable to make alignment over the protein sequence present in the complex (part present in the PDB file) rather than over the entire protein sequence.

12. Make sure that the contact amino acids are also conserved more than 30%. If the overall conservation is higher than conservation of the contact amino acids, it may suggest the presence of paralogs in your alignment.
13. The Rate4site tool can be downloaded from https://www.tau.ac.il/~itaymay/cp/rate4site.html. Please, refer to this website in order to obtain more details on the tool's usage.
14. These lists of contact residues will be instrumental in selecting the best docking model in Subheading 3.4.
15. The most conserved contact amino acids in human NSE1 are N88 and Q20. The most conserved contact amino acids in human NSE3 are G139, Q94, and L97.
16. Only one modeling project can be run at the time.
17. The I-TASSER server needs registration.
18. Optionally, assign template to guide I-TASSER modeling by clicking on *Option I* button (e.g., modeling of the Nse3 protein requires 3NW0 template assignment as there are MAGE paralogs structures present in PDB).
19. It takes roughly 24 h to complete one modeling project (of an average-size protein).
20. I-TASSER provides usually five models. Open the new models in COZOID tool, and align them with the original structure (3NW0) by clicking on the *Align Selected* button in *3D View Controls* (**step 2** in Subheading 3.4), and choose the most similar model.
21. Keep the order of the protein chains in the same way as in the original structure, i.e., receptor = chain A = Nse1 and ligand = chain B = Nse3.
22. On this page, tick corresponding amino acids in each list of sequences. Use the most conserved contact amino acids of corresponding yeast sequences (the most conserved human NSE1 contact residues N88 and Q20 correspond to yeast N81 and Q18, respectively; the most conserved human NSE3 contact residues G139, Q94, and L97 correspond to G64, R17, and I20, respectively).
23. The PyDock run takes usually several hours and provides 100 docking models.
24. PyDock calculates electrostatics, desolvation energy, and van der Waals contribution (recorded in the project8707.ene file). PyDock scoring function then uses these parameters to rank the docking models (recorded in the project8707.eneRST file).
25. Load all project files (from project8707_19.pdb to project8707_9996.pdb) at once.

26. For example, compare original human NSE1–NSE3 structure (3NW0) with configurations of the yeast Nse1–Nse3 proteins in the docking models of the highest PyDock rating. Projects nr. 8707_19 and 8707_553 were top ranked in project8707.eneRST file. You can hide the unwanted models by clicking on eye icon on the *Structures Overview* panel.
27. When ticking Q18, you will hide 44 models which miss this contact residue. To completely eliminate them, click on *Invert Active* on the *Structures Overview* panel; make sure that the original *3nw0* structure is not selected (i.e., click on the *3nw0* window while pressing CTRL key to deselect it if necessary), and click on the *Remove Active* on the same panel. To filter out more models (not fitting the original constraint criteria), you can repeat this step with another conserved residue (e.g., R17; you will remove another ten models).
28. Make sure that you have selected the original crystal structure (3NW0) as the primary structure.
29. The COZOID tool ranked the project nr. 19 at the first position (same as the PyDock ranking function).

Acknowledgments

Internal Masaryk University grant (MU/0822/2015) and the Czech MEYS - Projects CEITEC 2020 (LQ1601) are acknowledged for their financial support.

References

1. Huang SY (2014) Search strategies and evaluation in protein-protein docking: principles, advances and challenges. Drug Discov Today 19:1081–1096
2. Jiménez-García B, Pons C, Fernández-Recio J (2013) pyDockWEB: a web server for rigid-body protein-protein docking using electrostatics and desolvation scoring. Bioinformatics 29:1698–1699
3. Kundrotas PJ, Zhu Z, Janin J, Vakser IA (2012) Templates are available to model nearly all complexes of structurally characterized proteins. Proc Natl Acad Sci U S A 109:9438–9441
4. Das J et al. (2013) Cross-species protein interactome mapping reveals species-specific wiring of stress response pathways. Sci Signal 6:ra38
5. Valdar WS, Thornton JM (2001) Protein-protein interfaces: analysis of amino acid conservation in homodimers. Proteins 42:108–124
6. Mintseris J, Weng Z (2005) Structure, function, and evolution of transient and obligate protein-protein interactions. Proc Natl Acad Sci U S A 102:10930–10935
7. Caffrey DR, Somaroo S, Hughes JD, Mintseris J, Huang ES (2004) Are protein-protein interfaces more conserved in sequence than the rest of the protein surface? Protein Sci 13:190–202
8. Vo TV et al. (2016) A proteome-wide fission yeast interactome reveals network evolution principles from yeasts to human. Cell 164:310–323
9. Gandhi TK et al. (2006) Analysis of the human protein interactome and comparison with yeast, worm and fly interaction datasets. Nat Genet 38:285–293
10. Levy ED, Boeri Erba E, Robinson CV, Teichmann SA (2008) Assembly reflects evolution of protein complexes. Nature 453:1262–1265

11. Dey S, Ritchie DW, Levy ED (2018) PDB-wide identification of biological assemblies from conserved quaternary structure geometry. Nat Methods 15:67–72
12. Yu H et al. (2004) Annotation transfer between genomes: protein-protein interologs and protein-DNA regulogs. Genome Res 14:1107–1118
13. Andreani J, Guerois R (2014) Evolution of protein interactions: from interactomes to interfaces. Arch Biochem Biophys 554:65–75
14. Hopf TA et al. (2014) Sequence co-evolution gives 3D contacts and structures of protein complexes. Elife 3
15. Furmanová K et al. (2018) COZOID: contact zone identifier for visual analysis of protein-protein interactions. BMC Bioinformatics 19:125
16. Palecek JJ, Gruber S (2015) Kite proteins: a superfamily of SMC/Kleisin partners conserved across Bacteria, Archaea, and Eukaryotes. Structure 23:2183–2190
17. Doyle JM, Gao J, Wang J, Yang M, Potts PR (2010) MAGE-RING protein complexes comprise a family of E3 ubiquitin ligases. Mol Cell 39:963–974
18. Zabrady K et al. (2016) Chromatin association of the SMC5/6 complex is dependent on binding of its NSE3 subunit to DNA. Nucleic Acids Res 44:1064–1079
19. Hudson JJ et al. (2011) Interactions between the Nse3 and Nse4 components of the SMC5-6 complex identify evolutionarily conserved interactions between MAGE and EID families. PLoS One 6:e17270
20. Mayrose D, Graur N, Ben-Tal N, Pupko T (2004) Comparison of site-specific rate-inference methods for protein sequences: empirical Bayesian methods are superior. Mol Biol Evol 21:1781–1791
21. Zhang Y (2008) I-TASSER server for protein 3D structure prediction. BMC Bioinformatics 9(40)

Chapter 8

Path-LZerD: Predicting Assembly Order of Multimeric Protein Complexes

Genki Terashi, Charles Christoffer, and Daisuke Kihara

Abstract

Many important functions in a cell are carried out by protein complexes with more than two subunits. Similar to the folding of a single protein, multimeric protein complexes in general follow an energetically favored assembly path. Knowing the assembly path not only provides critical information about the molecular mechanism of the assembly but also serves as a foundation for artificial design of protein complexes, as well as development of drugs that interfere with complex formation. There are experimental approaches for determining the assembly path of a complex; however, such methods are resource intensive. We have recently developed a computational method, Path-LZerD, which predicts the assembly path of a complex by simulating the docking process of the complex. Here, we explain how to use the Path-LZerD software with examples.

Key words Protein docking, Assembly order, Multimeric protein complex, Protein–protein interaction, PPI, PPI network, Protein structure modeling, Structure prediction

1 Introduction

Many biological processes in a cell involve multimeric protein complexes. Having multiple subunits allows a protein complex to have a large structure and also to combine multiple different molecular functions in a coordinated fashion into an elaborate functional pipeline to be carried out by the complex. To understand the molecular mechanism of function of protein complexes, experimental methods, including X-ray crystallography and electron microscopy, as well as computational protein docking methods have been developed and extensively used. On the other hand, fewer studies have been conducted to investigate how a multimeric complex is formed, i.e., the assembly pathways of complexes. For single-chain proteins, many biophysical and computational studies have been conducted to elucidate the folding process of a protein

Stefan Canzar and Francisca Rojas Ringeling (eds.), *Protein-Protein Interaction Networks: Methods and Protocols*, Methods in Molecular Biology, vol. 2074, https://doi.org/10.1007/978-1-4939-9873-9_8,

chain into its native structures [1], but there have been far fewer studies of complex assembly pathways. Information about the assembly pathway is very helpful for in vitro reconstitution of a subcomplex of an entire multimeric protein complex, which is commonly performed in cases where solving the structure of the entire complex by experiment is difficult. Knowledge of assembly order is also useful for artificial design of protein complexes [2] and for designing drugs that interfere with a critical protein–protein interaction in a complex [3].

There are several experimental methods for elucidating the assembly steps of a protein complex. Reconstruction of stable intermediate subcomplexes can be detected by co-immunoprecipitation [4] and gel electrophoresis [5]. Recently, several types of mass spectrometry-based techniques have also been applied [6–11].

This article introduces software, Path-LZerD, for predicting assembly order of multimeric protein complexes [12]. There is an earlier work that predicts assembly order of a protein complex from an experimentally determined protein complex structure by examining the buried surface area of protein–protein interface of subunits (the subunits with a relatively large interface are predicted to dock earlier in the assembly process) [13, 14]. The difference between the earlier work and Path-LZerD is that Path-LZerD takes subunit structures as input, assembles them into protein complex models using a multimeric protein docking protocol, and predicts the assembly pathway of the complex from actually observed pathways in the docking process. Thus, Path-LZerD can be applied to complexes whose overall structure is not known. The algorithm of Path-LZerD and the prediction results were thoroughly discussed in our previous work [12]. Here, we focus on showing how to use the Path-LZerD software.

2 Materials

Path-LZerD is freely available to the academic community at our lab website, http://www.kiharalab.org/proteindocking/pathlzerd.php. From this link, the archived file (path_lzerd-master.zip) can be downloaded. After downloading the archived file, users can decompress the file by the command *unzip path_lzerd-master.zip*. Decompressing *path_lzerd-master.zip* creates a directory *path_lzerd-master*. This directory contains

1. LICENSE.txt: a copy of the GNU General Public License, Version 3, under which the Path-LZerD software is licensed.
2. path_lzerd.py: the main program of the Path-LZerD protocol. This program performs path order prediction.

3. score_pair.py: a component of path_lzerd.py. This script calls lzerd_pdbgen.py and SCORES.py.
4. lzerd_pdbgen.py: a component of score_pair.py. This script generates decoy docking models from the output file of the LZerD pairwise protein docking (A-B.out, A-C.out, etc.).
5. SCORES.py: a component of score_pair.py. This script evaluates pairwise docking decoys by external structure scoring programs.
6. shared.py: a component of path_lzerd.py.
7. PATHS.ini: the parameter file that defines environments (PATH) of external programs.
8. README.md: Readme file. It contains instructions for installing external programs and testing the example (test directory).
9. test: the directory that contains the test files.

Path-LZerD requires a pairwise protein docking program LZerD [15] and a multiple protein docking program Multi-LZerD [16]. These two programs are for modeling pairwise and multiple docking structures, respectively. They are available in the LZerD docking suite at the Kihara Lab website, http://www.kiharalab.org/proteindocking [17]. The LZerD pairwise docking program is available as a compressed file (lzerddistribution.tar.gz) from the LZerD section of the webpage (http://kiharalab.org/proteindocking/lzerd.php). Multi-LZerD is available as multilzerddistribution.tar.gz from the Multi-LZerD section (http://kiharalab.org/proteindocking/multilzerd.php). A version of the Multi-LZerD distribution specifically for use with Path-LZerD is available from the Path-LZerD website (http://kiharalab.org/proteindocking/pathlzerd.php).

You can decompress each archived file by the command *tar -zxf [archived file]*. All packages are intended to run on Linux machines.

Path-LZerD also requires the following external programs and python modules:

1. DFIRE (http://sparks-lab.org/index.php/Main/Downloads) [18].
2. GOAP (http://cssb.biology.gatech.edu/GOAP/index.html) [19].
3. ITScorePro (http://zoulab.dalton.missouri.edu/resources.html) [20].
4. SOAP-PP (https://salilab.org/SOAP/) [21].
5. OPUS-PSP (http://ma-lab.rice.edu/soft.php) [22].
6. ProFit (http://www.bioinf.org.uk/programs/profit/) [23].
7. Python modules: numpy, pandas, modeller, and mdt.

The first five programs are scoring functions for protein models. These are used for providing rankings to pairwise docking models generated for each pair of subunit structures. ProFit is a program for comparing protein structures and to compute root mean square deviation (RMSD) of the two structures. This program is used to identify the best (i.e., the lowest root-mean square deviation (RMSD) to the native structure) model among the models generated by the program Multi-LZerD. The identification of the lowest RMSD model is needed to perform one of the path prediction methods implemented in Path-LZerD, named the lowest RMSD method. See Subheading 3 below.

To install the required python modules, users can use the following commands with Anaconda:

```
> conda config --add channels salilab
> conda install modeller
> conda install -c salilab mdt
```

Once all external programs and python modules are installed, PATH.ini should be modified according to the environment of user's computer. This file specifies the location of the programs.

Example of PATH.ini:

```
[required]
lzerd_path: /home/user1/lzerddistribution/
multilzerd_path: /home/user1/multilzerd/
[optional]
goap_path: /home/user1/bin/goap-alone_long
itscorepro_cmd: /home/user1/bin/ITScorePro
soappp_cmd: /home/user1/bin/soap_pp/soap_pp.py
modeller_cmd: python
opuspsp_path: /home/user1/bin/OPUS_PSP/
```

The lines below [optional] specify paths of scoring functions to use. For example, if the user wants to use only GOAP score in Path-LZerD, only the *goap_path* needs to be specified.

To run Path-LZerD, users need protein structure files in the PDB format for subunits of a target complex. The input protein structure files should have a common file name and a suffix, ".pdb." Also, a prefix should be associated with the files to indicate the chain ID of the units. For example, PDB ID 1A0R has three subunits, chain B, G, and P. Three input files for 1A0R can be named as follows: B-1A0R.pdb, G-1A0R.pdb, and P-1A0R.pdb. In this case the common file name is "1A0R," and the prefixes that indicate the protein chain IDs are "B," "G," and "P". In addition, it is required that the chain ID is provided in each ATOM field. We suggest that the same chain ID be used as the prefix.

3 Methods

3.1 Overview of Path-LZerD

Path-LZerD uses a multiple-protein docking algorithm (Multi-LZerD) which assembles a protein complex structure from individual subunit structures. By analyzing the assembly pathways of predicted complex models, Path-LZerD predicts the assembly order of the protein complex.

The flowchart of Path-LZerD has four main steps (Fig. 1). In the first step, all pairwise combinations of input subunit structures are docked by a pairwise protein docking method, LZerD. LZerD uses a rotation invariant mathematical surface shape representation, the 3D.

Zernike Descriptors (3DZD) [24] for evaluating docking conformation. For each pair, typically 2000 to 6000 models (called decoys) are generated. In the second step, models of the entire complex are built by Multi-LZerD by combining the pairwise docking predictions generated in the first step. Multi-LZerD uses a genetic algorithm (GA) to explore the combinatorial space. GA generates a population of ~200 decoys of the complex at each

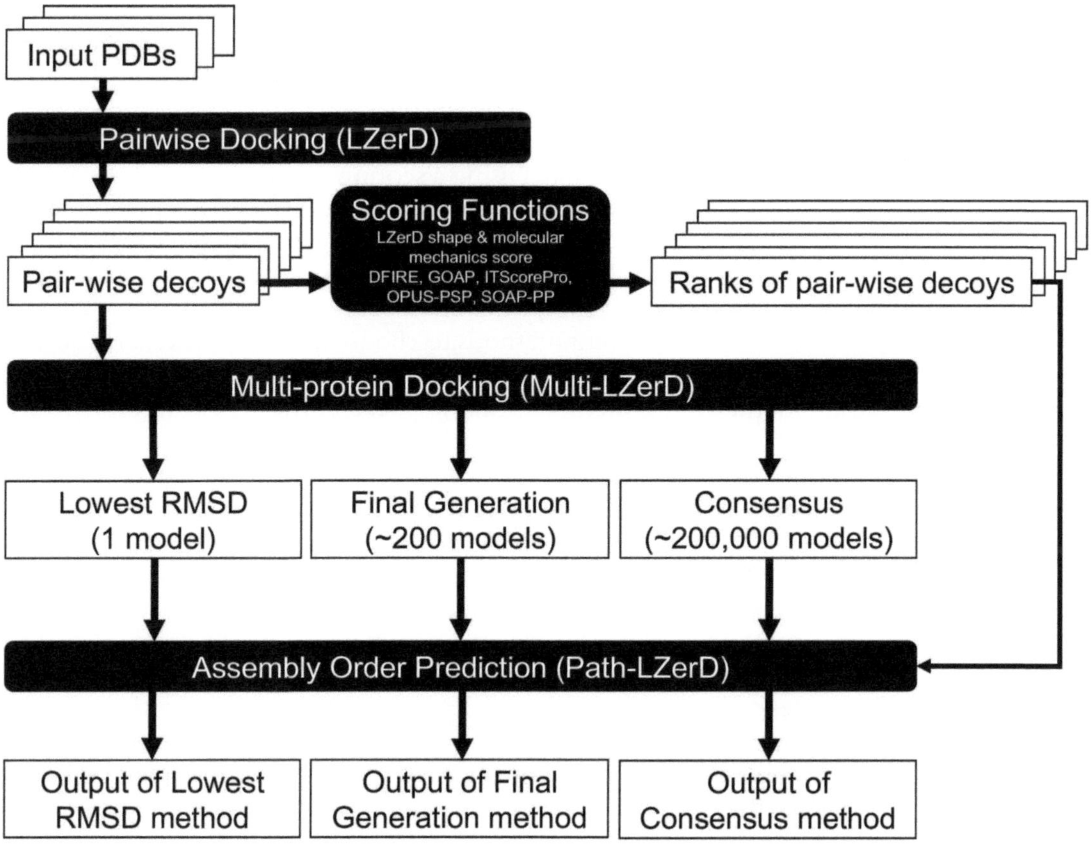

Fig. 1 Flowchart of Path-LZerD

"generation" and improves the decoy structures by assembling different combinations of pairwise decoys. This process is repeated for 3000 generations, and finally, about 200 models of the target protein complex will be generated in the final generation. In the third step, the pairwise decoys for each subunit pair were ranked by a scoring function which evaluates the binding energy of decoys. Path-LZerD uses seven scoring functions. Two scoring functions, a shape score and a molecular mechanics score, are originally equipped in LZerD and Multi-LZerD. The other five scoring functions are DFIRE [18], ITScorePro [20], GOAP [19], OPUS-PSP [22], and SOAP-PP [21] (for more information on scoring functions, *see* **Note 1**).

In the final step, the assembly order of the target complex is predicted by comparing the score ranks of the pairwise decoys that were assembled by Multi-LZerD to obtain the whole complex model. For example, if a model of an A-B-C complex is made up of the A-B decoy with rank 1 and the B-C decoy with rank 100 by a scoring function, then the complex is predicted to assemble A-B first, followed by AB bound with C (denoted as AB> ABC). Thus, a pair with a docking conformation that is energetically favorable relative to other docking conformations is predicted to dock earlier in the assembly process. Since a different scoring function may provide different ranking to decoys for each pair, the final assembly path prediction for a given protein complex model may be different when using a different scoring function.

As discussed above, an assembly path will be predicted for a protein complex model using a scoring function that ranks component pairwise decoys and moreover, Multi-LZerD generates many complex models across thousands of generations. Therefore, to predict the assembly pathway of a protein complex, we have several choices of which complex models to analyze. In Path-LZerD, users have three different methods to choose from (*see* **Note 2**). The first method is called the lowest RMSD (root mean square deviation) method. This method is only applicable when the correct structure of the complex is known. In the lowest RMSD method, among the ~200 complex models generated in the final generation of GA, the one with the lowest RMSD to the correct (native) structure is chosen, and the assembly path is predicted based on the lowest RMSD model. Thus, to identify the lowest RMSD model, users need to provide the native structure of the target complex. The next method is called the final generation method. This method uses all ~200 models from the final GA generation of Multi-LZerD regardless of the accuracy (i.e., RMSD to the native) of the models. Assembly order prediction is performed on each complex model, and then the most frequently occurring assembly order is selected as the predicted assembly order. The last choice, the consensus method, uses the complex models from the GA generation 1000 through the final generation. Total ~20,000 complex models are used to find the majority of the assembly order predictions. The latter two methods make the path prediction by majority vote from complex models generated; thus, they do not need the prior knowledge of the native structure of the target complex. For more details, please refer to the original paper of Path-LZerD [12].

3.2 Generating Multimeric Complex with Multi-LZerD

As discussed in the overview, the first step of Path-LZerD is to run Multi-LZerD to generate protein complex structure models.

Input files: protein structure files of subunits of the target protein complex in the PDB format. For example, for the case of 1a0r, they are B-1a0r.pdb, G-1a0r.pdb, and P-1a0r.pdb.

Output files: Output files for use by Path-LZerD: "lowest.ga.out," "consensus.ga.out," and "final.ga.out." "lowest.ga.out" Contains the decoy from the final generation with the lowest RMSD to the native structure. Similarly, "consensus.ga.out" and "final.ga.out" contain the decoys from the last 1000 generations and decoys from the final generation, respectively. Multi-LZerD also produces many other output files such as lists of pairwise docking decoys, e.g., "B-G.out" (this is the list of pairwise decoys for chain B and G); multiple docking decoys, e.g., "decoy-00001.pdb"; and lists of multiple docking decoys from intermediate Multi-LZerD generations, e.g., "01909-1a0r.ga.out" (this is a file of ~200 multiple docking from the 1909th generation).

These files will be generated in the directory where Multi-LZerD is run, e.g., the same directory as the input files. In Subheading 3, we use 1a0r, a three-chain complex with chains B, G, and P, to illustrate the path prediction procedure. The input files and the output files are available from the Path-LZerD website as Case 1.

The Multi-LZerD executables can be downloaded from the Path-LZerD webpage. To run Multi-LZerD, simply run the following script included in the Multi-LZerD package:

```
> $path_to_run_multilzerd_path.sh 1a0r B,G,P
```

For the example of 1a0r, if a native structure is available, it can be included by instead running.

```
> $path_to_run_multilzerd_path.sh 1a0r B,G,P $path_to_native
```

The Multi-LZerD script first computes LZerD docking decoys for each pair of subunits, e.g., BG, BP, and GP. Then, after clustering to remove redundant decoys, the decoys are combined by Multi-LZerD into complex decoys. Multi-LZerD saves each generation for later processing. After the required number of generations has been generated, decoy files are generated for the final generation. If a native structure was supplied, RMSDs between the decoys and the native structure are computed, and the lowest-RMSD decoy is saved to lowest.ga.out. The last 1000 generations are combined into consensus.ga.out, and the final generation is saved to final.ga.out.

Multi-LZerD is normally used with the default parameters. However, the number of generations computed (default 3000) can be set via the ML_NUM_GENERATIONS environment variable. The number of generations combined in the consensus can be set via the environment variable ML_CONSENSUS_SIZE.

3.3 Identifying the Lowest RMSD Model for the Lowest-RMSD Path Prediction Method

In Path-LZerD, as mentioned above, the assembly path can be predicted from the best generated model (i.e., the model with the lowest RMSD to native) if the native complex structure is known. To perform the lowest-RMSD method, simply include a PDB file containing the native structure as a third argument to *run_multilzerd_path.sh*; this will generate the file *lowest.ga.out*, which will then be used when Path-LZerD is run. If a native structure is not supplied to *run_multilzerd_path.sh*, the files *final.ga.out* and *consensus.ga.out* will still be generated. The lowest.ga.out file can be passed to Path-LZerD in the same way as the others, as detailed in Subheading 3.5.

run_multilzerd_path.sh Uses ProFit [23] to calculate the RMSD of each complex decoy to the native structure. ProFit will perform a sequence alignment of the decoys with the supplied native structure, so the RMSD calculation is resilient against differences, e.g., in residue numbering.

3.4 Using Multi-LZerD with Multiple Cores

For most inputs, Multi-LZerD run on a single processor core will require a large amount of time to complete. In the examples given in Table 4 of the Path-LZerD paper [12], the Multi-LZerD portion of the computation accounts for most of total computation time. For example, the three-chain 1a0r took 470.0 CPU-hours, the four-chain 3fh6 took 1214.5 CPU-hours, and the five-chain 1w88 took 1431.3 CPU-hours. To mitigate this problem, Multi-LZerD supports multithreaded computation. The executable will automatically attempt to use multiple threads. If a specific number of threads is desired, it can be set via the environment variable OMP_NUM_THREADS.

3.5 Path Prediction

The next step is to run Path-LZerD to make assembly path predictions using the output files (i.e., generated complex structure models) from Multi-LZerD. The program files of Path-LZerD are in the Path-LZerD package (*path_lzerd-master.zip*).

Before running the Path-LZerD script, users have to put the following files to the same directory. First, we explain how to run Path-LZerD on the sample data, which are provided in the *test* directory.

Input PDB files: Path-LZerD requires the same input PDB files that are used in the Multi-LZerD step. The input PDB files have to follow specific naming rules. Every PDB file name must start with a prefix indicating the chain ID. The prefix should be followed by a common name (such as PDB-ID “1A0R”) and a suffix “.pdb.” In the *test* directory, there are three input PDB files (*B-1A0R.pdb*, *G-1A0R.pdb*, and *P-1A0R.pdb*).

The output files of pairwise docking from LZerD runs: The prefixes of “*.out*” files represent the pair of chain IDs. For example, *B-G.out* is an output file of pairwise docking of chain B and G. These files are automatically generated by the Multi-LZerD program. In the *test* directory, there are three pairwise docking prediction files (*B-G.out*, *B-P.out*, and *G-P.out*).

The output file that contains multiple docking decoys from Multi-LZerD: a file with a “*.ga.out*” suffix is an output file of Multi-LZerD. In the *test* directory, there is a “*finalgene.ga.out*” that contains the data of 200 complex models from the final GA generation.

Once all required files are put in the same directory, the path of main program (*path_lzerd.py*), name id (“*1A0R*”), and “.ga.out” file should be specified in *run_test.sh*. In the test directory, *run_test.sh* was described as

```
#!/bin/bash

PATH_LZERD_HOME=".."

path_lzerd="$PATH_LZERD_HOME/path_lzerd.py"

cmd="$path_lzerd -p 1A0R -d . -g finalgene.ga.out -o finalgene_chain.csv"
echo $cmd
$cmd
```

After preparing all files, the Path-LZerD protocol will be executed by

```
bash ./run_test.sh
```

in the test directory.

In this example, *run_test.sh* generates six rank files by scoring functions for each protein pair. The file names are *B-G_dfire.txt*, *B-G goap.txt*, *B-G itscore.txt*, *B-G molmech.txt*, *B-G opuspsp.txt*, *and B-G soappp.txt for pairwise docking decoys of B and G chains. The predicted assembly order is recorded in finalgene_chain.csv*:

```
score,pathway,pdbid
dfirerank,BG>BGP,1a0r
soappprank,BG>BGP,1a0r
sumrank,BG>BGP,1a0r
goaprank,BG>BGP,1a0r
```

```
molmechrank,BG>BGP,1a0r
shapescorerank,BG>BGP,1a0r
opuspsprank,BG>BGP,1a0r
itscorerank,BG>BGP,1a0r
```

These files show path predictions using the different scoring functions using the final generation method. The first column is showing the type of scoring function, except "sumrank." The "sumrank" is the prediction made by using the sum of score rank by DFIRE, SOAPPP, GOAP, molecular mechanics (Molmec), LZerD shape score (shapescore), OPUSPSP, and ITScore. The second column is showing the predicted assembly order. For example, "BG>BGP" means chain B and G form a subcomplex first; then B-G complex bounds with chain P.

Users also can execute *path_lzerd.py* with modified parameters. Table 1 lists all available options in *path_lzerd.py*. In the next case studies section, we show three examples of assembly pathway predictions.

Table 1
Parameters in *path_lzerd.py*

Option	Reference
-h, --help	Show help message and exit
-p [PDBID], --pdbid [PDBID]	Name id of input PDB files. For example, when the input PDB files are *A-sample.pdb*, *B-sample.pdb*, and *C-sample.pdb*, the PDBID is set to "sample"
-d [TOP_DIR], --top_dir [TOP_DIR]	The directory where the input PDB files (*.pdb*), the pairwise docking prediction files (*.out*), and the output file of Multi-LZerD (*.ga.out*) are located
-g [.ga.out file] --ga_file [.ga.out file]	The output file of Multi-LZerD (*.ga.out*) Users can specify the output file (.ga.out) that is corresponding to the prediction method, from the final generation method, the lowest RMSD method, or the consensus method
-c CHAINS[;CHAINS ...], --same_chains CHAINS[;CHAINS ...]	CHAINS represent the same protein For example, "-c AB;CD" means chain A and B are same protein and chain C and D are same
-s {all, dfire, soappp, sum, molmech, opuspsp, shapescore, itscore, goap} [{all, dfire, soappp, sum, molmech, opuspsp, shapescore, itscore, goap} ...]] --scores {...}	Scoring functions for ranking of pairwise decoys. Default value is set to "all," which uses all scoring functions
-o [OUTPUT], --output [OUTPUT]	Output file name Default file name is *[PDBID]_path_lzerd.csv*

4 Case Studies

In this section, we show three examples of assembly order prediction by Path-LZerD. We describe details of commands that were used. In the command line, *$path_lzerd* denotes the path of the main program *path_lzerd.py*. For example, if the archived file *path_lzerd-master.zip* was decompressed in /home/user1/, *$path_lzerd* is set to */home/user1/path_lzerd-master/path_lzerd.py*.

4.1 Case 1: Transducin $\beta\gamma$ Dimer with Phosducin (PDB ID: 1a0r, Three Chains)

The first example is 1a0r (https://www.rcsb.org/structure/1A0R), which is the complex structure of transducin β subunit (chain B), γ subunit (chain G), and phosducin (chain P) [25]. Phosducin binds to the transducing βγ dimer in a regulatory fashion. The association between the transducin β and γ subunits is very strong [25, 26]. The transducin β and γ subunit forms a dimer first, then phosducin binds to the dimer. Thus, the assembly order is BG>BGP (Fig. 2). The input files and the output files of this example are provided as Case 1 on the Path-LZerD page.

For prediction of this example, the following commands were executed:

```
<Multi-LZerD part>
> $path_to_run_multilzerd_path.sh 1a0r B,G,P 1a0r.pdb
<Path-LZerD part>
```

Path-LZerD predicted assembly order from three output files (.ga.out) of MultiLZerD. In this example, we executed the following commands:

```
> $path_lzerd -p $id -d . -g finalgene.ga.out -o finalgene_chain.csv
> $path_lzerd -p $id -d . -g consensus.ga.out -o consensus_chain.csv
> $path_lzerd -p $id -d . -g lowest.ga.out -o lowest_chain.csv
```

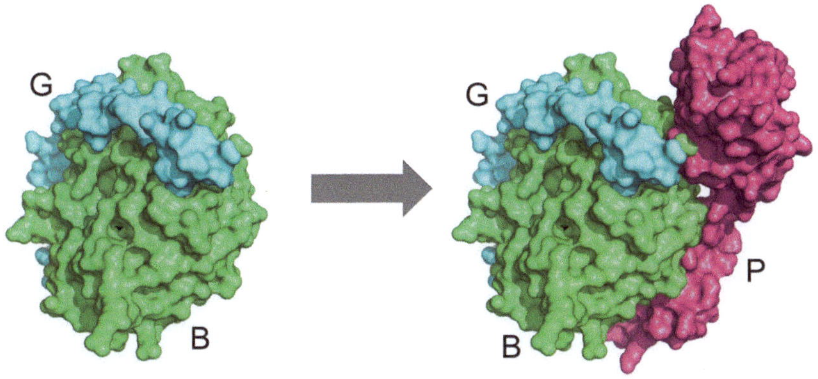

Fig. 2 Subunit marked B (green) and G (cyan) are β and γ subunits of G protein transducin, respectively. Subunit marked P is phosducin. The assembly pathway is BG $>$ BGP

The three commands above are for making predictions using the final generation method, the consensus method, and the lowest RMSD method. As mentioned in MultiLZerD part, *finalgene.ga.out*, *consensus.ga.out*, and *lowest.ga.out* correspond to the final generation method, consensus method, and lowest RMSD method, respectively. *finalgene.ga.out* Contains the 200 docking models of the final GA generation in Multi-LZerD process. *Consensus.ga.out* Contains a total of ~20,000 docking models. *lowest.ga.out* Contains the one docking model with the lowest RMSD to the correct (native) structure. From these three runs, three output files (*finalgene_chain.csv*, *consensus_chain.csv*, and *lowest_chain.csv*) are generated. In this example, these three files show the same results:

```
score,pathway,pdbid
dfirerank,BG>BGP,1a0r
soappprank,BG>BGP,1a0r
sumrank,BG>BGP,1a0r
goaprank,BG>BGP,1a0r
molmechrank,BG>BGP,1a0r
shapescorerank,BG>BGP,1a0r
opuspsprank,BG>BGP,1a0r
itscorerank,BG>BGP,1a0r
```

In the output files, all scoring functions predicted the correct assembly order of "BG>BGP."

4.2 Case 2: IVY Complex with Its Target HEWL (PDB ID: 1gpq, Four Chains)

1gpq (https://www.rcsb.org/structure/1GPQ) is the complex structure of inhibitor of vertebrate lysozyme (Ivy) from *E. coli* bound to hen egg white lysozyme C [27]. Ivy forms a homodimer (chains A and B, denoted as A and A′ in Fig. 3) and binds to lysozyme C (chains C and D, denoted as C and C′). Since Ivy is functional as homodimer [28], it needs to be formed first. Lysozyme C can function as a monomer or dimer [29, 30]. Thus, the known assembly order is AA′>AA′C>AA′CC′.

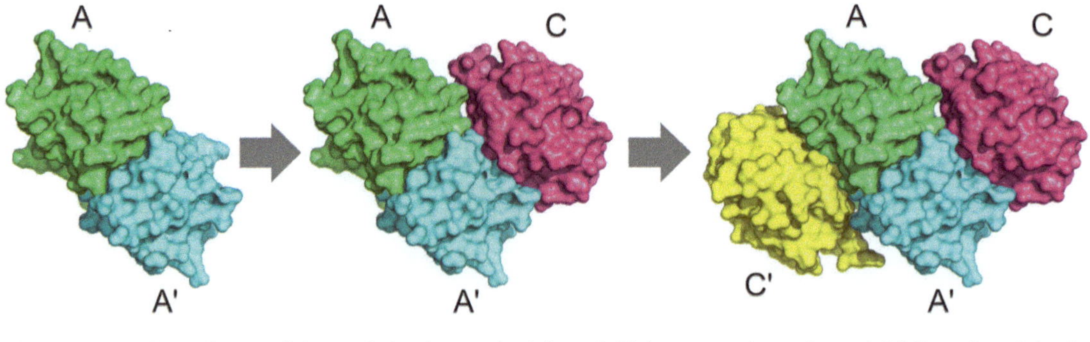

Fig. 3 Assembly pathway of 1gpq. Subunits marked A and A′ (green and cyan) are inhibitor of vertebrate lysozyme (IVY), and subunits marked C and C′ (magenta and yellow) are lysozyme C. The assembly pathway is AA′>AA′C>AA′CC′

The following commands were executed:

```
<Multi-LZerD part>
> $path_to_run_multilzerd_path.sh 1gpq A,B,C,D 1gpq.pdb
<Path-LZerD part>
```

For the Path-LZerD run of this example, we specified that chain A and B are the same and chain C and D are same by using "-c AB;CD" option:

```
> $path_lzerd -p $id -d . -g finalgene.ga.out -o finalgene_chain.csv -c AB;CD
> $path_lzerd -p $id -d . -g consensus.ga.out -o consensus_chain.csv -c AB;CD
> $path_lzerd -p $id -d . -g lowest.ga.out -o lowest_chain.csv -c AB;CD
```

The output files (*finalgene_chain.csv*, *consensus_chain.csv* and *lowest_chain.csv*) show the same results which are correct:

```
score,pathway,pdbid
dfirerank,AA>AAC>AACC,1gpq
soappprank,AA>AAC>AACC,1gpq
sumrank,AA>AAC>AACC,1gpq
goaprank,AA>AAC>AACC,1gpq
molmechrank,AA>AAC>AACC,1gpq
shapescorerank,AA>AAC>AACC,1gpq
opuspsprank,AA>AAC>AACC,1gpq
itscorerank,AA>AAC>AACC,1gpq
```

"AA" is corresponding to chains A and B. "CC" is corresponding chains C and D. All predictions show assembly order of "AA>AAC>AACC," which indicates that chain A and B form a homodimer first; then each of chains C and D binds to the homodimer.

4.3 Case 3: Pyruvate Dehydrogenase E1 with the Peripheral Subunit-Binding Domain of E2 (PDB ID: 1w88, Five Chains)

1w88 (https://www.rcsb.org/structure/1W88) Consists of a tetramer of pyruvate dehydrogenase E1 and the peripheral subunit-binding domain of dihydrolipoyl transacetylase (E2) [31]. Because the structure of E1 can be solved in the absence of E2 [32], the tetramer of E1 with two chains (A and C) of the α subunit and two β subunits (chain B and D) is expected to form before binding to the E2 subunit (chain I), thus the assembly order AA′BB′>AA′BB′I (Fig. 4).

The commands ran are.

```
<Multi-LZerD part>
> $path_to_run_multilzerd_path.sh 1w88 A,B,C,D,I 1w88.pdb
<Path-LZerD part>
```

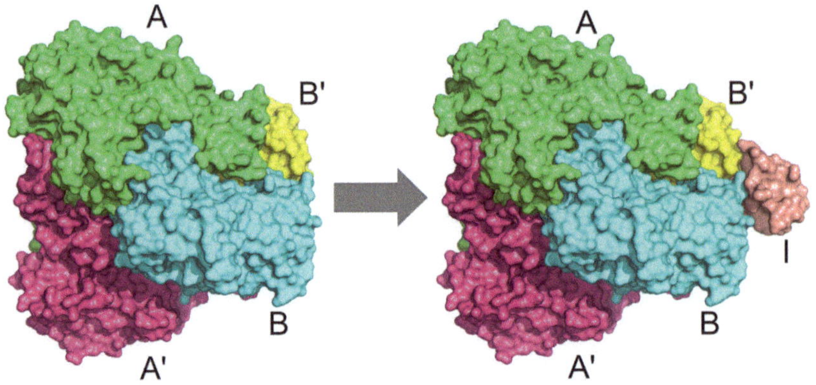

Fig. 4 Assembly pathway of 1w88. Subunits marked A and A′ (green and magenta) are pyruvate dehydrogenase E1 α subunit, subunits marked B and B′ (cyan and yellow) are E1 β subunit, and subunit I (salmon) is the peripheral subunit binding domain of E2. The assembly pathway is AA′BB′>AA′BB′I

In this example, we specified that chain A and C are same and chain B and D are same by using "-c AC;BD" option.

```
> $path_lzerd -p $id -d . -g finalgene.ga.out -o finalgene_chain.csv -c AC;BD
> $path_lzerd -p $id -d . -g consensus.ga.out -o consensus_chain.csv -c AC;BD
> $path_lzerd -p $id -d . -g lowest.ga.out -o lowest_chain.csv -c AC;BD
```

In this case, output results are not the same between the three methods. finalgene_chain.csv And consensus_chain.csv outputs were the same:

```
score,pathway,pdbid
dfirerank,BB>AI+BB>ABB+AI>AABBI,1w88
soappprank,BB>AI+BB>ABB+AI>AABBI,1w88
sumrank,BB>AI+BB>ABB+AI>AABBI,1w88
goaprank,BB>AI+BB>ABB+AI>AABBI,1w88
molmechrank,BB>AI+BB>ABB+AI>AABBI,1w88
shapescorerank,BB>AI+BB>ABB+AI>AABBI,1w88
opuspsprank,BB>ABB>ABB+AI>AABBI,1w88
itscorerank,BB>AI+BB>ABB+AI>AABBI,1w88
```

All the scoring functions made wrong prediction. Many scoring functions predicted an incorrect assembly order of BB>AI +BB>ABB+AI>AABBI. This was because many of the complex models generated by Multi-LZerD had a large RMSD with incorrect topologies. Particularly, placing E2 (chain I) in the correct position was difficult in the docking because chain I is very small (49 residues) relative to the other subunits (E1 α, A and A′: 368 residues; E1 β, B and B′, 324 residues).

On the other hand, results of lowest RMSD model method (*lowest_chain.csv*) show different predictions:

```
score,pathway,pdbid
dfirerank,BB>ABB>AABB>AABBI,1w88
soappprank,BB>ABB>AABB>AABBI,1w88
sumrank,BB>ABB>AABB>AABBI,1w88
goaprank,BB>BBI>ABBI>AABBI,1w88
molmechrank,BB>ABB>ABBI>AABBI,1w88
shapescorerank,BB>ABB>AABB>AABBI,1w88
opuspsprank,BB>ABB>AABB>AABBI,1w88
itscorerank,BB>ABB>AABB>AABBI,1w88
```

The lowest-RMSD model method predicted the correct assembly order AABB>AABBI with all but two (GOAP and molecular mechanics) scoring functions. As this example shows, if the native structure is available, the lowest RMSD method can often exclude incorrect conclusions that originate from incorrect docking models.

5 Notes

1. *Scoring Functions*

   In Path-LZerD, the pairwise docking decoys are ranked by a scoring function, and the scores are then used for the binding energy rank comparison. These scoring functions were originally designed for modeling and evaluating protein structure predictions. Each scoring function is designed from a different idea. As shown in the third case, path prediction results can be different by different scoring functions. Table 2 briefly summarizes seven scoring functions.

2. *Accuracy of Path-LZerD*

   According to the original Path-LZerD paper (S3, S4, and S5 Table) [12], the three methods, the lowest RMSD, the final generation, and the consensus methods, successfully predicted 11, 10, and 10 out of 21 (52.4 and 47.6%) assembly paths, respectively (here the highest value was shown among the results by different scoring functions). When combinations of scoring functions and the methods are concerned, the lowest RMSD method with OPUS-PSP showed the highest accuracy (52.4%, 11 out of 21 paths). Therefore, the lowest RMSD method with OPUS-PSP showed the highest performance, although the other combinations showed similar results.

   The benchmark study of Path-LZerD showed that the path prediction accuracy was higher for cases when complex structures were accurately predicted. Also, the path prediction accuracy was higher for smaller complexes, when the number of chains of target complexes was less than five.

Table 2
Scoring functions used in Path-LZerD

Name	Label used in Path-LZerD	Description
LZerD shape score	*shapescore*	Shape-based scoring function used in the LZerD pairwise docking program
LZerD molecular mechanics-based score	*molmech*	A linear combination of van der Waals, electrostatics, hydrogen and disulfide bond, solvation, and knowledge-based contact potential terms
DFIRE	*dfire*	A distance-dependent atom contact potential that considers 167 atom types
GOAP	*goap*	An orientation dependent statistical potential that combines an orientation-dependent term and DFIRE
ITScorePro	*itscore*	A distance-dependent atom contact potential based on 20 atom types
OPUS-PSP	*opuspsp*	An orientation-dependent statistical potential that considers orientation-specific packing interactions of side chains
SOAP-PP	*soappp*	A statistical potential for protein–protein interaction that considers atom pair distances based on 158 atom types, bond orientation, and relative solvent-accessible surface area

Acknowledgments

We thank Tunde Aderinwale for testing this software. This work was partly supported by the National Institute of General Medical Sciences of the NIH (R01GM123055) and the National Science Foundation (DMS1614777).

References

1. Englander SW, Mayne L (2014) The nature of protein folding pathways. Proc Natl Acad Sci U S A 111(45):15873–15880. https://doi.org/10.1073/pnas.1411798111
2. Bale JB, Gonen S, Liu Y, Sheffler W, Ellis D, Thomas C, Cascio D, Yeates TO, Gonen T, King NP, Baker D (2016) Accurate design of megadalton-scale two-component icosahedral protein complexes. Science 353 (6297):389–394. https://doi.org/10.1126/science.aaf8818
3. Shin WH, Christoffer CW, Kihara D (2017) In silico structure-based approaches to discover protein-protein interaction-targeting drugs. Methods 131:22–32. https://doi.org/10.1016/j.ymeth.2017.08.006
4. Kennedy KA, Gachelet EG, Traxler B (2004) Evidence for multiple pathways in the assembly of the Escherichia coli maltose transport complex. J Biol Chem 279(32):33290–33297. https://doi.org/10.1074/jbc.M403796200
5. Mizushima S, Nomura M (1970) Assembly mapping of 30S ribosomal proteins from E. coli. Nature 226(5252):1214
6. Hernandez H, Robinson CV (2007) Determining the stoichiometry and interactions of macromolecular assemblies from mass spectrometry. Nat Protoc 2(3):715–726. https://doi.org/10.1038/nprot.2007.73
7. Davis JH, Tan YZ, Carragher B, Potter CS, Lyumkis D, Williamson JR (2016) Modular assembly of the bacterial large ribosomal

subunit. Cell 167(6):1610–1622.e1615. https://doi.org/10.1016/j.cell.2016.11.020

8. Mulder AM, Yoshioka C, Beck AH, Bunner AE, Milligan RA, Potter CS, Carragher B, Williamson JR (2010) Visualizing ribosome biogenesis: parallel assembly pathways for the 30S subunit. Science 330(6004):673–677. https://doi.org/10.1126/science.1193220
9. Talkington MW, Siuzdak G, Williamson JR (2005) An assembly landscape for the 30S ribosomal subunit. Nature 438 (7068):628–632. https://doi.org/10.1038/nature04261
10. Sharon M, Witt S, Glasmacher E, Baumeister W, Robinson CV (2007) Mass spectrometry reveals the missing links in the assembly pathway of the bacterial 20 S proteasome. J Biol Chem 282(25):18448–18457. https://doi.org/10.1074/jbc.M701534200
11. Heck AJ (2008) Native mass spectrometry: a bridge between interactomics and structural biology. Nat Methods 5(11):927–933. https://doi.org/10.1038/nmeth.1265
12. Peterson LX, Togawa Y, Esquivel-Rodriguez J, Terashi G, Christoffer C, Roy A, Shin WH, Kihara D (2018) Modeling the assembly order of multimeric heteroprotein complexes. PLoS Comput Biol 14(1):e1005937. https://doi.org/10.1371/journal.pcbi.1005937
13. Marsh JA, Hernandez H, Hall Z, Ahnert SE, Perica T, Robinson CV, Teichmann SA (2013) Protein complexes are under evolutionary selection to assemble via ordered pathways. Cell 153(2):461–470. https://doi.org/10.1016/j.cell.2013.02.044
14. Levy ED, Boeri Erba E, Robinson CV, Teichmann SA (2008) Assembly reflects evolution of protein complexes. Nature 453 (7199):1262–1265. https://doi.org/10.1038/nature06942
15. Venkatraman V, Yang YD, Sael L, Kihara D (2009) Protein-protein docking using region-based 3D Zernike descriptors. BMC Bioinformatics 10:407. https://doi.org/10.1186/1471-2105-10-407
16. Esquivel-Rodriguez J, Yang YD, Kihara D (2012) Multi-LZerD: multiple protein docking for asymmetric complexes. Proteins 80 (7):1818–1833. https://doi.org/10.1002/prot.24079
17. Esquivel-Rodriguez J, Filos-Gonzalez V, Li B, Kihara D (2014) Pairwise and multimeric protein-protein docking using the LZerD program suite. Methods Mol Biol 1137:209–234. https://doi.org/10.1007/978-1-4939-0366-5_15
18. Zhou H, Zhou Y (2002) Distance-scaled, finite ideal-gas reference state improves structure-derived potentials of mean force for structure selection and stability prediction. Protein Sci 11(11):2714–2726
19. Zhou H, Skolnick J (2011) GOAP: a generalized orientation-dependent, all-atom statistical potential for protein structure prediction. Biophys J 101(8):2043–2052. https://doi.org/10.1016/j.bpj.2011.09.012
20. Huang SY, Zou X (2011) Statistical mechanics-based method to extract atomic distance-dependent potentials from protein structures. Proteins 79(9):2648–2661. https://doi.org/10.1002/prot.23086
21. Dong GQ, Fan H, Schneidman-Duhovny D, Webb B, Sali A (2013) Optimized atomic statistical potentials: assessment of protein interfaces and loops. Bioinformatics 29 (24):3158–3166. https://doi.org/10.1093/bioinformatics/btt560
22. Lu M, Dousis AD, Ma J (2008) OPUS-PSP: an orientation-dependent statistical all-atom potential derived from side-chain packing. J Mol Biol 376(1):288–301. https://doi.org/10.1016/j.jmb.2007.11.033
23. McLachlan AD (1982) Rapid comparison of protein structures. Acta Cryst A38:871–873
24. Kihara D, Sael L, Chikhi R, Esquivel-Rodriguez J (2011) Molecular surface representation using 3D Zernike descriptors for protein shape comparison and docking. Curr Protein Pept Sci 12(6):520–530
25. Loew A, Ho YK, Blundell T, Bax B (1998) Phosducin induces a structural change in transducin beta gamma. Structure 6(8):1007–1019
26. Dingus J, Hildebrandt JD (2012) Synthesis and assembly of G protein betagamma dimers: comparison of in vitro and in vivo studies. Subcell Biochem 63:155–180. https://doi.org/10.1007/978-94-007-4765-4_9
27. Abergel C, Monchois V, Byrne D, Chenivesse S, Lembo F, Lazzaroni JC, Claverie JM (2007) Structure and evolution of the ivy protein family, unexpected lysozyme inhibitors in Gram-negative bacteria. Proc Natl Acad Sci U S A 104(15):6394–6399. https://doi.org/10.1073/pnas.0611019104
28. Monchois V, Abergel C, Sturgis J, Jeudy S, Claverie JM (2001) Escherichia coli ykfE ORFan gene encodes a potent inhibitor of C-type lysozyme. J Biol Chem 276 (21):18437–18441. https://doi.org/10.1074/jbc.M010297200
29. Maroufi B, Ranjbar B, Khajeh K, Naderi-Manesh H, Yaghoubi H (2008) Structural studies of hen egg-white lysozyme dimer: comparison with monomer. Biochim Biophys Acta 1784(7–8):1043–1049. https://doi.org/10.1016/j.bbapap.2008.03.010

30. Cegielska-Radziejewska R, Lesnierowski G, Kijowski J (2008) Properties and application of egg white lysozyme and its modified preparations-a review. Pol J Food Nutr Sci 58:5–10

31. Frank RA, Titman CM, Pratap JV, Luisi BF, Perham RN (2004) A molecular switch and proton wire synchronize the active sites in thiamine enzymes. Science 306(5697):872–876. https://doi.org/10.1126/science.1101030

32. Kato M, Wynn RM, Chuang JL, Tso SC, Machius M, Li J, Chuang DT (2008) Structural basis for inactivation of the human pyruvate dehydrogenase complex by phosphorylation: role of disordered phosphorylation loops. Structure 16(12):1849–1859. https://doi.org/10.1016/j.str.2008.10.010

Chapter 9

Embedding Alternative Conformations of Proteins in Protein–Protein Interaction Networks

Farideh Halakou, Attila Gursoy, and Ozlem Keskin

Abstract

While many proteins act alone, the majority of them interact with others and form molecular complexes to undertake biological functions at both cellular and systems levels. Two proteins should have complementary shapes to physically connect to each other. As proteins are dynamic and changing their conformations, it is vital to track in which conformation a specific interaction can happen. Here, we present a step-by-step guide to embedding the protein alternative conformations in each protein–protein interaction in a systems level. All external tools/websites used in each step are explained, and some notes and suggestions are provided to clear any ambiguous point.

Key words Protein binding, Protein–protein interaction, PPI network, Structural PPI network, Protein conformational change, Protein alternative conformations

1 Introduction

Protein–protein interaction (PPI) networks are fundamentally used for understanding the biological relationships behind disease phenotypes [1–3]. These networks use a node and edge demonstration in which nodes represent proteins and edges represent the physical binding of proteins. Usefulness of this traditional representation aside, it lacks the molecular details contained in physical binary interactions [4, 5]. The structural details of protein bindings are necessary to understand the cellular processes and protein functions [6–8]. Embedding structural information to PPI networks enables us to study the binding regions of proteins and to find the incompatibility of structures between binding partners [9, 10].

In structural PPI networks, it is common to investigate just one protein structure for each node. However, as proteins are dynamic and changing their conformation continually, it is necessary to track and investigate these conformational changes in the networks [11, 12]. Each protein might have multiple alternative

Stefan Canzar and Francisca Rojas Ringeling (eds.), *Protein-Protein Interaction Networks: Methods and Protocols*, Methods in Molecular Biology, vol. 2074, https://doi.org/10.1007/978-1-4939-9873-9_9,

conformations based on the environmental factors. So it can bind to different proteins from time to time as shown in Fig. 1.

As binary protein interactions need two proteins of complementary shapes to come together, conformational changes can make/disrupt the interactions in PPI networks. So structural PPI networks are not only static but also changing their topology from time to time [13, 14].

Figure 2 shows a schematic representation of a PPI network enriched with protein alternative conformations. Using this representation, we can see which alternative conformation (s) contributed in each PPI in the network.

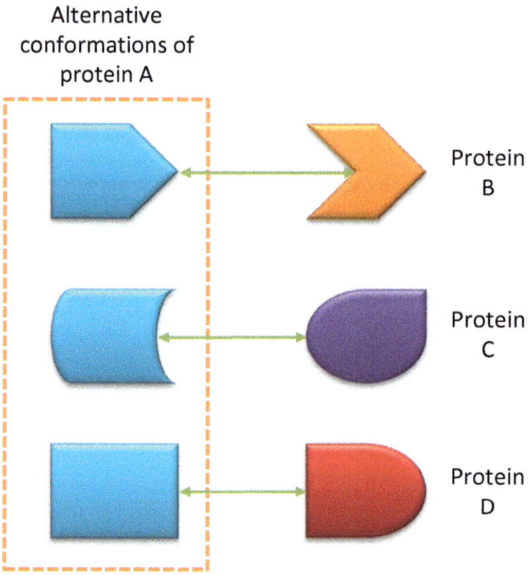

Fig. 1 Schematic representation of protein conformational changes. Protein A changes its conformation in time and finds new binding partners in each conformation

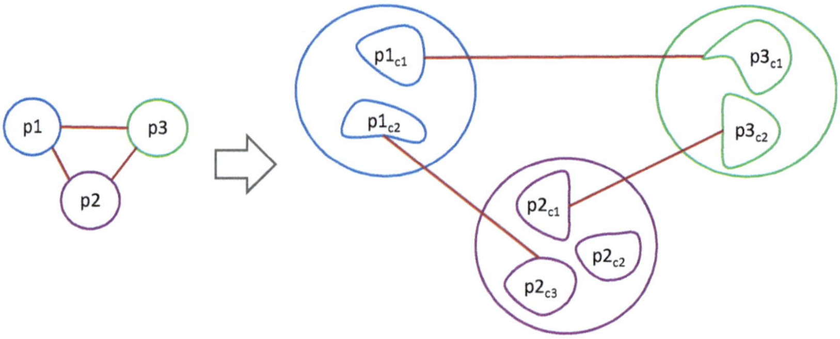

Fig. 2 Structural PPI network in traditional representation and the new representation enriched with protein alternative conformations on the left and right, respectively. Shapes inside each node show the alternative conformations of that protein

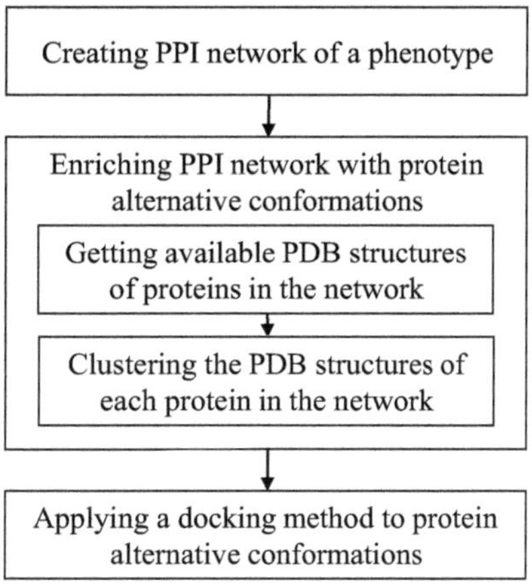

Fig. 3 Steps of PPI network creation enriched with protein alternative conformations

We have already shown the impact of inspecting protein alternative conformations in protein docking predictions and structural network creation [15]. Here, we give a step-by-step illustration of structural PPI network creation enriched with alternative conformations of proteins for a phenotype of interest. The steps of the structural network creation are shown in Fig. 3.

The process starts with creating the traditional PPI network for the phenotype of interest. There are many tools and databases which can be used in this step. We will explain GUILDify web server [16] which we used for this purpose. Then, all available structures for the proteins in the network should be downloaded from Protein Data Bank (PDB). As PDB is a redundant database, the structures for each protein should be clustered to remove the repetitive similar structures. This step gives the available alternative conformations of proteins in the network. To see the possible interactions between these alternative conformations, a protein docking tool is needed. We will describe PRISM as a protein docking tool. All these steps will be explained in detail in the following sections.

2 Materials

2.1 GUILDify

GUILDify [16] is a web server in which we can get a scored PPI network by free text search or by providing a list of genes. GUILDify has two steps, i.e., BIANA [17] and GUILD [18]. BIANA is a tool which integrates the biological data from different publicly available databases. GUILDify queries BIANA to find the proteins associated with the user-provided keywords. Usually, BIANA brings an unweighted massive PPI network. In the next step, GUILD is applied on the PPI network to score the nodes based on their relevance using the prioritization algorithms. As the final network is scored, it is possible to extract the high-scored region, which contains the most relevant proteins, for structural analysis and protein docking. GUILDify is freely accessible through http://sbi.imim.es/web/GUILDify.php.

2.2 Clustering PDB Structures

To get the alternative conformations of the proteins in the PPI network, it is necessary to cluster the PDB structures corresponding to each protein, because as mentioned earlier, many structures deployed in PDB are redundant. We cluster the PDBs based on sequence identity and structural similarities using agglomerative complete linkage clustering method. Each cluster shows an alternative conformation of a protein, so a PDB structure is selected for each cluster as the representative to feed into the docking step. Proteins having less than 30 residues are eliminated from the clustering process. Two PDB structures will be clustered together if they have more than 95% sequence identity and smaller than 2 Å RMSD value. The clustering code and other necessary scripts are written in Python and are available through https://github.com/ku-cosbi/ppi-network-alternative.

2.3 Protein Docking

To see if two proteins, connected with an edge in the network, can bind to each other or not, we need a protein docking tool. We used PRISM [19–21] to model the potential protein complexes in the multi-conformational PPI network. PRISM has a template interface dataset, consisted of 22,604 interfaces, extracted from PDB complex structures. It tries to find the two sides of a template interface on the surfaces of two proteins. If it can find such two protein surfaces, it models their complex structure using that template interface. PRISM consisted of four phases, i.e., surface extraction of target proteins, structural alignment of template interfaces with target surfaces, transforming the target surfaces to the similar template interface, and flexible refinement of the predicted protein complex. PRISM can get a list of PPIs, i.e., a network as input which is needed in this study. PRISM webserver is accessible through http://cosbi.ku.edu.tr/prism/.

3 Methods

This section describes in detail the steps to create a multi-conformational PPI network.

3.1 Creating the Traditional PPI Network Using GUILDify

If you don't have a PPI network beforehand, you can use GUILDify to easily create a phenotype specific network (*see* **Note 1**). If you have the PPI network in advance, you can skip this section.

3.1.1 Input

To use the GUILDify webserver, we need to provide keywords or gene names. When the keywords are quoted together, it means that the entries in BIANA will be retrieved if their description matches the whole quoted keywords, e.g., "breast cancer lung metastasis" will retrieve the entries having all four words. Otherwise, each keyword will be searched separately. In the case of providing a list of genes, they should be separated by semicolon (;), e.g., SNAI1; SNUPN;KPNB1;KPNA2.

3.1.2 Using GUILDify

To create the network, we can launch GUILDify webserver through http://sbi.imim.es/web/GUILDify.php. Then, we need to enter the keywords in the text box and select the species. By clicking the "Search in BIANA Knowledge base," the server will start the searching process and find the related genes. It will show us the gene names found in BIANA in a table in which we can deselect the genes we don't want to have in the final PPI network. After selecting the gene names in the table, we can click the "GUILDify" button which starts the process of prioritizing the gene products in the network. This process can take 30 min to several hours based on the network size and the number of queued/running jobs on the server.

3.1.3 Output

When the prioritization process ends, the results page (Fig. 4) provides the list of proteins in the network with their GUILD scores.

In this page, we can download the whole PPI network by clicking on the "Download Interactome" link. As mentioned earlier, normally this network is very big and not manageable for

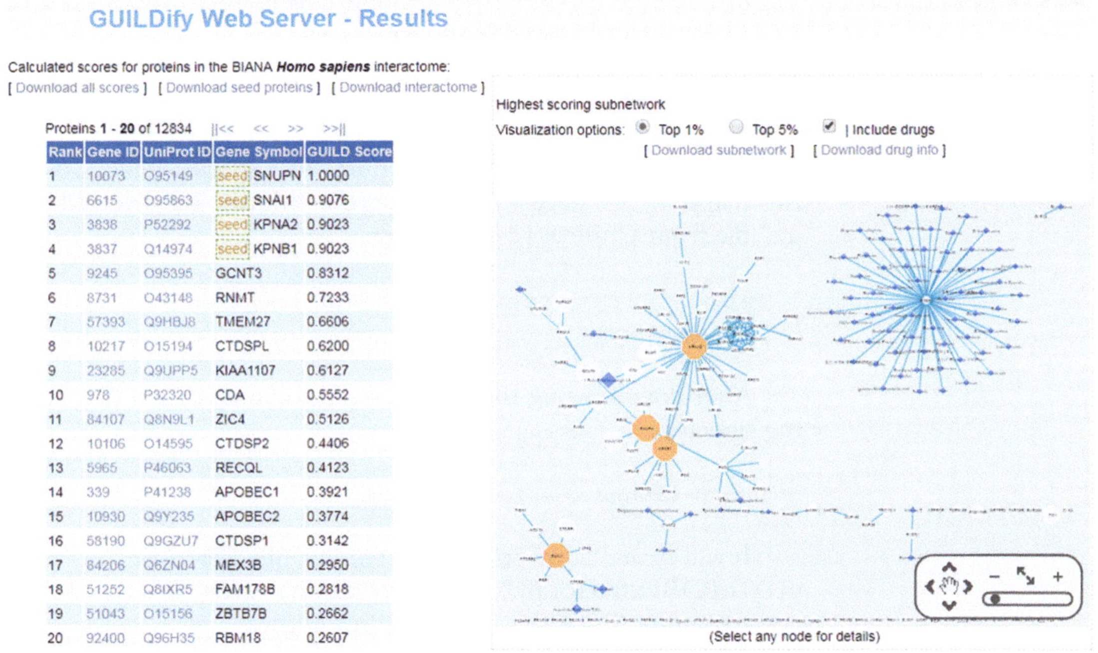

Fig. 4 Screenshot of GUILDify results page for genes list SNAI1;SNUPN;KPNB1;KPNA2. All network gene IDs are listed on the left, and the visualization of the prioritized network is shown on the right

structural analysis. So we can instead download the highest-scored subnetwork by selecting top 1% or top 5% option in the results page and clicking "Download subnetwork." The downloaded network/ subnetwork is a plain text file in which each line shows an interaction in the format of "BIANA_ID interaction BIANA_ID" (*see* **Note 2**).

We can also download the scores and seeds, which are the related genes found in BIANA to create the network, in the results page. By downloading the scores, we can see each BIANA ID is equivalent to which gene as it shows their corresponding gene ID and gene symbol.

We have saved the network file as "subnetwork.sif.5" and the scores file as "guild_scores.txt." In the next step, we need gene IDs for PDB clustering. So to convert the network file from BIANA IDs to gene IDs, we need to run the "IdMapping.py" file like (in a Linux system):

```
> python IdMapping.py
```

Network and scores file should be in the same directory with "IdMapping.py." By running this script, it will produce two files, i.e., "PPI_Network.txt" and "Network_Gene_Ids.txt," which represent the PPI network as gene IDs and list gene IDs of all network nodes, respectively (*see* **Note 3**).

3.2 Clustering Protein Conformations

In this section, we will explain how to get the available structures of the proteins in the PPI network and clustering them to remove repetitive calculations in the docking step.

3.2.1 Preparing PDB Files for Clustering

To get the PDB files of the proteins in the PPI network, the mappings between gene IDs and PDB IDs are needed to find out which PDB structures are the products of a specific gene. To get this mapping, we need to download the "idmapping_selected.tab.gz" file from UniProt [22] website:

```
> wget ftp://ftp.uniprot.org/pub/databases/uniprot/current_release/knowledgebase/idmapping/idmapping_selected.tab.gz
```

After downloading this file, we need to unzip it into the working directory:

```
> gunzip idmapping_selected.tab.gz
```

It will create "idmapping_selected.tab" file. Then we run "geneIDToPDBExtractor.py" script to extract gene IDs and their corresponding PDB IDs from UniProt entries.

```
> python geneIDToPDBExtractor.py
```

It creates "geneIDToPDBMapping.txt" file which has all gene IDs and PDB IDs match results like:

```
7531 2BR9:A; 3UAL:A; 3UBW:A; 6EIH:A

7533 2C63:A; 2C63:B; 2C63:C; 2C63:D; 2C74:A; 2C74:B

22629 5YQG:A; 5YQG:B; 5YQG:C; 5YQG:D

5520 3DW8:B; 3DW8:E
```

We need to extract the gene IDs in our network from this file and download their protein products for clustering. To do so, we can run "PdbDownloader.py" code.

```
> python PdbDownloader.py
```

It extracts the network gene Ids from "geneIDToPDBMapping.txt" file and puts them in "usedGeneIDToPDBMapping.txt" file. In addition, it downloads the corresponding PDB files into "pdbFiles" folder (*see* **Note 4**). Using this mapping file and the downloaded PDB files, we can run the clustering code in the next step.

It should be noted that there are many genes which have no available structure in PDB. So the number of genes in "usedGeneIDToPDBMapping.txt" is smaller than that of "Network_Gene_Ids.txt."

3.2.2 Running Clustering Code

As mentioned earlier, PDB structures corresponding to a gene ID might be very similar to each other. So there is no need to consider them as different conformations and spend time for docking them in the next step.

Clustering code needs TM-align for PDB structure comparisons. We download TM-align from https://zhanglab.ccmb.med.umich.edu/TM-align to "externalTools/TMalign."

```
> wget zhanglab.ccmb.med.umich.edu/TM-align/TMalignc.tar.gz
```

After downloading TM-align, we need to unzip it and install it:

```
> tar xvzf TMalignc.tar.gz

> g++ -static -O3 -ffast-math -lm -o TMalign TMalign.cpp
```

After installing TM-align, we can run the clustering code. To cluster the similar PDB structures corresponding to a gene ID, we run "mainProgram.py" code.

```
> python mainProgram.py
```

This code extracts the monomer structures from PDB files and stores them as separate files. Then, using TM-align [23] tool, it compares the PDB structures corresponding to a gene ID and put the similar ones in a cluster (*see* **Note 5**). Finally, it creates "clusteredGeneIDToPDBMapping.txt" file in which each line shows a gene ID, number of alternative conformations, and representative PDB IDs for each conformation, like:

```
3688 3 3T9K_A,3VI4_D,3G9W_D
4651 3 3AU4_A,2LW9_A,2LW9_B
6386 1 1OBZ_B
2316 6 4M9P_A,2K7Q_A,2J3S_A,3HOP_B,2AAV_A,2K7P_A
3146 3 2LY4_A,2YRQ_A,2RTU_A
```

As PDB structures do not necessarily cover the complete protein sequence, it is possible to have some clusters representing different parts of a protein. These clusters are still important because they can contribute to different interactions in PPI networks.

3.3 Protein Docking Using PRISM

Having the PPI network and proteins' alternative conformations, we need to use a protein docking tool to see which protein conformations contribute in the PPIs present in the network. Here we explain PRISM web server for protein docking.

3.3.1 Preparing PDB-PDB Interaction List for PRISM

In the previous step, we got the alternative conformations of proteins in our network by clustering their corresponding PDB structures. Here, for each PPI in "PPI_Network.txt," we need to dock their potential interactions using their alternative conformations which are listed in "clusteredGeneIDToPDBMapping.txt." To do so, we need to run "PpiNetworkToPDBNetwork.py":

```
> python PpiNetworkToPDBNetwork.py
```

It will create "PDB_Network.txt" file with all PDB-PDB interactions. These interactions will be submitted to PRISM to model their complex structures.

3.3.2 Submitting Interactions to PRISM and Getting the Predictions

In PRISM web server, there is a "Network" tab in which we can submit a maximum of ten interactions as one job. The docking process will start by clicking "Submit" button and would take several minutes to several hours depending on the size of PDB structures, number of interactions, and number of jobs running on server.

Finishing the docking process, the results will be shown for each interaction (Fig. 5).

Results:

Target1	Target2	Interface	Energy	Structure
2w83C	2o61B	1ft9AB	-39.12	View
4e4vA	2f3iA	3bh7AB	-34.44	View
3gutC	4e4vA	3gwrAB	-30.45	View
2f3iA	4e4vA	2zrtBC	-20.66	View
3qfvB	1aoxA	3g17AB	-14.61	View
3qfvB	1aoxA	1zkpBD	-10.26	View
4e4vA	1nfiC	1x8dAB	-9.94	View
4e4vA	2o61A	1f9aAF	-9.3	View
4e4vA	3gutC	3s29CD	-7.61	View

Fig. 5 Sample screenshot of PRISM results page. Each row shows a binary interaction between two PDB IDs and the interface name which is used to bind the PDB structures, plus binding free energy of the predicted complex structure

For each interaction, PRISM may find no/many binding mechanisms using different interface structures. The results are sorted based on the binding free energy of the predicted complex structures. The smaller the energy value, the better the predicted complex structure. By clicking "View" button, the predicted complex structure will be shown (Figs. 5 and 6).

By clicking "Contacts of Interface Residues," one-to-one binding of the residues on the interface will be shown as a text file.

3.3.3 Getting the Alternative Conformations Interactions

Having the protein docking results, we can check for each interaction in our PPI network, which alternative conformations contributed to. To do so, we need to download all PRISM predictions and extract the predictions corresponding to our network. In PRISM web server, we can go to "Predictions" tab and click "Download All Predictions" at the bottom of the page. Then we store this text file as "PrismPredictions.txt" and run "PpiNetworkPrismPredictions.py" code:

```
> python PpiNetworkPrismPredictions.py
```

It will create "PPI_Network_Prism_Predictions.txt" file in which the alternative conformations and PRISM predictions, if they exist, are listed for each PPI separately. So we can see for each PPI in the network, which alternative conformations contributed to and which ones did not. An example can be as follows:

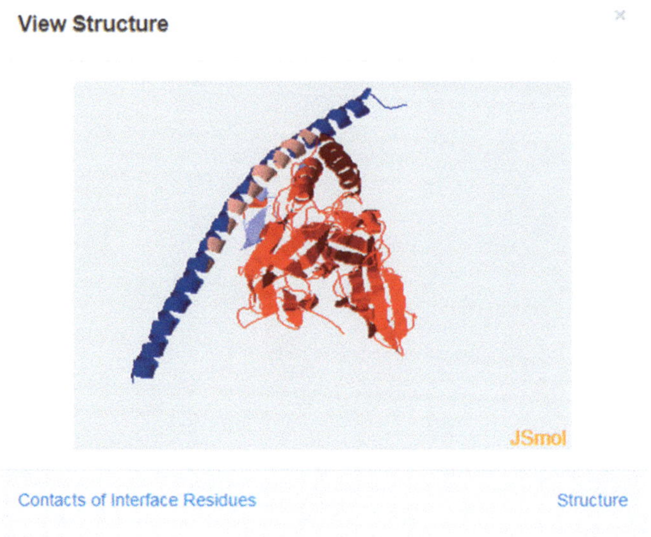

Fig. 6 PRISM prediction for 2w83C-2o61B interaction. Two PDB structures are colored in red and blue. The interacting residues are colored in pink and purple

```
5970-3838
2O61A 1NFIC 3GUTC
1QGKB 4E4VA
1nfiC 4e4vA 3dcgAC -3.97 2018-12-31 21:00:55
3gutC 4e4vA 3gwrAB -30.45 2018-12-31 21:00:55
```

This shows that for PPI interaction between gene IDs 5970 and 3838, PRISM could find two possible complexes using alternative conformations 1nfiC and 3gutC for gene ID 5970 and 4e4vA for gene ID 3838. Although both genes have one more alternative conformation, it seems that they could not contribute to this specific interaction (*see* **Note 6**).

4 Notes

1. GUILDify is just a sample tool to create a PPI network. You can use any tool that you prefer.
2. If we select *Homo sapiens* for species, GUILDify final network will include the interactions with drugs too. BIANA ID of drugs starts with "DB," and they are put in the first lines of the network text file. So if we don't need them, we can just simply remove those lines.
3. In some rare cases, GUILDify scores file does not have the gene IDs of the nodes. In these cases, the code skips the interactions without gene IDs and does not put them in the output files.

4. A few PDB structures may be available just as .cif files instead of .pdb. These structures are skipped in this step.
5. Minimum residue number for a PDB structure to be investigated in the clustering process, minimum sequence similarity, and maximum RMSD values for clustering PDB structures can be changed inside "mainProgram.py."
6. For the two predictions shown, the binding free energy scores differ considerably. So all the docking results need to be structurally analyzed to make sure they are correct predictions.

References

1. Vinayagam A, Zirin J, Roesel C, Hu Y, Yilmazel B, Samsonova AA, Neumuller RA, Mohr SE, Perrimon N (2014) Integrating protein-protein interaction networks with phenotypes reveals signs of interactions. Nat Methods 11(1):94–99. https://doi.org/10.1038/nmeth.2733
2. Hu L, Huang T, Liu XJ, Cai YD (2011) Predicting protein phenotypes based on protein-protein interaction network. PLoS One 6(3): e17668. https://doi.org/10.1371/journal.pone.0017668
3. Carter H, Hofree M, Ideker T (2013) Genotype to phenotype via network analysis. Curr Opin Genet Dev 23(6):611–621. https://doi.org/10.1016/j.gde.2013.10.003
4. Yang JS, Campagna A, Delgado J, Vanhee P, Serrano L, Kiel C (2012) SAPIN: a framework for the structural analysis of protein interaction networks. Bioinformatics 28(22):2998–2999. https://doi.org/10.1093/bioinformatics/bts539
5. Mosca R, Ceol A, Aloy P (2013) Interactome3D: adding structural details to protein networks. Nat Methods 10(1):47–53. https://doi.org/10.1038/nmeth.2289
6. Guven Maiorov E, Keskin O, Gursoy A, Nussinov R (2013) The structural network of inflammation and cancer: merits and challenges. Semin Cancer Biol 23(4):243–251. https://doi.org/10.1016/j.semcancer.2013.05.003
7. Acuner-Ozbabacan ES, Engin BH, Guven-Maiorov E, Kuzu G, Muratcioglu S, Baspinar A, Chen Z, Van Waes C, Gursoy A, Keskin O, Nussinov R (2014) The structural network of Interleukin-10 and its implications in inflammation and cancer. BMC Genomics 15(Suppl 4):S2. https://doi.org/10.1186/1471-2164-15-S4-S2
8. Engin HB, Guney E, Keskin O, Oliva B, Gursoy A (2013) Integrating structure to protein-protein interaction networks that drive metastasis to brain and lung in breast cancer. PLoS One 8(11):e81035. https://doi.org/10.1371/journal.pone.0081035
9. Tuncbag N, Kar G, Gursoy A, Keskin O, Nussinov R (2009) Towards inferring time dimensionality in protein-protein interaction networks by integrating structures: the p53 example. Mol BioSyst 5(12):1770–1778. https://doi.org/10.1039/B905661K
10. Kar G, Gursoy A, Keskin O (2009) Human cancer protein-protein interaction network: a structural perspective. PLoS Comput Biol 5 (12):e1000601. https://doi.org/10.1371/journal.pcbi.1000601
11. Gerstein M, Echols N (2004) Exploring the range of protein flexibility, from a structural proteomics perspective. Curr Opin Chem Biol 8(1):14–19. https://doi.org/10.1016/j.cbpa.2003.12.006
12. Boehr DD, Nussinov R, Wright PE (2009) The role of dynamic conformational ensembles in biomolecular recognition. Nat Chem Biol 5 (11):789–796. https://doi.org/10.1038/nchembio.232
13. Goh CS, Milburn D, Gerstein M (2004) Conformational changes associated with protein-protein interactions. Curr Opin Struct Biol 14 (1):104–109. https://doi.org/10.1016/j.sbi.2004.01.005
14. Ozgur B, Ozdemir ES, Gursoy A, Keskin O (2017) Relation between protein intrinsic normal mode weights and pre-existing conformer populations. J Phys Chem B 121 (15):3686–3700. https://doi.org/10.1021/acs.jpcb.6b10401
15. Halakou F, Kilic ES, Cukuroglu E, Keskin O, Gursoy A (2017) Enriching traditional protein-protein interaction networks with alternative conformations of proteins. Sci Rep 7(1):7180. https://doi.org/10.1038/s41598-017-07351-0

16. Guney E, Garcia-Garcia J, Oliva B (2014) GUILDify: a web server for phenotypic characterization of genes through biological data integration and network-based prioritization algorithms. Bioinformatics 30 (12):1789–1790. https://doi.org/10.1093/bioinformatics/btu092
17. Garcia-Garcia J, Guney E, Aragues R, Planas-Iglesias J, Oliva B (2010) Biana: a software framework for compiling biological interactions and analyzing networks. BMC Bioinformatics 11:56. https://doi.org/10.1186/1471-2105-11-56
18. Guney E, Oliva B (2012) Exploiting protein-protein interaction networks for genome-wide disease-gene prioritization. PLoS One 7(9): e43557. https://doi.org/10.1371/journal.pone.0043557
19. Ogmen U, Keskin O, Aytuna AS, Nussinov R, Gursoy A (2005) PRISM: protein interactions by structural matching. Nucleic Acids Res 33 (Web Server issue):W331–W336
20. Tuncbag N, Gursoy A, Nussinov R, Keskin O (2011) Predicting protein-protein interactions on a proteome scale by matching evolutionary and structural similarities at interfaces using PRISM. Nat Protoc 6(9):1341–1354. https://doi.org/10.1038/nprot.2011.367
21. Baspinar A, Cukuroglu E, Nussinov R, Keskin O, Gursoy A (2014) PRISM: a web server and repository for prediction of protein-protein interactions and modeling their 3D complexes. Nucleic Acids Res 42 (Web Server issue):W285–W289. https://doi.org/10.1093/nar/gku397
22. The UniProt Consortium (2017) UniProt: the universal protein knowledgebase. Nucleic Acids Res 45(D1):D158–D169. https://doi.org/10.1093/nar/gkw1099
23. Zhang Y, Skolnick J (2005) TM-align: a protein structure alignment algorithm based on the TM-score. Nucleic Acids Res 33 (7):2302–2309. https://doi.org/10.1093/nar/gki524

Chapter 10

Informed Use of Protein–Protein Interaction Data: A Focus on the Integrated Interactions Database (IID)

Chiara Pastrello, Max Kotlyar, and Igor Jurisica

Abstract

Protein–protein interaction data is fundamental in molecular biology, and numerous online databases provide access to this data. However, the huge quantity, complexity, and variety of PPI data can be overwhelming, and rather than helping to address research problems, the data may add to their complexity and reduce interpretability. This protocol focuses on solutions for some of the main challenges of using PPI data, including accessing data, ensuring relevance by integrating useful annotations, and improving interpretability. While the issues are generic, we highlight how to perform such operations using Integrated Interactions Database (IID; http://ophid.utoronto.ca/iid).

Key words Protein interactions, Interaction networks, Databases

1 Introduction

Protein–protein interactions (PPIs) are a widely used resource in biomedicine. Their importance comes from their involvement in almost all cellular processes like growth, metabolism, and repair. It is therefore paramount, for researchers studying molecular mechanisms behind any biological process, to use knowledge of PPIs and investigate how they form and break complexes and signaling cascades (pathways). At the present time, even if incomplete, PPI data has led to a number of discoveries in molecular biology, among which are drug discovery, identification of disease genes [1], and prediction of gene function [2].

PPIs improve relevance of models and interpretability of almost any analysis of gene or protein sets (for brevity referred to as protein sets from now on) regardless of the research aim (e.g., prognostic/predictive signatures, drug target identification, drug mechanism of action) or experimental methods (i.e., high-throughput gene expression, proteomic data). PPIs can enhance analysis in multiple ways—providing a better understanding of the functionality of the

Stefan Canzar and Francisca Rojas Ringeling (eds.), *Protein-Protein Interaction Networks: Methods and Protocols*, Methods in Molecular Biology, vol. 2074, https://doi.org/10.1007/978-1-4939-9873-9_10,

protein set, the relationships within the protein set, and the relationship between the protein set and the wider proteome. The functionality of a protein set, and its other properties are typically determined through pathway or Gene Ontology [3] enrichment analysis; these analyses can be enhanced with PPIs, for example, by augmenting the protein set with its highly connected interaction partners, to provide a biologically relevant model. Relationships within a protein set can be identified through topology analysis; for example, a highly connected protein set may correspond to a protein complex, while a linearly connected set may be part of a signaling pathway. Similarly, topology analysis can help clarify the relationships between the protein set and the wider proteome; for example, many members of the set may be connected to the same interaction partners, possibly regulating their function. These types of analysis can provide a better understanding of a protein set, improve its predictive or diagnostic value, generate new hypothesis, and help formulate and prioritize the next investigative steps [4–6].

Carrying out these analyses requires access to comprehensive, reliable, up-to-date, and annotated PPIs, which are relevant to the given problem. While there are more than 25 online PPI databases [7], determining which ones to use, when, and how, is not straightforward. PPI databases can be categorized by multiple criteria: interaction type (physical or functional, stable or transient), species, data source, annotation details and variety, and specializations, such as immune system or extracellular matrix interactions [8, 9]. One of the first criteria in choosing a database may be interaction type; many databases provide only physical interactions (e.g., IntAct [10], MINT [11], DIP [12]), while some also provide genetic [13] or functional interactions [14]. Another key criterion is species—most databases primarily have human PPIs (e.g., BioGRID [13], IntAct [10], DIP [12]), some have only human [15], and a few are devoted to nonhuman species [16]. PPI databases can also be divided based on how they acquire PPI data; *primary databases* obtain PPIs by curating journal articles that report experimentally detected PPIs [10, 12, 13], *secondary databases* obtain PPIs from multiple primary databases [17, 18], and some databases also include computationally predicted PPIs [14, 19].

This protocol shows the steps needed to retrieve interactions from the Integrated Interactions Database (IID; http://ophid.utoronto.ca/iid [19]), covering key strengths and limitations of general PPI searches. IID integrates PPIs from major curated databases with computationally predicted PPIs, covers interactions across 18 species, and supports diverse query filters to cover different biological contexts, such as tissue and disease. Figure 1 shows the outline of a common PPI search.

Fig. 1 Outline of a PPI search. A set of proteins (**a**) is used to search a PPI database (**b**) to retrieve PPIs for the protein of interest and their annotations (**c**). The output file can then be visualized as a network (**d**), where visualization features are used to show annotation data (**e**)

2 Methods

2.1 Retrieve Interactions

Interactions can be retrieved from IID programmatically (through an API), through a plugin in NAViGaTOR [20, 21] (a visualization software; http://ophid.utoronto.ca/navigator) or directly from http://ophid.utoronto.ca/iid. In IID, paste the names or IDs of protein(s) of interest into the input window, selecting a species (if not human) and choosing submit or download. It is suggested to use less than 1000 IDs as larger numbers can result in hundreds of thousands of PPIs and longer response times. IDs may include any combination of UniProt, Entrez, gene symbol, separated by tabs, commas, or newline characters. If IDs are not recognized as valid (e.g., are not from the selected species, are not supported, are invalid or deprecated), then a descriptive error message is returned (*see* **Note 1** for common problems with this step). If PPIs are found, a data table is returned, with each row corresponding to a unique interaction. IID also allows users to retrieve just the interactions among query proteins or interactions of query proteins. For example, we typed MLH1 and MSH2 in the search tab of IID and next discuss the output network.

2.2 PPI Annotation

In most applications of PPIs, there is a need to understand cellular processes in a very specific context, such as particular tissues, disease states, or subcellular localizations. However, most of the PPIs returned from a database may not occur in the required context—for example, most human genes are not expressed in all tissues [22]. It is helpful to know which annotation is important for the disease or molecular mechanism being studied and select it in the filter options (if available) when searching in a PPI database. IID allows users to select any number of contexts among 133 tissues, 91 diseases, and 13 cellular localizations and druggability and to combine the selections in different ways—enabling a huge number of annotation combinations. Within each context type (e.g., tissue), users can specify whether returned PPIs can be any of the selected annotations (e.g., present in either the kidney or liver) or must be in all selected contexts (e.g., present in kidney and liver). If multiple annotation types are selected (e.g., tissues and cellular localizations), the context types will be combined (*see* **Note 2** for other databases with PPI annotations). In our example, we selected human as organism, focusing only on the lung and ovary as tissues.

2.3 Output

Different PPI databases return results in many possible formats; the most frequent is the tab-separated ASCII format. Running the search in IID will return one PPI (with its annotations) per line. The output can be used for further analysis in any software (e.g., R, Excel) or for visualization and network analysis (e.g., using NAViGaTOR or Cytoscape). The output file we obtained in our example is provided at http://iid.ophid.utoronto.ca/static/supplementary_files/supplementaryFile1.txt.

2.4 Network Analysis

Figure 2 shows one possible annotated graph resulting from the output file obtained from IID. PPIs obtained from one database can be further analyzed—as mentioned above—and annotations integrated with the ones provided by other software. In our example, we loaded the output file from IID into NAViGaTOR and further annotated interactors of our proteins of interest with Gene Ontology annotations. Topology analyses focus on identifying key nodes (proteins) in the network and major network components. IID provides topology measures for the results of a query (degree, clustering coefficient, and betweenness centrality). Other network analysis-specific software, such as NAViGaTOR, allow performing similar and additional functions to quickly identify important network features and provide a better understanding of query proteins. Using degree and finding graph articulation points, for example (further described in **Note 3**), NAViGaTOR can help identify query proteins and interaction partners that are likely to have the strongest impact on phenotype, such as lethality and synthetic lethality. For further possibilities of network analyses and tools, we refer the reader to a recent review [21].

1. Upon opening NAViGaTOR it is possible to select a file to load from the Open dialog. A wizard will guide the user through the

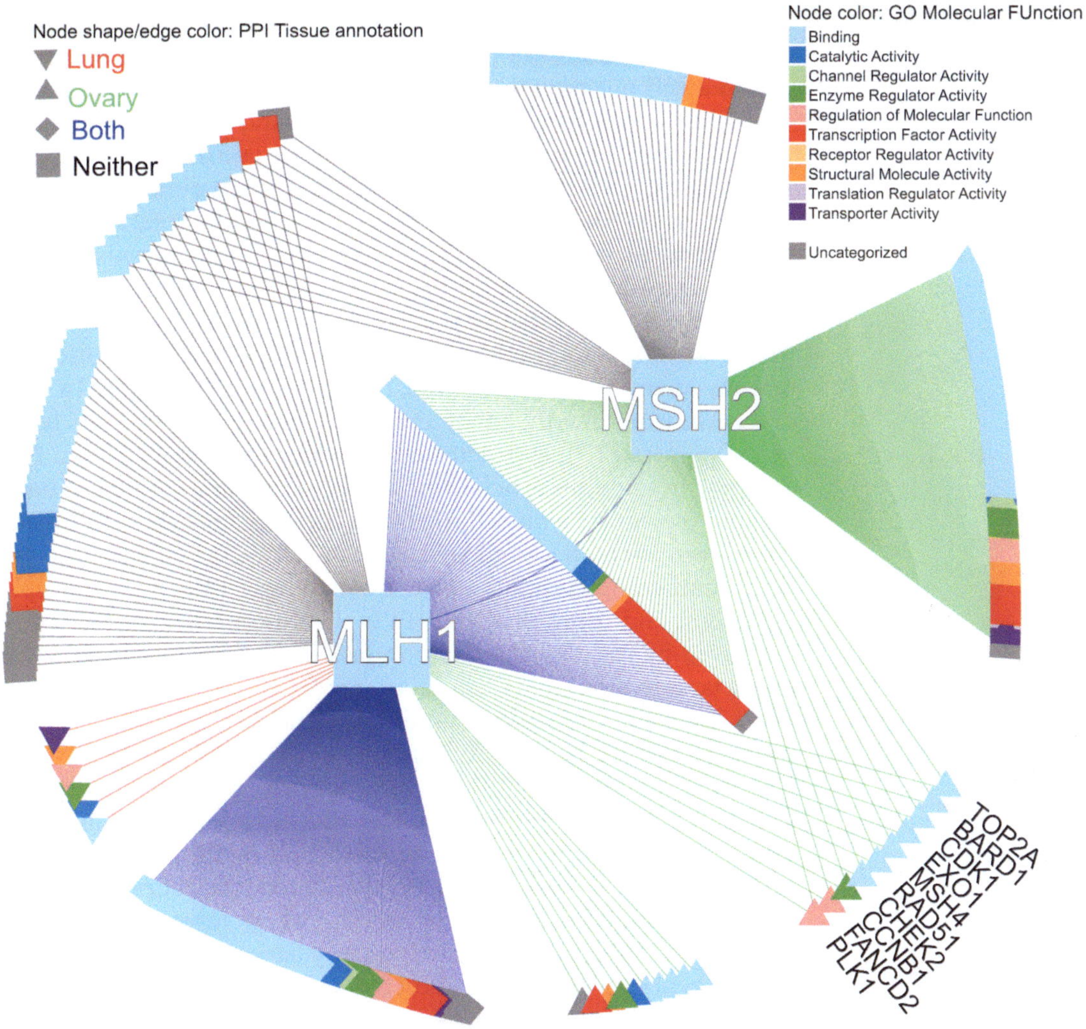

Fig. 2 Network representing PPIs obtained in our example search in IID. Colors represent Gene Ontology Molecular Function annotations, while node shapes and edge colors show PPIs belonging to different tissues (as per legend). Node size is proportional to node degree

steps needed to open a tab-separated file, at the end of which the PPI graph will be visible. The wizard also allows the user to import any annotation linked to the proteins or the PPIs.

2. The analysis tab allows the user to calculate topological features (such as degree, centrality) and identify key nodes (such as articulation points, labeled as "bi-connected components" in NAViGaTOR).
3. The appearance tab lets the user visually annotate any imported annotation (e.g., nodes can be colored based on presence in a specific tissue, and edge thickness can be proportional to the number of studies describing the PPI or its confidence level).

2.5 Potential Errors

Data returned by PPI databases may have two types of errors: missing interactions (i.e., interactions of query proteins that happen in a cell but are missing from the data) and false interactions (i.e., interactions that are reported in the data, but do not occur in cells). The rate of missing interactions, *false-negative rate = missing PPIs of query proteins/all PPIs of query proteins*, is difficult to assess, as there is no standard approach for estimating the total number of PPIs involving a set of proteins. However, for many nonhuman species, the rate of missing PPIs is very high, as almost no interactions have been detected or predicted. For human, and a few other well-studied species, the rate may still be quite high due to several reasons: many proteins have not been thoroughly studied by PPI detection experiments, many detection methods have high false-negative rates [23], and detected PPIs may be absent from a PPI database because the articles where they are reported have not been curated (or curators missed the interactions by mistake). PPI databases have adopted several strategies for reducing missing interactions: improving curation efficiency, including PPIs from multiple databases, and using predicted PPIs. Curation efficiency has been improved by dividing the curation task between multiple database projects and developing curation standards and quality controls [24]. Inclusion of PPIs from multiple databases is achieved in two ways: by storing PPIs from multiple-source databases (e.g., IID) and by using the PSICQUIC [25] service (e.g., IntAct), which searches multiple PPI databases. IID also includes computationally predicted interactions from state-of-the-art methods and interactions predicted by orthology [26]. Predicted interactions are especially important for nonhuman species, where most proteins have no detected PPIs. False interactions may have two causes: errors in PPI detection or prediction and errors in curation (i.e., an interaction reported in an article is incorrectly recorded in a database [27]). The rate of false interactions, *false discovery rate = false PPIs/identified PPIs*, is difficult to assess as well. Estimated false discovery rates of several high-throughput studies ranged from 35% to 83% [28]. Estimated false discovery rates for predicted PPIs ranged from 60% to 87% [29]. An investigation of curation errors found a 45% error rate [27]. Importantly, not all curation databases have such a high error rate, and curation process improvements reduced the error substantially [24]. Some PPI databases provide a degree of control over the false discovery rate from detection or prediction. IID enables users to filter interactions based on the numbers of studies or methods that identified them. Other databases (e.g., IntAct [10], iRefWeb [30], HAPPI-2 [31]) provide confidence scores for filtering interactions. Currently, there is no standard way to calculate confidence scores, but scores typically consider the number of studies that detected the interactions, as well as the number and types of detection methods. Filtering by number of studies (or scores that incorporate them) also reduces the false discovery rate from curation, since the same error is unlikely to occur in curation of multiple studies. However, an important drawback of such filtering is that few interactions are left and thus drastically increasing false-negative rate. For example, only about 15% of human experimentally detected PPIs are reported in more than one study. False discovery rate related to curation errors can be reduced by filtering on the number of databases that curated an interaction, but this also loses many interactions. Fortunately, primary databases from the IMEx consortium [24] have implemented systematic approaches for reducing curation errors including automated syntactic and semantic checking [32] and cross-curation [24].

3 Notes

1. Data retrieval from PPI databases can run into several common problems. The first usually being too many input IDs. To solve this, it is important to check database limitations before usage and, if needed, reduce the input list. It is key to check that the database being used allows to search for the ID at hand (or to use a tool to convert to another type of ID, e.g., UniProt https://www.uniprot.org/uploadlists/). All PPI databases sometimes fail to recognize input IDs—often due to formatting, unofficial IDs, uncommon IDs, and incorrect species. Formatting problems are typically simple issues such as the type of delimiters used between IDs (e.g., commas instead of carriage returns). The problem of unofficial gene names and symbols, and slight variations in spelling, is quite common since nonstandard nomenclature is widespread in literature. All databases have limitations on the types of IDs that are recognized. Some databases (e.g., IID) do not recognize input IDs that are not from the selected species, while other databases (e.g., IntAct) do not require a species to be selected and instead return PPIs from all species where the IDs are found. PPI databases may return interactions that are quite different than what a user intended. Some databases may return not only physical PPIs but also functional (STRING) or genetic interactions (BioGRID). It is fundamental to thoroughly read the description of a database before using it, to make sure the output is the one expected.
2. Several databases other than IID also enable filtering PPIs by context. HIPPIE [33], MyProteinNet [34], and TissueNet [35], allow filtering by tissue, based on whether the interacting proteins or their encoding genes are expressed in the tissue. HIPPIE [33] and MyProteinNet [34] support filtering by cellular localizations based on annotations of interacting

protein pairs. HIPPIE [33] also enables filtering by disease using a similar approach as for cellular localization.

3. Important nodes are identified through degree, centrality, and articulation status. Degree is the number of interactions that a protein has in the network, and it has been associated to the biological importance of the protein—proteins of high degree (hubs) tend to be conserved across species and might have an impact on phenotype [36]. Of note, it has been proposed that proteins associated to specific diseases tend to have higher degree because they (and their interactions) are studied more [37]. Moreover, it is important to keep in mind that a high number of connections of a hub protein could be condition-specific and consequently not happen unless the condition is met [38]. Node articulation status indicates whether removal of the node will disconnect the network—there will no longer be paths (sequences of edges) between certain nodes in the network. Nodes that disconnect the network can be essential for survival [4]. Querying a PPI database with just a few proteins will often return hundreds or thousands of interactions. Making sense of this network can be difficult and usually requires experience with visualization and network analysis software. However, a few PPI databases are starting to help with this task by providing intuitive topological and enrichment analysis functionality. A detailed review of network analyses tools and functionalities has recently been published [21].

Acknowledgments

The work was supported in part by the Canada Research Chair Program (CRC #225404), Krembil Foundation, Ontario Research Fund (GL2-01-030 and #34876), Natural Sciences Research Council (NSERC #203475), Canada Foundation for Innovation (CFI #225404, #30865), and IBM.

References

1. Navlakha S, Kingsford C (2010) The power of protein interaction networks for associating genes with diseases. Bioinformatics 26:1057–1063
2. Warde-Farley D, Donaldson SL, Comes O, Zuberi K, Badrawi R, Chao P, Franz M, Grouios C, Kazi F, Lopes CT et al (2010) The GeneMANIA prediction server: biological network integration for gene prioritization and predicting gene function. Nucleic Acids Res 38:W214–W220
3. Ashburner M, Ball CA, Blake JA, Botstein D, Butler H, Cherry JM, Davis AP, Dolinski K, Dwight SS, Eppig JT et al (2000) Gene ontology: tool for the unification of biology. Nat Genet 25:25–29
4. Przulj N, Wigle DA, Jurisica I (2004) Functional topology in a network of protein interactions. Bioinformatics 20:340–348
5. Barabási A-L, Gulbahce N, Loscalzo J (2011) Network medicine: a network-based approach to human disease. Nat Rev Genet 12:56–68

6. Wachi S, Yoneda K, Wu R (2005) Interactome-transcriptome analysis reveals the high centrality of genes differentially expressed in lung cancer tissues. Bioinformatics 21:4205–4208
7. Kotlyar M, Pastrello C, Rossos A, Jurisica I (2018) Protein–protein interaction databases. In: Ranganathan S, Nakai K, Schönbach C and Gribskov M (eds.), Encyclopedia of Bioinformatics and Computational Biology 1:988–996. Oxford: Elsevier
8. Breuer K, Foroushani AK, Laird MR, Chen C, Sribnaia A, Lo R, Winsor GL, Hancock REW, Brinkman FSL, Lynn DJ (2013) InnateDB: systems biology of innate immunity and beyond—recent updates and continuing curation. Nucleic Acids Res 41:D1228–D1233
9. Chautard E, Fatoux-Ardore M, Ballut L, Thierry-Mieg N, Ricard-Blum S (2011) MatrixDB, the extracellular matrix interaction database. Nucleic Acids Res 39:D235–D240
10. Orchard S, Ammari M, Aranda B, Breuza L, Briganti L, Broackes-Carter F, Campbell NH, Chavali G, Chen C, del-Toro N et al (2014) The MIntAct project—IntAct as a common curation platform for 11 molecular interaction databases. Nucleic Acids Res 42:D358–D363
11. Licata L, Briganti L, Peluso D, Perfetto L, Iannuccelli M, Galeota E, Sacco F, Palma A, Nardozza AP, Santonico E et al (2012) MINT, the molecular interaction database: 2012 update. Nucleic Acids Res 40:D857–D861
12. Salwinski L, Miller CS, Smith AJ, Pettit FK, Bowie JU, Eisenberg D (2004) The database of interacting proteins: 2004 update. Nucleic Acids Res 32:D449–D451
13. Chatr-aryamontri A, Oughtred R, Boucher L, Rust J, Chang C, Kolas NK, O'Donnell L, Oster S, Theesfeld C, Sellam A et al (2017) The BioGRID interaction database: 2017 update. Nucleic Acids Res 45:D369–D379
14. Szklarczyk D, Morris JH, Cook H, Kuhn M, Wyder S, Simonovic M, Santos A, Doncheva NT, Roth A, Bork P et al (2017) The STRING database in 2017: quality-controlled protein-protein association networks, made broadly accessible. Nucleic Acids Res 45:D362–D368
15. Keshava Prasad TS, Goel R, Kandasamy K, Keerthikumar S, Kumar S, Mathivanan S, Telikicherla D, Raju R, Shafreen B, Venugopal A et al (2009) Human protein reference database—2009 update. Nucleic Acids Res 37: D767–D772
16. Brandão MM, Dantas LL, Silva-Filho MC (2009) AtPIN: Arabidopsis thaliana protein interaction network. BMC Bioinformatics 10:454
17. Prieto C, De Las Rivas J (2006) APID: Agile Protein Interaction DataAnalyzer. Nucleic Acids Res 34:W298–W302
18. Cowley MJ, Pinese M, Kassahn KS, Waddell N, Pearson JV, Grimmond SM, Biankin AV, Hautaniemi S, Wu J (2012) PINA v2.0: mining interactome modules. Nucleic Acids Res 40: D862–D865
19. Kotlyar M, Pastrello C, Malik Z, Jurisica I (2019) IID 2018 update: context-specific physical protein–protein interactions in human, model organisms and domesticated species. Nucleic Acids Res 47:D581–D589
20. Brown KR, Otasek D, Ali M, McGuffin MJ, Xie W, Devani B, Toch IL, Jurisica I (2009) NAViGaTOR: Network Analysis, Visualization and Graphing Toronto. Bioinformatics 25:3327–3329
21. Hauschild A-C, Pastrello C, Rossos AEM, Jurisica I (2018) Visualization of biomedical networks. In: Reference module in life sciences. Elsevier
22. Emig D, Kacprowski T, Albrecht M (2011) Measuring and analyzing tissue specificity of human genes and protein complexes. EURASIP J Bioinform Syst Biol 2011:5
23. Braun P, Tasan M, Dreze M, Barrios-Rodiles-M, Lemmens I, Yu H, Sahalie JM, Murray RR, Roncari L, de Smet AS et al (2009) An experimentally derived confidence score for binary protein-protein interactions. Nat Methods 6:91–97
24. Orchard S, Kerrien S, Abbani S, Aranda B, Bhate J, Bidwell S, Bridge A, Briganti L, Brinkman FSL, Brinkman F et al (2012) Protein interaction data curation: the International Molecular Exchange (IMEx) consortium. Nat Methods 9:345–350
25. Aranda B, Blankenburg H, Kerrien S, Brinkman FSL, Ceol A, Chautard E, Dana JM, De Las Rivas J, Dumousseau M, Galeota E et al (2011) PSICQUIC and PSISCORE: accessing and scoring molecular interactions. Nat Methods 8:528–529
26. Brown KR, Jurisica I (2007) Unequal evolutionary conservation of human protein interactions in interologous networks. Genome Biol 8:R95
27. Cusick ME, Yu H, Smolyar A, Venkatesan K, Carvunis A-R, Simonis N, Rual J-F, Borick H, Braun P, Dreze M et al (2009) Literature-curated protein interaction datasets. Nat Methods 6:39–46
28. Hart GT, Ramani AK, Marcotte EM (2006) How complete are current yeast and human protein-interaction networks? Genome Biol 7:120

29. Kotlyar M, Pastrello C, Pivetta F, Lo Sardo A, Cumbaa C, Li H, Naranian T, Niu Y, Ding Z, Vafaee F et al (2015) In silico prediction of physical protein interactions and characterization of interactome orphans. Nat Methods 12:79–84
30. Turinsky AL, Razick S, Turner B, Donaldson IM, Wodak SJ (2014) Navigating the global protein-protein interaction landscape using iRefWeb. Methods Mol Biol 1091:315–331
31. Chen JY, Pandey R, Nguyen TM (2017) HAPPI-2: a comprehensive and high-quality map of human annotated and predicted protein interactions. BMC Genomics 18:182
32. Montecchi-Palazzi L, Kerrien S, Reisinger F, Aranda B, Jones AR, Martens L, Hermjakob H (2009) The PSI semantic validator: a framework to check MIAPE compliance of proteomics data. Proteomics 9:5112–5119
33. Alanis-Lobato G, Andrade-Navarro MA, Schaefer MH (2017) HIPPIE v2.0: enhancing meaningfulness and reliability of protein-protein interaction networks. Nucleic Acids Res 45:D408–D414
34. Basha O, Flom D, Barshir R, Smoly I, Tirman S, Yeger-Lotem E (2015) MyProteinNet: build up-to-date protein interaction networks for organisms, tissues and user-defined contexts. Nucleic Acids Res 43:W258–W263
35. Barshir R, Basha O, Eluk A, Smoly IY, Lan A, Yeger-Lotem E (2013) The TissueNet database of human tissue protein-protein interactions. Nucleic Acids Res 41:D841–D844
36. He X, Zhang J (2006) Why do hubs tend to be essential in protein networks? PLoS Genet 2: e88
37. Ideker T, Sharan R (2008) Protein networks in disease. Genome Res 18:644–652
38. Han J-D, Bertin N, Hao T, Goldberg DS, Berriz GF, Zhang LV, Dupuy D, Walhout AJM, Cusick ME, Roth FP et al (2004) Evidence for dynamically organized modularity in the yeast protein–protein interaction network. Nature 430:88–93

Chapter 11

Generation and Interpretation of Context-Specific Human Protein–Protein Interaction Networks with HIPPIE

Gregorio Alanis-Lobato and Martin H. Schaefer

Abstract

High-throughput techniques for the detection of protein–protein interactions (PPIs) have enabled a systems approach for the study of the living cell. However, the increasing amount of protein interaction data, the varying quality of these measurements, and the lack of context information make it difficult to construct meaningful and reliable protein networks.

The Human Integrated Protein–Protein Interaction rEference (HIPPIE) is a web tool that integrates and annotates experimentally supported human PPIs from a heterogeneous set of data sources. In HIPPIE, one can query for the interactors of one or more proteins and generate high-quality and context-specific networks. This chapter highlights HIPPIE's most important features and exemplifies its functionality through a proposed use case.

Key words Protein–protein interactions, Protein network, Context-specific networks, Network biology, Systems biology

1 Introduction

To date, more than 38,000 studies have been performed to detect human protein–protein interactions (PPIs). On the one hand, the resulting PPI network data is scattered over a large number of databases. Moreover, the experimental techniques employed to measure PPIs have largely different quality [1]. This raises the question of how to retrieve reliable and comparable sets of PPIs.

HIPPIE is a versatile web tool to query and analyze the human interactome. It automatically retrieves PPIs from several source databases, examines the evidence supporting each of the interactions, and transforms this evidence into a confidence score. By doing so, HIPPIE helps researchers to distinguish between true PPIs and likely false experimental measurements.

PPIs are under tight physiological control and happen only under certain conditions. Experimental techniques to detect PPIs

Stefan Canzar and Francisca Rojas Ringeling (eds.), *Protein-Protein Interaction Networks: Methods and Protocols*, Methods in Molecular Biology, vol. 2074, https://doi.org/10.1007/978-1-4939-9873-9_11,

are often performed under conditions that poorly resemble the physiological context in which the interaction happens in the cell (e.g., in vitro or human proteins expressed in yeast). To address the problem of missing context information, HIPPIE implements methods that assign functional and spatial information to PPIs and help the user to construct networks specific to a particular cellular context. This information is inferred from various attributes of the interacting proteins.

In the following, we describe the basic functionally of HIPPIE and provide a walk-through example of how to query the web tool for a protein of interest, apply filters to generate context-specific networks, and apply basic network algorithms to highlight relevant interactions.

1.1 Protein Interaction Retrieval

HIPPIE aggregates experimentally detected protein interactions using the PSICQUIC web service [2]. PSICQUIC is an effort of the Human Proteome Organization-Proteomics Standards Initiative (HUPO-PSI) to facilitate programmatic access to the molecular interactions reported by 33 providers [2]. We update HIPPIE's set of PPIs annually.

HIPPIE focuses on the retrieval of interactions annotated with the PSI-MI categories Association (MI, 0914), Physical Association (MI, 0915), Direct Interaction (MI, 0407), and Colocalization (MI:0403) [3, 4]. The PPI corpus includes data from IntAct [5], MINT [6], BioGRID [7], HPRD [8], DIP [9], BIND [10], and MIPS [11]. We integrate these interactions by mapping database-specific protein identifiers to Entrez gene IDs and then to their corresponding UNIPROT IDs, UNIPROT accessions, and gene symbols (*see* Subheading 2.3).

1.2 Protein Interaction Annotation and Enrichment Analysis

HIPPIE allows to construct tissue-, function-, subcellular localization-, and disease-specific protein networks. To make this possible, we annotate PPIs with gene expression information, Gene Ontology (GO) terms, and MeSH disease headings. To generate a context-specific network, proteins that are not expressed in the tissue of interest, or that are not annotated with the selected GO term or MeSH heading, are excluded from the result.

GTEx is our reference gene expression dataset [12]. It provides gene expression quantifications in 53 healthy tissues from postmortem samples. If the median expression of a gene over samples in a tissue is at least 1 RPKM, we consider that its gene product is present in that tissue [13]. Regarding the GO and MeSH terms, we take into account the hierarchical structure of these ontologies to annotate PPIs. Given a GO term or MeSH disease heading, we associate an interaction with the function or disease when both interactors are annotated with the given term/heading or with its children in the ontology hierarchy [14].

When a user queries a network in HIPPIE, she can carry out functional and disease-enrichment analyses of the constituent proteins of the resulting network via third-party tools. For the former, we use the web services provided by PANTHER [15] and for the latter the ones provided by GS2D [16]. In both cases, the statistically overrepresented GO terms or diseases associated with the network proteins are determined.

1.3 Edge Directionality and Interaction Effects

Result networks generated by HIPPIE can also be annotated with inferred PPI orientation reflecting the directionality of information flow in cellular signaling pathways. The inference process requires a predefined set of source and sink proteins. The user can provide these sets or employ the HIPPIE defaults, which correspond to proteins annotated with GO terms *receptor* (sources) and *sequence-specific DNA binding transcription factor activity* (sinks). All pairwise shortest paths between source-sink pairs are computed along the output network, and the direction of the path determines the direction of the edge. We do not assign directions to PPIs with conflicting orientations [14]. Directionality inference can be performed over the unweighted protein network or using HIPPIE confidence scores as edge weights (*see* Subheading 1.4).

To infer whether a protein activates or inhibits the activity of its interaction partners, we incorporate image data from a genome-wide RNAi knockdown screening into HIPPIE [17]. Furthermore, expert-curated PPI directions and effects from the KEGG database [18] can be used instead of the inferred ones.

1.4 HIPPIE Confidence Score

For each PPI in HIPPIE, we compute a confidence score that serves as an indicator of the amount and quality of the evidence supporting the interaction. This score is a weighted sum of three subscores:

$$S = w_s s_s + w_o s_o + w_t s_t$$

with $w_s + w_o + w_t = 1$ and each subscore s_i being a saturating function of the form

$$s_i(n) = \frac{2}{1 + e^{-\alpha_i n}} - 1$$

where α_i controls the steepness of the function.

For each subscore, parameter n has a specific meaning. When $i = s$, n is the number of studies in which the interaction was reported (i.e., its number of associated PubMed IDs). When $i = o$, n is the number of species in which orthologs of the interacting proteins also interact. Finally, when $i = t$, n is the sum of the reliability scores of the different techniques used to measure the PPI. These reliability scores, ranging from 0 to 10, were assigned to each technique by experts in PPI detection methods [4] and can be downloaded from HIPPIE's website (*see* Subheading 2.4).

To determine the best combination of values for parameters α_i and w_i, we performed a grid search with step 0.1 in the ranges [0, 3] and [0, 1], respectively. Roughly speaking, we successively removed each study, rescored the remaining dataset, and identified those parameters that would score the overlap between the two datasets highest (*see* [4] for more details).

1.5 Comparison with Other Web Tools

There are several other resources that integrate PPI data from different databases, allowing for the construction, visualization, and analysis of protein interaction networks. One of the earliest ones is the STRING database [19], which records functional associations between genes and scores gene pairs based on the probability that they participate in the same KEGG pathway [20]. UniHI stores experimental and predicted interactions between genes, proteins, and drugs and provides phenotype-based filters and enrichment analyses for the queried networks [21]. iRefWeb consolidates PPI data from different databases and organisms, scores interactions with an evidence-/homology-based metric and offers filters by this score, organism, and interaction type, to name a few [22]. HAPPI focuses on human physical and functional protein interactions that are scored based on the reliability of the experimental or computational platform used to measure or predict them, respectively [23]. InWeb_InBioMap collects human PPIs from eight sources and scores them based on their reproducibility in different publications [24].

The accumulation of tissue- and cell-specific gene and protein expression profiles has sparked the exploration of networks that form in different conditions [25]. In this framework, TissueNet is among the first resources to enable tissue-sensitive views of PPI networks [26]. In its current version, it uses the MyProteinNet tool [27] to construct a high-quality network with interactions from four databases that can be filtered based on gene and protein expression across tens of human tissues [26]. GIANT employs a Bayesian methodology to predict context-specific networks based on gene expression profiles in hundreds of human tissues and cell types [28]. Finally, the IID resource comprises experimentally detected PPIs, interologs, and predicted interactions and permits the identification of tissue-dependent networks that are conserved in two or more organisms [29].

HIPPIE differs from all the abovementioned tools in two main aspects: it incorporates only experimentally determined data for both PPIs and gene-tissue associations (*see* Subheadings 1.1 and 1.2), and it has a transparent, robust, and comprehensive confidence scoring system (*see* Subheading 1.4). These two points, in combination with protein- and interaction-based filters, enrichment analyses, directionality, and effect inference, allow for the generation of high-quality and meaningful protein networks that the users can examine in the light of their appropriate functional, tissue, and biomedical context.

1.6 HIPPIE Size

The first version of HIPPIE [4] comprised 73,137 PPIs. This number grew almost fourfold in HIPPIE v2.0 [30], for a total of 287,357 interactions and then to 340,629 in its most recent release (v2.1). The latter numbers are in agreement with current estimates of the size of the human interactome, which lie in the range between 154,000 [31, 32] and 650,000 [33] PPIs, excluding interactions between splice variants. However, only 27% of the interactions in HIPPIE v2.1 have high confidence scores ($\geq$0.72), which indicates that many human PPIs remain to be determined.

1.7 Limitations

One limitation of HIPPIE is that it contains only interaction information on human proteins. While we envision for further updates incorporating interactomes of other model organisms, for now, HIPPIE is limited to queries and analyses of human interactions.

An inherent property of PPI networks is a strong study bias toward disease proteins [34]: proteins involved in disease are studied more often for interaction partners and hence form larger networks in public PPI databases. This can lead to misinterpretation of biological network data [34]. In future updates of HIPPIE, we will offer to query optional bias-reduced networks, as well as tools to deal with the inherent bias in existing networks.

2 Materials

2.1 Hardware Requirements

HIPPIE is a web tool that can be accessed by any standard computer connected to the Internet. Queries and operations are solved on the server side, which means that HIPPIE does not have major memory requirements. In terms of hard disk space, the most common query results do not require more than a couple of available megabytes.

2.2 Software Requirements

HIPPIE is compatible with the most popular web browsers. It has been tested with Chrome (version $\geq$61), Firefox (version $\geq$55), Opera (version $\geq$47), Safari (version $\geq$11), and Internet Explorer (version $\geq$11). The link to the web tool is http://cbdm.uni-mainz.de/hippie/.

2.3 Valid Protein Identifiers

HIPPIE supports four different protein identifiers: UniProt IDs (e.g., HD_HUMAN), UniProt accessions (e.g., P42858), gene symbols (e.g., HTT), and Entrez gene IDs (e.g., 3064).

2.4 Results Download

Single and multiple protein queries in HIPPIE result in a table with a button that allows to export it as a tab-separated file. It is also possible to graphically represent these tables through a network visualization that can be downloaded as a PNG, JPEG, or JSON file (the latter file format enables export to other graph analysis and visualization tools such as Cytoscape [35]). Full HIPPIE datasets and scores assigned to experimental techniques (*see* Subheading 1.4) are available under the HIPPIE's Download tab.

2.5 REST Web Services

Users can also access HIPPIE via its REST web service, making it easy to integrate query results into bioinformatics pipelines. REST requests to HIPPIE are based on the following template:

`http://cbdm-01.zdv.uni-mainz.de/~mschaefer/hippie/queryHIPPIE.php?`**`proteins`**`=xxx,xxx;xxx|xxx&`**`layers`**`=xxx&`**`conf_thres`**`=xxx&`**`out_type`**`=xxx.`

The parts of the template that are highlighted in boldface correspond to parameters required for REST requests:

- `proteins` = One or more proteins of interest separated by ",", ";" or "|" (mandatory).
- `layers` = 0 to query interactions within the input set or 1 to query interactions between the input set and HIPPIE (optional, default = 1).
- `conf_thres` = Only protein interactions with confidence scores above this threshold, which ranges between 0 and 1, are considered (optional, default = 0).
- `out_type` = The query output format. *Browser* shows the list of interactions in a table in HIPPIE, viz shows a network visualization, *mitab* generates a MITAB file, and *conc_file* generates a simple tab-separated text file (optional, default = *conc_file*).

Note that when not specified, parameters take their default values.

3 Proposed Use Case

3.1 The TP53 Interactome

HIPPIE allows querying for interactions of single proteins by different types of identifiers from the start page or the network query page, which offers a large variety of filter, analysis, or visualization options. If we, for example, query HIPPIE for interaction partners of the protein TP53 by entering its name into the search field on the start page, we retrieve 1043 interaction partners in a tabular format. If we instead enter TP53 into the search field in the network query tab without changing any parameters, we retrieve a graphical representation of the network surrounding TP53 (Fig. 1a). It becomes immediately apparent that for proteins with a high number of interaction partners (such as TP53, which is among the proteins with the largest number of binding partners),

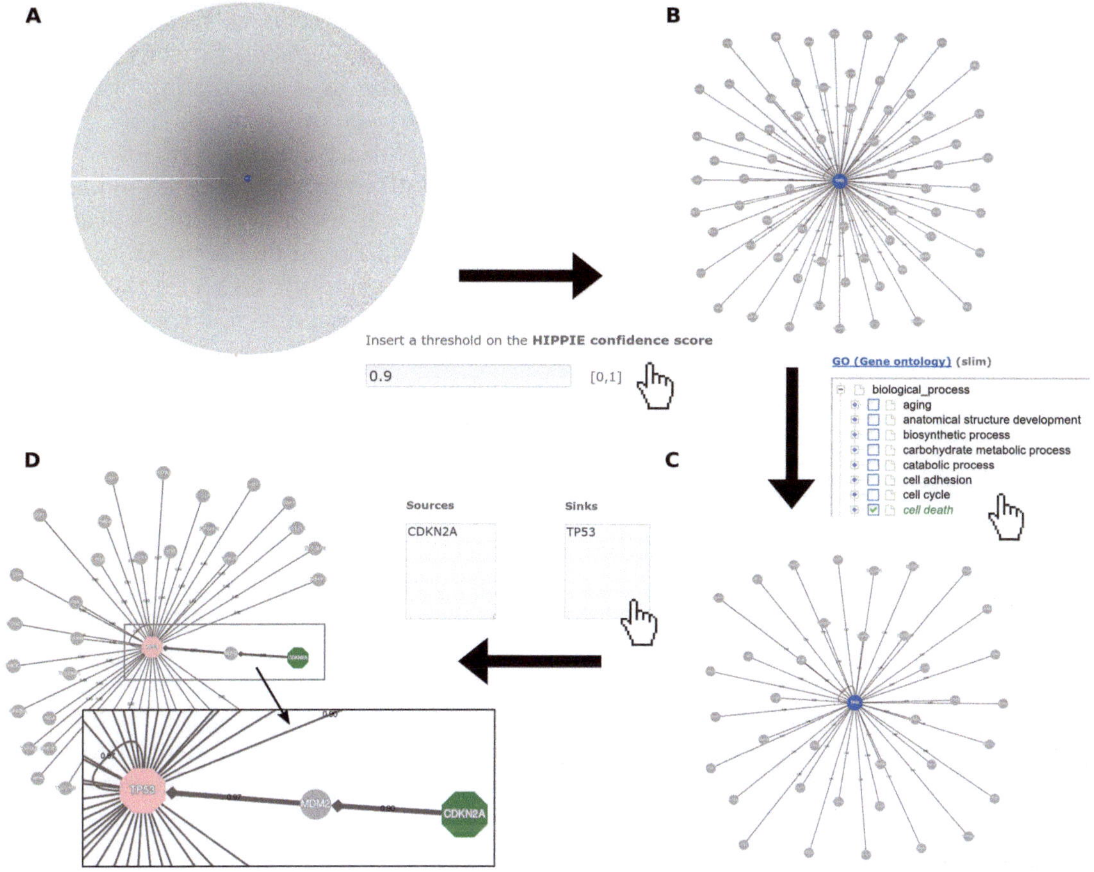

Fig. 1 Subsequent application of filters and network analyses with HIPPIE. (**a**) For genes with many interaction partners, the output is a large "hairball network." Here, the TP53 interactome is shown. (**b**) Confidence filtering leads to a largely reduced network. (**c**) Only showing TP53 binding partners involved in the "cell death" GO term further reduces the network. (**d**) Shortest path analysis can identify important interactions. Here, computing the shortest path between CDKN2A and TP53 correctly highlights the regulatory relation between CDKN2A and MDM2 and MDM2 and TP53, respectively

it is difficult to extract any useful biological insights into the function and regulation of the protein from the resulting hairball network (Fig. 1a) without applying any filters. One easy way of filtering is to remove interactions which are supported only by weak experimental evidence. In the case of TP53, this reduces the resulting network to 467 PPIs when the default "high-confidence" filter of 0.72 is applied and further down to 69 PPIs when the even more stringent cutoff of 0.9 on the confidence score is applied. The resulting network is concise enough for visual inspection and allows for the easy identification of interactions important for TP53 biology (Fig. 1b).

3.2 Function-Specific Network

TP53 is a master regulator of apoptosis and cellular senescence upon different stimuli [36] but has a multifaceted biology including other molecular functions, such as facilitating DNA repair by holding the cell cycle upon detection of DNA damage [37]. One of HIPPIE's major strengths is the creation of context-specific networks that allows to filter out proteins from a result network that are not associated with a particular function.

Given our TP53 example, one could be interested in identifying interaction partners of TP53 that are involved in cell death and to exclude other cellular functions. If the TP53 network is reduced to its cell death interactome by selecting the GO term "cell death" from the GO tree view menu in the functional filter category of the "network query" tab, the remaining network has 38 PPIs (Fig. 1c) including many known regulators of TP53-initiated apoptosis, such as TP63 or TP53BP2.

3.3 Regulation of TP53

TP53 is embedded into regulatory pathways, which tightly control TP53 activation [38]. The canonical pathway activating TP53 upon oncogenic signaling is mediated by the E3 ubiquitin ligase MDM2 (by ubiquitination, which targets TP53 for degradation). One regulator of MDM2 is CDKN2A (by binding and sequestration in the nucleolus). One application of HIPPIE is to infer directionality of signal flow by applying shortest path algorithms. If we query HIPPIE with TP53 and CDKN2A (as we know that it is a top-level regulator of TP53) and apply the same filters as mentioned above (cell death + 0.9 confidence score), we retrieve a network of 39 interactions. We can then add CDKN2A to the sources and TP53 to the sinks to predict signal flow between CDKN2A and TP53. Doing so correctly highlights CDKN2A as a regulator of MDM2 and MDM2 as the mediator of TP53 activation (Fig. 1d).

Acknowledgments

HIPPIE has been developed and is maintained in the lab of Prof. Miguel Andrade. He also gave valuable feedback on this book chapter. We also thank the Zentrum für Datenverarbeitung of the Johannes Gutenberg Universität for their help in the maintenance of the web server that hosts HIPPIE.

References

1. von Mering C, Krause R, Snel B, Cornell M, Oliver SG, Fields S, Bork P (2002) Comparative assessment of large-scale data sets of protein–protein interactions. Nature 417:399–403
2. del-Toro N, Dumousseau M, Orchard S, Jimenez RC, Galeota E, Launay G, Goll J, Breuer K, Ono K, Salwinski L et al (2013) A new reference implementation of the PSICQUIC web service. Nucleic Acids Res 41:W601–W606
3. Hermjakob H, Montecchi-Palazzi L, Bader G, Wojcik J, Salwinski L, Ceol A, Moore S, Orchard S, Sarkans U, von Mering C et al (2004) The HUPO PSI's molecular interaction format—a community standard for the representation of protein interaction data. Nat Biotechnol 22:177–183
4. Schaefer MH, Fontaine J-F, Vinayagam A, Porras P, Wanker EE, Andrade-Navarro MA (2012) HIPPIE: integrating protein

interaction networks with experiment based quality scores. PLoS One 7:e31826

5. Kerrien S, Aranda B, Breuza L, Bridge A, Broackes-Carter F, Chen C, Duesbury M, Dumousseau M, Feuermann M, Hinz U et al (2012) The IntAct molecular interaction database in 2012. Nucleic Acids Res 40: D841–D846
6. Licata L, Briganti L, Peluso D, Perfetto L, Iannuccelli M, Galeota E, Sacco F, Palma A, Nardozza AP, Santonico E et al (2012) MINT, the molecular interaction database: 2012 update. Nucleic Acids Res 40:D857–D861
7. Chatr-aryamontri A, Breitkreutz B-J, Oughtred R, Boucher L, Heinicke S, Chen D, Stark C, Breitkreutz A, Kolas N, O'Donnell L et al (2015) The BioGRID interaction database: 2015 update. Nucleic Acids Res 43: D470–D478
8. Keshava Prasad TS, Goel R, Kandasamy K, Keerthikumar S, Kumar S, Mathivanan S, Telikicherla D, Raju R, Shafreen B, Venugopal A et al (2009) Human protein reference database—2009 update. Nucleic Acids Res 37: D767–D772
9. Salwinski L, Miller CS, Smith AJ, Pettit FK, Bowie JU, Eisenberg D (2004) The database of interacting proteins: 2004 update. Nucleic Acids Res 32:D449–D451
10. Isserlin R, El-Badrawi RA, Bader GD (2011) The biomolecular interaction network database in PSI-MI 2.5. Database 2011:baq037
11. Pagel P, Kovac S, Oesterheld M, Brauner B, Dunger-Kaltenbach I, Frishman G, Montrone C, Mark P, Stumpflen V, Mewes H-W et al (2005) The MIPS mammalian protein-protein interaction database. Bioinformatics 21:832–834
12. The GTEx Consortium (2013) The genotype-tissue expression (GTEx) project. Nat Genet 45:580–585
13. Mortazavi A, Williams BA, McCue K, Schaeffer L, Wold B (2008) Mapping and quantifying mammalian transcriptomes by RNA-Seq. Nat Methods 5:621–628
14. Schaefer MH, Lopes TJS, Mah N, Shoemaker JE, Matsuoka Y, Fontaine J-F, Louis-Jeune C, Eisfeld AJ, Neumann G, Perez-Iratxeta C et al (2013) Adding protein context to the human protein-protein interaction network to reveal meaningful interactions. PLoS Comput Biol 9:e1002860
15. Mi H, Poudel S, Muruganujan A, Casagrande JT, Thomas PD (2016) PANTHER version 10: expanded protein families and functions, and analysis tools. Nucleic Acids Res 44: D336–D342
16. Andrade-Navarro MA, Fontaine JF (2016) Gene set to diseases (GS2D): disease enrichment analysis on human gene sets with literature data. Genom Comput Biol 2:e33
17. Suratanee A, Schaefer MH, Betts MJ, Soons Z, Mannsperger H, Harder N, Oswald M, Gipp M, Ramminger E, Marcus G et al (2014) Characterizing protein interactions employing a genome-wide siRNA cellular phenotyping screen. PLoS Comput Biol 10: e1003814
18. Kanehisa M, Goto S (2000) KEGG: Kyoto encyclopedia of genes and genomes. Nucleic Acids Res 28:27–30
19. Szklarczyk D, Franceschini A, Wyder S, Forslund K, Heller D, Huerta-Cepas J, Simonovic M, Roth A, Santos A, Tsafou KP et al (2015) STRING v10: protein-protein interaction networks, integrated over the tree of life. Nucleic Acids Res 43:D447–D452
20. von Mering C, Jensen LJ, Hooper SD, Krupp M, Foglierini M, Jouffre N, Huynen MA, Bork P (2005) STRING: known and predicted protein-protein associations, integrated and transferred across organisms. Nucleic Acids Res 33:D433–D437
21. Kalathur RKR, Pinto JP, Hernández-Prieto MA, Machado RSR, Almeida D, Chaurasia G, Futschik ME (2014) UniHI 7: an enhanced database for retrieval and interactive analysis of human molecular interaction networks. Nucleic Acids Res 42:D408–D414
22. Turner B, Razick S, Turinsky AL, Vlasblom J, Crowdy EK, Cho E, Morrison K, Donaldson IM, Wodak SJ (2010) iRefWeb: interactive analysis of consolidated protein interaction data and their supporting evidence. Database 2010:baq023
23. Chen JY, Pandey R, Nguyen TM (2017) HAPPI-2: a comprehensive and high-quality map of human annotated and predicted protein interactions. BMC Genomics 18:182
24. Li T, Wernersson R, Hansen RB, Horn H, Mercer J, Slodkowicz G, Workman CT, Rigina O, Rapacki K, Staerfeldt HH et al (2017) A scored human protein–protein interaction network to catalyze genomic interpretation. Nat Methods 14:61–64
25. Yeger-Lotem E, Sharan R (2015) Human protein interaction networks across tissues and diseases. Front Genet 6:257
26. Basha O, Barshir R, Sharon M, Lerman E, Kirson BF, Hekselman I, Yeger-Lotem E (2017) The TissueNet v.2 database: a quantitative view of protein-protein interactions across human tissues. Nucleic Acids Res 45:D427–D431

27. Basha O, Flom D, Barshir R, Smoly I, Tirman S, Yeger-Lotem E (2015) MyProteinNet: build up-to-date protein interaction networks for organisms, tissues and user-defined contexts. Nucleic Acids Res 43:W258–W263
28. Greene CS, Krishnan A, Wong AK, Ricciotti E, Zelaya RA, Himmelstein DS, Zhang R, Hartmann BM, Zaslavsky E, Sealfon SC et al (2015) Understanding multicellular function and disease with human tissue-specific networks. Nat Genet 47:569–576
29. Kotlyar M, Pastrello C, Sheahan N, Jurisica I (2016) Integrated interactions database: tissue-specific view of the human and model organism interactomes. Nucleic Acids Res 44: D536–D541
30. Alanis-Lobato G, Andrade-Navarro MA, Schaefer MH (2017) HIPPIE v2.0: enhancing meaningfulness and reliability of protein–protein interaction networks. Nucleic Acids Res 45:D408–D414
31. Hart GT, Ramani AK, Marcotte EM (2006) How complete are current yeast and human protein-interaction networks. Genome Biol 7:120
32. Venkatesan K, Rual J-F, Vazquez A, Stelzl U, Lemmens I, Hirozane-Kishikawa T, Hao T, Zenkner M, Xin X, Goh K-I et al (2009) An empirical framework for binary interactome mapping. Nat Methods 6:83–90
33. Stumpf MP, Thorne T, de Silva E, Stewart R, An HJ, Lappe M, Wiuf C (2008) Estimating the size of the human interactome. Proc Natl Acad Sci 105:6959–6964
34. Schaefer MH, Serrano L, Andrade-Navarro MA (2015) Correcting for the study bias associated with protein–protein interaction measurements reveals differences between protein degree distributions from different cancer types. Front Genet 6:260
35. Shannon P (2003) Cytoscape: a software environment for integrated models of biomolecular interaction networks. Genome Res 13:2498–2504
36. Meek DW (2015) Regulation of the p53 response and its relationship to cancer. Biochem J 469:325–346
37. Williams AB, Schumacher B (2016) p53 in the DNA-damage-repair process. Cold Spring Harb Perspect Med 6:a026070
38. Hollstein M, Hainaut P (2010) Massively regulated genes: the example of TP53. J Pathol 220:164–173

Chapter 12

Explore Protein–Protein Interactions for Cancer Target Discovery Using the OncoPPi Portal

Andrey A. Ivanov

Abstract

Protein–protein interactions (PPIs) control all functions and physiological states of the cell. Identification and understanding of novel PPIs would facilitate the discovery of new biological models and therapeutic targets for clinical intervention. Numerous resources and PPI databases have been developed to define a global interactome through the PPI data mining, curation, and integration of different types of experimental evidence obtained with various methods in different model systems. On the other hand, the recent advances in cancer genomics and proteomics have revealed a critical role of genomic alterations in acquisition of cancer hallmarks through a dysregulated network of oncogenic PPIs. Deciphering of cancer-specific interactome would uncover new mechanisms of oncogenic signaling for therapeutic interrogation. Toward this goal our team has developed a high-throughput screening platform to detect PPIs between cancer-associated proteins in the context of cancer cells. The established network of oncogenic PPIs, termed the OncoPPi network, is available through the OncoPPi Portal, an interactive web resource that allows to access and interpret a high-quality cancer-focused network of PPIs experimentally detected in cancer cell lines integrated with the analysis of mutual exclusivity of genomic alterations, cellular co-localization of interacting proteins, domain-domain interactions, and therapeutic connectivity. This chapter presents a guide to explore the OncoPPi network using the OncoPPi Portal to facilitate cancer biology.

Key words Cancer, Oncogenic signaling, High-throughput PPI screening, Target discovery, Data integration, Web resource

1 Introduction

Protein–protein interactions (PPIs) dictate the signal transduction and regulate diverse physiological programs, including cell death and survival. Discovery and understanding of novel PPIs may uncover new pathways and mechanisms of how cell functions are regulated. Indeed, over the past decades, the PPIs have emerged as promising therapeutic targets in proliferative disease, including cancer, cardiovascular disease, diabetes, and neurodegenerative disorders [1–5]. Recent advances in high-throughput screening and proteomic technologies have enabled large-scale PPI studies that

Stefan Canzar and Francisca Rojas Ringeling (eds.), *Protein-Protein Interaction Networks: Methods and Protocols*, Methods in Molecular Biology, vol. 2074, https://doi.org/10.1007/978-1-4939-9873-9_12,

resulted in the discovery of thousands of PPIs involved in hundreds of metabolic and signaling pathways [6–11]. Consequently, a growing number of PPIs detected with various methods under different physiological conditions in different model systems (in most cases in normal human HEK293T cells or yeast) are now available in large PPI databases, providing a general landscape of human interactome [12–17].

However, in cancer cells, genomic alterations, including gene mutations, deletions, or amplifications, lead to a rewired network of PPIs that promotes the acquisition of cancer hallmarks [18, 19]. Discovery of those cancer-associated PPIs could lead to new biological models for oncogenic signaling and enable new strategies for cancer therapeutic development. Indeed, emerging pharmacological and clinical data suggest a highly promising role for cancer-specific PPIs as druggable cancer targets [4, 19–21]. Accordingly, a number of powerful bioinformatics approaches have been developed to predict oncogenic PPIs based on mRNA expression data analysis [22, 23] or through the analysis of mutual exclusivity of genomic alterations [24, 25] or the analysis of genomic dependencies in loss-of-function screens [26–28]. On the other hand, experimental detection of cancer-associated PPIs in a cancer cell environment remains a challenge, and resources focused on physical oncogenic PPIs are limited [29–31].

Recently, our team has developed a time-resolved fluorescence energy transfer (TR-FRET)-based high-throughput screening (HTS) platform to establish a network of cancer-associated PPIs determined in cancer cells [32]. The characterization of ~3500 PPIs tested for a set of lung cancer-related proteins combined with extensive statistical analysis resulted in a network of high-confidence direct PPIs, termed OncoPPi (version 1). Evaluation of PPIs reported in public databases revealed that more than 85% of the OncoPPi interactions are novel. Furthermore, the validation of newly discovered PPIs with conventional low-throughput methods, such as the affinity pull-down assays, revealed that at least 80% of OncoPPi PPIs can be confirmed as true-positive interactions. Detailed analysis of OncoPPi network led to a discovery of new models to regulate oncogenic pathways, for example, through MKK3-MYC, NSD3-MYC, or STK11-CDK4 PPIs [32–34].

To enable streamlined and integrated analysis of PPI datasets, the OncoPPi Portal has been developed [35]. The OncoPPi Portal is a web-based resource that integrates the network of experimentally detected cancer-associated PPIs with cancer genomic, pharmacological, and protein structural data. This chapter will describe the detailed procedures to explore and annotate the cancer-related PPIs using the OncoPPi Portal to facilitate discovery of new oncogenic programs and support cancer research.

2 Materials

2.1 The OncoPPi Portal

The OncoPPi Portal [35] provides an interactive user-friendly interface to explore the cancer-associated PPI networks generated in the PPI high-throughput screening (HTS) experiments to uncover novel new biological models and potential targets for therapeutic interrogation in cancer. The portal is freely available online at http://oncoppi.emory.edu.

2.2 OncoPPi Datasets

The following datasets are available through the OncoPPi Portal: (a) The OncoPPi network of cancer-related protein–protein interactions established with the time-resolved fluorescence energy transfer (TR-FRET) high-throughput screening experiments performed in cancer cell lines [32]. This is the major part of the Portal and the main focus of this chapter. (b) The PPI network established for master regulator transcription factor MYC using NanoLuc-based protein-fragment complementation assay (NanoPCA) performed in lung and colon cancer cell lines [36] (*see* **Note 1**). (c) The therapeutic connectivity network dataset connects experimentally determined cancer-related PPIs with currently approved drugs, as described before [35]. (d) The Portal also provides the graphical interface for the Mining Essentiality Data to Identify Critical Interactions (MEDICI) algorithm that allows to estimate the essentiality of 7900 PPIs in 206 different cancer cell lines as described previously [37].

2.3 External Resources

To facilitate the data mining and annotations of experimentally determined cancer-associated PPIs, the OncoPPi links individual genes, PPIs, and small molecules with a number of external resources, including the general protein annotation servers such as Human Genome Organization (HUGO) Gene Nomenclature Committee (HGCN) [38], Ensembl [39], UniProt [40], and Gene [41]; literature database PubMed [41]; cancer-focused resources such as Cancer Target Discovery and Development (CTD^2) Dashboard [42], TumorPortal [43], and cBioPortal [44, 45]; pharmacological databases such as Cancer Therapeutics Response Portal (CTRP) [46], PubChem [47], DrugBank [48], and Genomics of Drug Sensitivity in Cancer (GDSC) server; and [49] general PPI databases such as String [15], GeneMania [17], and IntAct [13] (Fig. 1, Table 1).

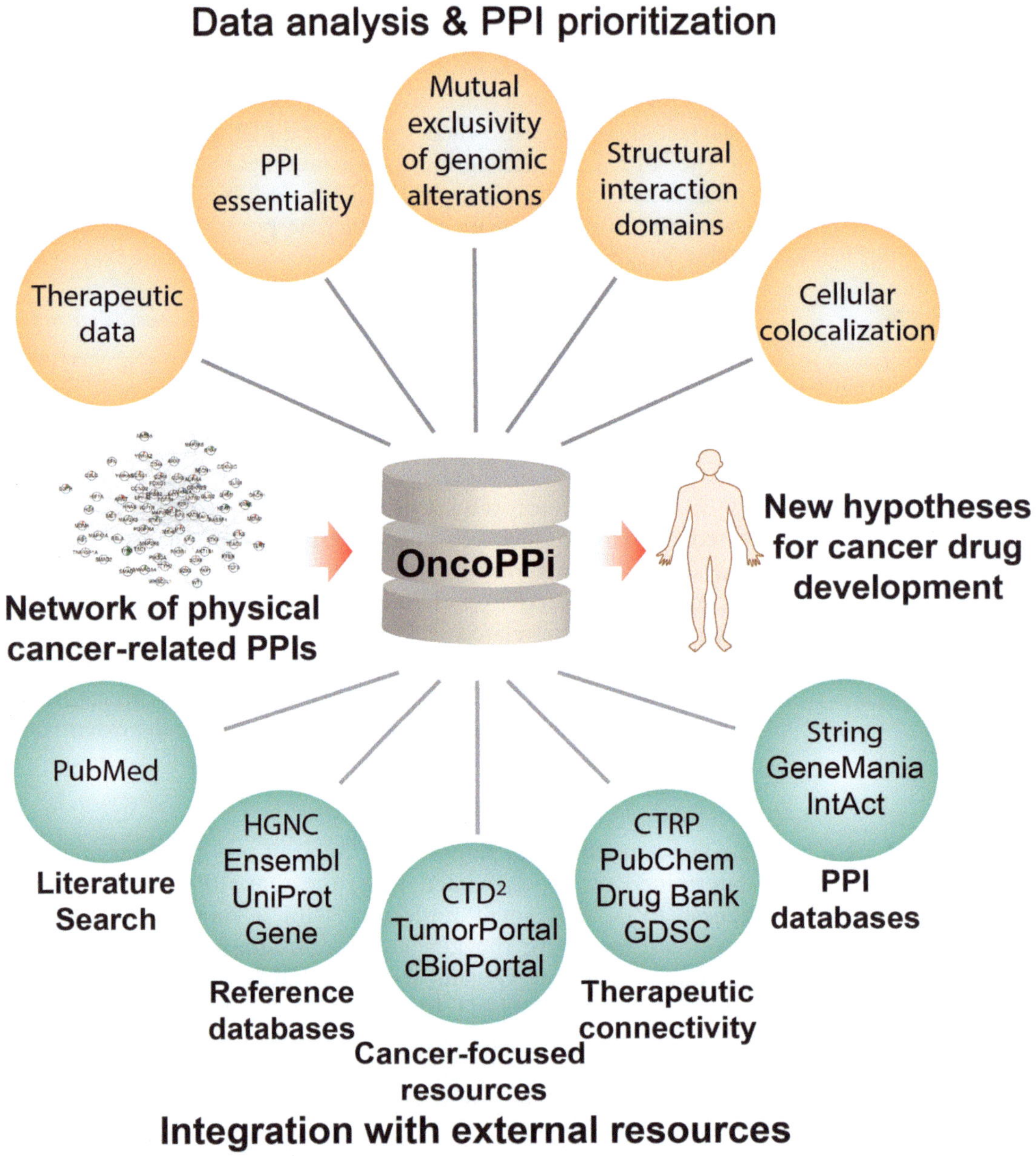

Fig. 1 OncoPPi Portal overview. The OncoPPi Portal provides a web-based interface to explore, visualize, and export the network of experimentally determined cancer-associated PPIs. To facilitate prioritization of PPIs for further biological studies, the PPI network is integrated with genomic, pharmacological, and structural data. A direct connection of PPIs and individual proteins with external resources enables detailed protein annotations and efficient data mining. Together, the OncoPPi Portal provides a framework to generate new hypotheses and biological models for cancer target discovery

Table 1
External resources integrated with the OncoPPi network

Database	Link	Description	Reference
HGNC	https://www.genenames.org	Human Genome Organization (HUGO) Gene Nomenclature Committee (HGCN) database	[38]
Ensembl	https://ensembl.org	A genome browser that annotates genes, computes multiple alignments, predicts regulatory function, and collects disease data	[39]
UniProt	https://www.uniprot.org	Resource of protein sequence and functional information	[40]
Gene	https://www.ncbi.nlm.nih.gov/gene	Gene database. Integrates information from different species and provides nomenclature, reference sequences, pathways, variations, phenotypes, and others	[41]
PubMed	https://www.ncbi.nlm.nih.gov/pubmed	Biomedical literature database	[41]
CTD^2 Dashboard	https://ctd2-dashboard.nci.nih.gov/dashboard	The Cancer Target Discovery and Development Dashboard combines and provides access to cancer-focused observations generated by the members (centers) of the CTD^2 network	[42]
TumorPortal	http://www.tumorportal.org	Resource to explore and analyze mutations in cancer genes across 21 cancer types	[43]
cBioPortal	http://www.cbioportal.org	Provides visualization and analysis of large-scale cancer genomic datasets	[44, 45]
CTRP	https://portals.broadinstitute.org/ctrp.v2.1	Cancer therapeutics response portal links genetic, lineage, and other cellular features of cancer cell lines to small-molecule sensitivity	[46]
PubChem	https://pubchem.ncbi.nlm.nih.gov	Provides information on chemical properties and biological activities of chemical molecules	[47]
DrugBank	https://www.drugbank.ca	Drug and drug target database	[48]
GDSC	https://www.cancerrxgene.org	Genomics of drug sensitivity in Cancer server helps to identify molecular features of cancers that predict response to anticancer drugs	[49]
String	https://string-db.org	Database of known and predicted PPIs	[15]
GeneMania	https://genemania.org	Physical and functional PPI database	[17]

(continued)

Table 1 (continued)

Database	Link	Description	Reference
IntAct	https://www.ebi.ac.uk/intact	Molecular interaction database	[13]
Pfam	https://pfam.xfam.org	Database of protein families and structural domains	[59]
3DID	https://3did.irbbarcelona.org	Database of three-dimensional interacting domains	[57]

3 Methods

3.1 Browsing the PPI Networks

Network of cancer-related PPIs. The OncoPPi Portal enables visualization and browsing of experimentally determined networks of PPIs between cancer-associated proteins. Currently, two PPI datasets are available to explore: (a) the recently published OncoPPi v1 dataset built on the TR-FRET-based PPI screening experiments performed in lung cancer H1299 cells [32] and (b) a focused set of PPIs detected in lung H1299 and colon cancer HCT116 cell lines for the master regulator transcription factor MYC using the NanoPCA assay (*see* **Note 2**) [36].

A user can access both datasets through the Networks folder located on the front page of the Portal. For example, to explore the OncoPPi network, a user can select the OncoPPi set from the Networks menu, and the OncoPPi network will appear for the analysis. A total of 3486 PPIs tested in the TR-FRET high-throughput PPI screening are available through the Portal along with the corresponding statistical characteristics. That includes the fold-over-control (FOC) values, permutation test p-values, and q-values [32]. In contrast to other public PPI databases, the OncoPPi Portal provides experimental data for both positive and negative interactions that allows to identify the PPIs tested and not detected under the indicated experimental conditions. The negative PPIs are defined as PPIs that demonstrate the FOC < 1.2 and/or p-value >0.05.

As described previously [32], three levels of positive PPIs can be defined:

1. Statistically significant PPIs (SS-PPIs) are characterized by the FOC ≥ 1.2 and p-value <0.05.
2. High-confidence PPIs (HC-PPIs) are characterized by the FOC ≥ 1.5, p-value <0.05, and q-value <0.01.
3. The OncoPPi network includes all HC-PPIs plus SS-PPIs previously detected in other assays.

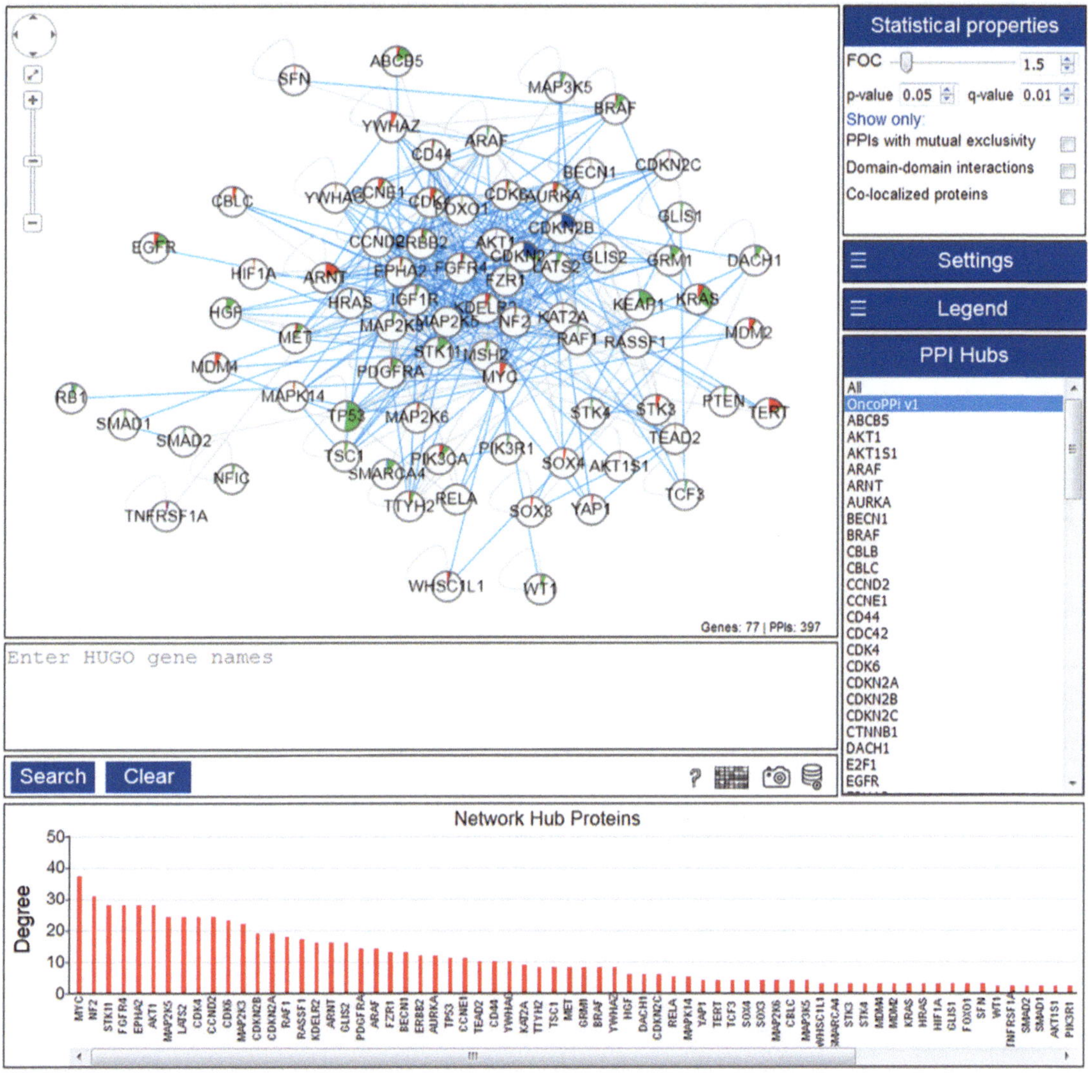

Fig. 2 The main window with the PPI network. The OncoPPi Portal allows the user to explore the network of PPIs experimentally determined in cancer cells. The PPIs supported by mutual exclusivity of genomic alterations are highlighted with blue lines. The frequencies of gene mutations, amplification, or deletions in lung cancer patients are indicated inside the corresponding circles with green, red, and blue sectors, respectively. The Statistical Properties panel allows to adjust the thresholds of the FOC, *p*-value, and *q*-value and filter the PPIs based on the mutual exclusivity of genomic alterations, cellular co-localization, and domain-domain interactions. The distribution of the hub protein degree in selected PPI network is visualized with the interactive bar graph located under the network window

User interface. The page for PPI network visualization and analysis consists of the main window with the PPI network (Fig. 2), Statistical Properties panel that allows to control statistical parameters and filters for network visualization, the Settings panel that provides a control for network layouts, the PPI Hubs panel that lists all proteins available for the analysis, as well as the predefined OncoPPi network version 1 set of PPIs. There is a Search panel located under the main network window. The Tool panel located under the Search panel allows to export the network-associated data either in tabular or graphical format, as well as to visualize the corresponding FOC values as a heatmap. It also provides a link to the OncoPPi Portal manual. The interactive Network Hub Proteins graph located at the bottom of the page shows the distribution of protein degree in currently selected PPI network.

PPI hub proteins. The OncoPPi Portal provides several ways to visualize the PPI network for the analysis. For example, the SS-PPI network can be generated by selecting "All" from the PPI Hubs menu and setting the FOC threshold to 1.2, p-value threshold to 0.05, and q-value to 1 in the "Statistical Properties" panel. The generated network of SS-PPIs includes 798 PPIs detected for 80 proteins. Changing the settings to FOC = 1.5, p-value = 0.05, and q-value = 0.01 will result in a network of 348 HC-PPIs determined for 72 proteins. The OncoPPi network is a predefined set that includes HC-PPIs and SS-PPIs supported by external experimental evidence [32]. This network is shown by default and contains 397 PPIs experimentally determined for 77 proteins. The OncoPPi network can be also visualized by selecting the "OncoPPi v1" from the PPI Hubs menu.

Detailed PPI network analyses, biological studies, and gene knockout experiments have revealed that deletions of proteins that are involved in many PPIs, known as hub proteins, are likely to be more lethal for the cells compared to non-hub protein deletions [50, 51]. Therefore, identification of hub proteins in the cancer-associated PPI network may facilitate a discovery of novel critical regulators of tumorigenesis as potential therapeutic targets. Accordingly, the analysis of the degree of protein connectivity is implemented as an interactive bar graph shown under the PPI network. The Network Hub Proteins graph allows an immediate visualization and ranking of the major hub proteins and identification of their binding partners for further analysis. Specifically, the X-axis of the graph shows the proteins included in the given network, and the Y-axis shows the number of protein binding partners (node degree) identified with the selected statistical parameters. The protein order along the X-axis and data shown on the graph changes dynamically when the network and statistical thresholds are changed. Furthermore, the subnetwork of binding partners of the hub proteins can be immediately accessed by clicking on the corresponding bar on the graph. The same hub protein PPI subnetwork can be also shown by selecting the corresponding protein name from the PPI Hubs panel on the right side of the page.

Search for specific proteins. The Search panel allows the user to query the network for single or multiple proteins. A selection of multiple proteins results in a subnetwork of PPIs identified for this particular set of proteins that meet the specified statistical

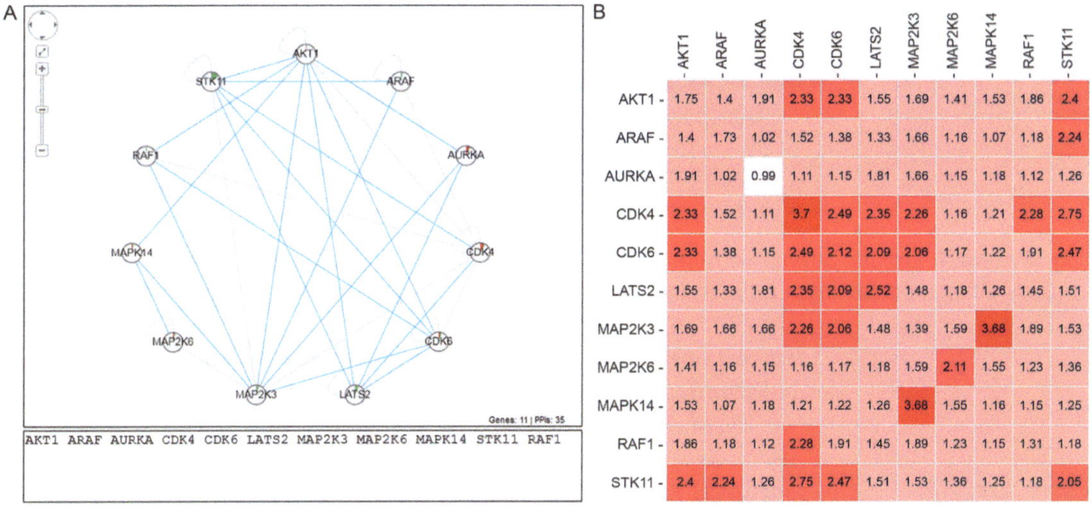

Fig. 3 The Search panel allows the user to query the network for single or multiple proteins. (**a**) A network of PPIs between different kinases included in the OncoPPi network is shown as an example. (**b**) A relative strength of the PPIs included in the network can be visualized using the heatmap tool. The heatmap shows the FOC values associated with the currently selected PPIs. The white color indicates that the PPI was not detected in the corresponding assay, and the increasing intensity of the red color reflects the increase in the PPI signal

parameters and filters. The standard HUGO names [38] should be used in the queries; however, if there are common gene names, for instance, PRAS40, MKK3, or LKB1, the corresponding HUGO names (e.g., AKT1S1, MAP 2K3, STK11) will be suggested to select. If a single protein name is used for the query, the corresponding protein hub subnetwork will be shown. The same network can be generated by selecting a protein from the PPI Hubs panel. In contrast, if multiple protein names are included in the query, the interactions within just this protein set will be shown. As an example, a query for PPIs between different protein kinases "AKT1 ARAF AURKA CDK4 CDK6 LATS2 MAP 2K3 MAP 2K6 MAPK14 STK11 RAF1" with the specific statistical parameters FOC = 1.5, *p*-value = 0.01, and *q*-value = 0.01 will result in a network of PPIs shown in Fig. 3a. While some of these interactions, e.g., MAPK14-MAP 2K3/6, are well-known, the STK11-CDK4 connectivity was not previously reported and has been discovered and validated based on the OncoPPi network analysis [32].

Determine the relative PPI strength. While the network view allows a user to explore the network structure and connectivity between the proteins, a relative strength of the PPIs included in the network can be visualized using the heatmap tool (Fig. 3b). The heatmap can be accessed through the heatmap icon located at the bottom of the network panel, and it shows the FOC values associated with the currently selected network. The white color on the heatmap indicates that the PPI was not detected in the

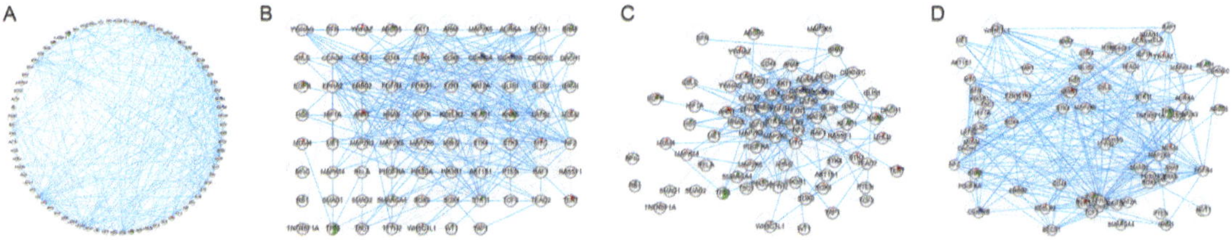

Fig. 4 Network layouts. Several network layouts available for users to automatically arrange the network proteins through the Settings/Layout option, including (**a**) circle, (**b**) grid, (**c**) springy, and (**d**) random layouts

corresponding assay. In contrast, the increasing intensity of the red color reflects the increase in the PPI signal.

Network layouts. There are several network layouts available for users to automatically arrange the network proteins through the Settings/Layout option. For example, the Springy layout tends to place proteins with a higher degree of connectivity in the middle of the network, while proteins with less number of binding partners are placed on the network periphery (Fig. 4). A Circle layout places all proteins in a circle. This layout can help to visually identify proteins with high number of PPIs based on the density of the PPI edges. Grid layout arranges the proteins in an ordered grid that helps to see all network proteins and find a specific protein of interest within the network. Lastly, a Random layout can be used to put proteins in random positions that can be helpful for illustration purposes. Besides the predefined layout, a user can rearrange proteins within the network by dragging each protein node manually.

3.2 PPI Prioritization

Analysis of mutual exclusivity of genomic alterations. Analysis of large genomic datasets, such as TCGA [52], revealed that genes involved in the same biological pathways often demonstrate mutual exclusivity (ME) of genomic alterations in cancer patient samples [24, 25]. Accordingly, ME analysis has become a widely used approach to discover functionally connected proteins and to associate those proteins with specific oncogenic pathways. Since the OncoPPi v1 network has been constructed for lung cancer-associated proteins, all 3486 PPIs tested in the original HT PPI screening have been annotated with the data of ME of genomic alterations observed in lung adenocarcinoma and lung squamous cell carcinoma patients [32]. A total of 2516 PPIs with ME were identified, and 257 PPIs out of the 397 (65%) in the OncoPPi network were predicted to be functionally connected based on the ME analysis.

The PPIs that demonstrate ME of genomic alterations are highlighted in the network with blue lines. In addition, the PPIs that are not supported by the ME analysis can be hidden by checking the "PPIs with mutual exclusivity" option in the Settings panel (Fig. 2).

Cellular co-localization of interacting proteins. The interacting proteins are expected to share a common compartment in the cell. Therefore, experimentally determined protein cellular localization can provide additional support for the observed PPIs and may serve as an important factor for PPI prioritization for further studies. Toward this goal, each protein in the OncoPPi library was annotated with its cellular localization compartments based on the analysis of GO annotations [53, 54] and recently reported cellular localization profiling studies [55]. Based on this analysis, a total of 353 out of 397 PPIs (89%) included in the OncoPPi network are supported by potential cellular co-localization. The PPIs supported by cellular co-localization data can be visualized in the network using the "Co-localized proteins" option in the Settings panel.

Analysis of domain-domain interactions. Often, a presence of specific structural domains and binding motifs determines the overall specificity and binding affinity of the interacting proteins [56–58]. Furthermore, the presence of such complementary structural domains in two different proteins may indicate that the proteins can be potential binding partners. Thus, analysis of protein structural interacting domains provides an effective approach to further support the PPIs determined in HTS experiments. On the other hand, the analysis of PPIs between proteins that lack known interacting domains may provide a new evidence; indeed their structural domains can be involved in a complex formation. Therefore, each PPI included in the OncoPPi set has been annotated with the domain-domain interaction analysis. Specifically, each protein has been annotated with the structural domain information available through the Pfam database [59]. Then, the interacting domains observed in co-crystallized protein–protein complexes have been identified for each protein–protein pair using the 3DID server [57]. A total of 180 out of 397 OncoPPi interacting proteins (45%) share the known interacting domains. These PPIs can be highlighted by checking the "Domain-domain interactions" option in the Settings panel.

As a working example, the ME analysis, cellular co-localization, and domain-domain interactions will be applied to prioritize the PPIs experimentally detected for neurofibromin 2 (NF2).

1. First, select the OncoPPi v1 from the PPI Hubs panel. The OncoPPi network will appear.
2. The Network Hub Proteins graph indicates that the NF2 is the second largest hub in the network after MYC.
3. Click on the NF2 bar to visualize the NF2 PPI subnetwork. From the Settings panel, select Circle layout to rearrange the NF2 direct binding partners. The NF2 subnetwork contains 204 PPIs detected for 30 proteins.

4. Apply the "PPIs with mutual exclusivity" filter to hide the PPIs that are not supported by ME analysis. This will result in a new network of 133 PPIs.
5. Selection of "Domain-domain interactions" will further prioritize 64 PPIs that contain potentially interacting structural domains.
6. Select the "Co-localized proteins" option to identify the PPIs supported by cellular co-localization analysis. This will result in a network of 59 PPIs detected for 22 proteins.
7. Click on the NF2 protein node to highlight NF2 binding partners: FGFR4, EPHA2, EGFR, BECN1, and RAF1.
8. Type these protein names along with NF2 itself in Search panel to visualize the network of prioritized NF2 binding partners.
9. Click on the heatmap icon to explore the FOC values calculated from the TR-FRET PPI assay. The heatmap indicates that all three receptor tyrosine kinases that appear in the network (EGFR, FGFR, and EPHA2) as well as the autophagy regulator BECN1 demonstrate a significant value of FOC > 2, and the NF2-EPHA2 PPI is characterized by the highest FOC value of 5.79.

Previous data indicated a role of NF2 in regulation of activity of different receptor tyrosine kinases, including EGFR, through protein–protein interactions [60, 61]. However, the PPIs between NF2 and EPHA2 or FGFR4, as well as BECN1 and RAF1, have not been previously observed. Thus, the network analysis and integration of structural, cancer genomic, and cellular data suggest novel roles for NF2 in regulation of oncogenic signaling through its PPIs with newly identified binding partners.

3.3 Protein Annotation and Integration with External Resources

General information. To facilitate the data mining and PPI prioritization for detailed biological studies, the OncoPPi networks are integrated with different types of information, including biochemical, biophysical, genomic, and clinical data. This detailed information is available in the "Protein Information" window accessible via a user's right-click on a protein in the network (Fig. 5).

To summarize the protein connectivity data for the currently selected network, the "Protein Information" panel includes (a) the standard gene symbol and complete protein name, (b) the protein degree (number of its binding partners) in the currently selected network, and (c) the list of all binding partners detected for the given protein in the network. Furthermore, the PPIs reported for a given protein in external databases can be directly accessed through links to the protein pages in String, IntAct, and GeneMania datasets [13, 15, 17].

The Protein Information window also shows the major cellular localization sites and known structural domains for each protein.

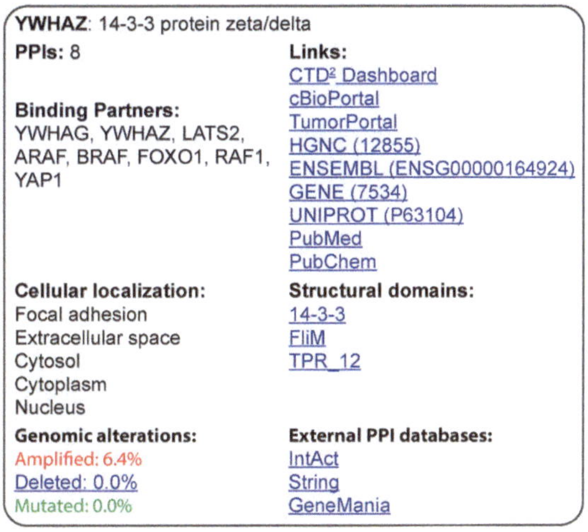

Fig. 5 Protein Information window. The Protein Information window is available on a right-click on a protein name in the network. It provides detailed information about the selected protein. It includes the standard gene name, the complete protein name, the number of protein binding partners detected with the given statistical thresholds, and a complete list of the binding proteins. In addition, the cellular localization, structural domains linked with the Pfam server, and percent of genomic alterations in lung adenocarcinoma patients are provided. To facilitate further data mining and protein annotations, the Protein Information panel links every protein with the major general PPI and cancer-specific external resources

Detailed description of each structural domain can be further accessed through the direct links to the corresponding pages on the Pfam web server. To provide direct access to general genomic and proteomic annotations, every protein is linked to the HGNC, Ensembl, Gene, and UniProt databases. Furthermore, the recently published literature related to a selected protein can be accessed through a direct link to PubMed.

Genomic alterations. Genomic alterations play a critical role in regulation of oncogenic signaling and tumorigenesis through a rewired network of protein–protein interactions [43, 62–64]. Thus, the analysis of PPIs between the proteins altered in cancer patients may facilitate the discovery of novel PPIs with therapeutic potential. Since the OncoPPi network was established for a set of lung cancer-associated genes [32], every gene in the library is annotated with a percentage of lung adenocarcinoma patient samples with gene DNA amplifications, deletions, or mutations. This information is provided in the Protein Information panel, and the frequency of these alterations is visualized with red, blue, and green sectors inside the protein node circle in the network. In addition, detailed information on genomic alterations in different tumor types and their impact on cancer patient survival can be accessed and explored though direct links to cBioPortal

[44, 45] and TumorPortal [43]. Furthermore, a direct link to the CTD^2 Dashboard [42] allows a user to get the most recent experimental cancer biology data generated for the protein of interest by the network of leading groups in cancer biology, clinical oncology, genomic, and bioinformatics fields.

3.4 Therapeutic Connectivity

The role of individual proteins as a biological target, their druggability, and available small molecule modulators of their activity can be gathered through the PubChem database [47] that is linked with the OncoPPi proteins through the Protein Information panel.

To facilitate the discovery of new tumor vulnerability and PPI dependency, the OncoPPi Portal provides an interface to link the PPIs between known oncogenes and tumor suppressors with currently approved drugs. The OncoPPi protein-drug connectivity network can be visualized through the therapeutic connectivity page of the OncoPPi Portal. Currently, a total of 42 protein-drug interactions have been identified for the 32 approved drugs and 14 OncoPPi network proteins, including BRAF, RAF1, CD44, CDK4, CDK6, EGFR, EPHA2, ERBB2, FGFR4, IGF1R, KEAP1, MET, PDGFRA, and PIK3CA. A user can get detailed chemical, biological, and pharmacological data associated with each compound, as well as its chemical structure, through the Compound Information window that is available on a right-click on the compound node in the network. For example, the most recent studies that utilized the drug in different cancer models can be accessed through a direct link to corresponding CTD^2 Dashboard server [42]. The Cancer Therapeutic Response Portal [46] and Genomics of Drug Sensitivity in Cancer project [49] servers allow to explore the drug sensitivity of cancer cell lines with different genomic backgrounds. PubChem [47] and DrugBank [48] databases provide detailed pharmacological and clinical information for the drugs.

There are several options to query the protein-drug connectivity. To query for all PPIs and PDIs for a specific protein, a user can use the HUGO gene Search panel located on the left side under the network window. For example, a search for “BRAF” will show BRAF PPIs with KRAS, HRAS, and 14-3-3 proteins included in the OncoPPi network as well as its association with the approved inhibitors dabrafenib, vemurafenib, sorafenib, and regorafenib. A user can also use the drug Search panel located on the right side under the network window to query the connectivity for a specific drug or several drugs. For instance, a search for lapatinib and neratinib will show that both compounds can inhibit EGFR and lapatinib is also approved as an ERBB2 inhibitor.

To illustrate the application of the OncoPPi Portal to predict new tumor dependencies, the protein-drug interactions determined for CDK4 PPI hub will be analyzed. The following steps should be performed for the analysis: (a) select CDK4 from the

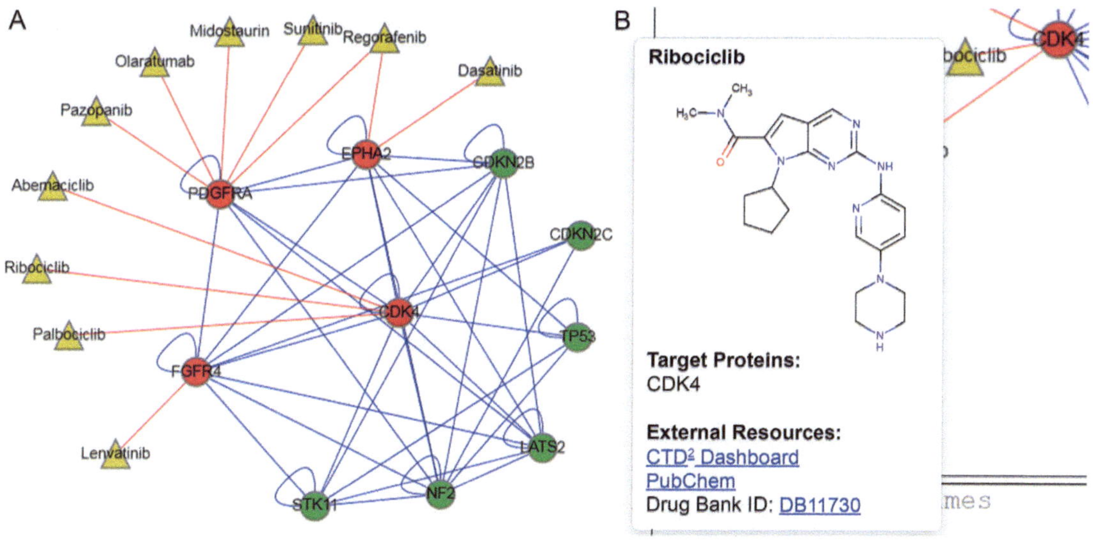

Fig. 6 Analysis of therapeutic connectivity. (**a**) Drug-PPI network determined for CDK4 kinase. Circles represent proteins, including known oncogenes (red) and tumor suppressors (green). Approved drugs are shown as yellow triangles. Blue lines indicate protein–protein interactions. Protein-drug interactions are represented by red lines. (**b**) Therapeutic Information window that is available on a right-click on the drug name shows the drug name, chemical structure, and main protein target for the drug as well as the direct links to CTD^2 Dashboard, PubChem, and DrugBank databases

Proteins panel; (b) select All from the Drugs panel; (c) click Generate Network button to visualize the CDK4 hub linked with approved drugs (Fig. 6a). The resulting network indicates three CDK4/CDK6 inhibitors currently approved as drugs: palbociclib, abemaciclib, and ribociclib. The chemical structure can be viewed in the Therapeutic Information window that is available on a right-click on a drug name. It also links the compound to external resources, inducing DrugBank, PubChem, and CDT^2 Dashboard servers for detailed exploration (Fig. 6b). The generated network shows that CDK4 interacts with several known oncogenes, including RAF1, FGFR4, PDGFRA, and EPHA2, that also have known inhibitors approved as drugs. This observation suggests that a combination of these inhibitors with CDK4 approved drugs can be beneficial for cancer patients with upregulated CDK4 interactions with RAF1, FGFR4, PDGFRA, or EPHA2. On the other hand, CDK4 binds to several tumor suppressors, including its known CDK inhibitors CDKN2B and CDKN2C, as well as TP53, LATS2, NF2, and STK11. Recently, it was shown that indeed CDK4 binds to STK11 in lung cancer cells at endogenous levels and that the CDK4 inhibitor, palbociclib, disrupts the CDK4/STK11 PPI [32]. Moreover, STK11 silenced cells showed increased sensitivity to palbociclib compared to the parental H1792 lung cancer cells [32]. Thus, the analysis of the OncoPPi protein-drug connectivity data revealed an enhanced CDK4 dependency for cancer cells that lack STK11. Similarly, one can expect that LATS2 mutations may also lead to an enhanced sensitivity of cancer cells to CDK4 inhibition, and the CDK4/LATS2 PPI can also be regulated by palbociclib and other CDK4 inhibitors. Together, exploration and analysis of OncoPPi protein-drug connectivity provides a valuable resource to discover new mechanisms of tumor dependencies.

3.5 Data Export and the OncoPPi API

All data associated with the OncoPPi PPI sets can be exported in a tabular comma-separated (csv) format for further analysis and exploration. To export the data, a user can click on the "Export PPI data" icon located at the bottom of the page. In addition, the "Save Image" icon allows a user to save currently shown PPI network as an image file.

The application programming interface (API) is a powerful tool to exchange the data between different servers and databases. To provide a rapid programming access to the OncoPPi data, the OncoPPi Portal has implemented the Representational State Transfer (REST) API that allows to access, query, and export all the data associated with the PPIs within the OncoPPi networks.

The base URL for the OncoPPi API is www.oncoppi.emory.edu/api?request.

The terms that can be used to access the data are summarized in Table 2.

For example, the following request will return a JSON object with statistical parameters and annotations available for MYC-STK11 PPIs:

Table 2
OncoPPi Portal API terms

Term	Description
output	Specifies the output format as json or csv
portal	Specifies the network to access, e.g., oncoppi or myc_nanopca
ppi	Indicates the protein pair to query. For example, MYC-STK11
ppi_hub	Specifies the protein PPI hub to query. Can be either a single HUGO symbol, e.g., CDK4, or all to access the whole dataset
foc	Sets the threshold for the fold-over-control value
pvalue	Sets the threshold for the *p*-value
qvalue	Sets the threshold for the *q*-value
me	Limits the query to PPIs between genes with mutually exclusive genomic alterations
ddi	Limits the query to PPIs between proteins with predicted domain-domain interactions
coloc	Limits the query to PPIs between proteins with predicted cellular co-localization

http://webappl.pharm.emory.edu/PPI_Portal/api/?request=portal:oncoppi~ppi:MYC-STK11~foc:1.5~pvalue:0.05~qvalue:0.01~output:json.

The following request will generate a JSON object that contains all MYC PPIs determined with the FOC ≥ 1.5, *p*-value <0.05, and *q*-value <0.01:

http://webappl.pharm.emory.edu/PPI_Portal/api/?request=portal:oncoppi~ppi_hub:myc~foc:1.5~pvalue:0.05~qvalue:0.01~output:json.

The next example can be used to extract all OncoPPi network PPIs between genes with mutually exclusive genomic alterations, interacting structural domains, predicted cellular co-localization, FOC ≥ 1.5, *p*-value <0.05, and *q*-value <0.01:

http://webappl.pharm.emory.edu/PPI_Portal/api/?request=portal:oncoppi~ppi_hub:all~foc:1.5~pvalue:0.05~qvalue:0.01~me~ddi~coloc~output:json.

The same information can be directly saved as a comma-separated csv file using the output:csv term instead of output:json.

4 Notes

1. It should be noted that both the TR-FRET and NanoPCA assays are designed to detect the physical PPIs between over-expressed tagged proteins. On the other hand, both assays demonstrate a high sensitivity and can be used to detect even weak PPIs in live cells (NanoPCA) or cell lysates (TR-FRET) at concentrations close to the physiological level.
2. The OncoPPi is a dynamic resource, expanding in several directions, including the gene library, the number of tested cancer-associated PPIs, the PPI screening technologies, and the tumor models for the PPI screening. The upcoming versions of the Portal will include new datasets, options for users to import external PPI sets for the analysis, and new tools for network analysis and PPI prioritization.

Acknowledgments

The results published here are in part based upon data generated by TCGA Research Network: http://cancergenome.nih.gov. This research was supported in part by the National Cancer Institute of the NIH (Cancer Target Discovery and Development Network grants U01CA168449 and U01CA217875), Emory University Research Committee grant, Winship Cancer Institute (NIH 5P30CA138292), and Fadlo R. Khuri Translational Research Award of the Winship Cancer Institute, Emory University.

References

1. Petta I, Lievens S, Libert C, Tavernier J, De Bosscher K (2016) Modulation of protein-protein interactions for the development of novel therapeutics. Mol Ther 24(4):707–718
2. Ballatore C, Brunden KR, Trojanowski JQ, Lee VM, Smith AB 3rd, Huryn DM (2011) Modulation of protein-protein interactions as a therapeutic strategy for the treatment of neurodegenerative tauopathies. Curr Top Med Chem 11(3):317–330
3. Morelli X, Bourgeas R, Roche P (2011) Chemical and structural lessons from recent successes in protein-protein interaction inhibition (2P2I). Curr Opin Chem Biol 15(4):475–481
4. Scott DE, Bayly AR, Abell C, Skidmore J (2016) Small molecules, big targets: drug discovery faces the protein-protein interaction challenge. Nat Rev Drug Discov 15 (8):533–550
5. Arkin MR, Tang Y, Wells JA (2014) Small-molecule inhibitors of protein-protein interactions: progressing toward the reality. Chem Biol 21(9):1102–1114
6. Huttlin EL, Ting L, Bruckner RJ, Gebreab F, Gygi MP, Szpyt J et al (2015) The BioPlex network: a systematic exploration of the human Interactome. Cell 162(2):425–440
7. Rual JF, Venkatesan K, Hao T, Hirozane-Kishikawa T, Dricot A, Li N et al (2005) Towards a proteome-scale map of the human protein-protein interaction network. Nature 437(7062):1173–1178
8. Luck K, Sheynkman GM, Zhang I, Vidal M (2017) Proteome-scale human interactomics. Trends Biochem Sci 42(5):342–354
9. Vo TV, Das J, Meyer MJ, Cordero NA, Akturk N, Wei X et al (2016) A proteome-wide fission yeast interactome reveals network evolution principles from yeasts to human. Cell 164(1–2):310–323
10. Rolland T, Tasan M, Charloteaux B, Pevzner SJ, Zhong Q, Sahni N et al (2014) A proteome-scale map of the human interactome network. Cell 159(5):1212–1226
11. Stelzl U, Worm U, Lalowski M, Haenig C, Brembeck FH, Goehler H et al (2005) A human protein-protein interaction network: a resource for annotating the proteome. Cell 122 (6):957–968
12. Schaefer CF, Anthony K, Krupa S, Buchoff J, Day M, Hannay T et al (2009) PID: the pathway interaction database. Nucleic Acids Res 37 (Database issue):D674–D679
13. Orchard S, Ammari M, Aranda B, Breuza L, Briganti L, Broackes-Carter F et al (2014) The MIntAct project—IntAct as a common curation platform for 11 molecular interaction databases. Nucleic Acids Res 42(Database issue):D358–D363
14. Chatr-Aryamontri A, Breitkreutz BJ, Oughtred R, Boucher L, Heinicke S, Chen D et al (2015) The BioGRID interaction database: 2015 update. Nucleic Acids Res 43(Database issue):D470–D478
15. Szklarczyk D, Franceschini A, Wyder S, Forslund K, Heller D, Huerta-Cepas J et al (2015) STRING v10: protein-protein interaction networks, integrated over the tree of life. Nucleic Acids Res 43(Database issue): D447–D452
16. Kanehisa M, Furumichi M, Tanabe M, Sato Y, Morishima K (2017) KEGG: new perspectives on genomes, pathways, diseases and drugs. Nucleic Acids Res 45(D1):D353–D361
17. Warde-Farley D, Donaldson SL, Comes O, Zuberi K, Badrawi R, Chao P et al (2010) The GeneMANIA prediction server: biological network integration for gene prioritization and predicting gene function. Nucleic Acids Res 38.(Web Server issue:W214–W220
18. Hanahan D, Weinberg RA (2011) Hallmarks of cancer: the next generation. Cell 144 (5):646–674
19. Ivanov AA, Khuri FR, Fu H (2013) Targeting protein-protein interactions as an anticancer strategy. Trends Pharmacol Sci 34 (7):393–400
20. Nero TL, Morton CJ, Holien JK, Wielens J, Parker MW (2014) Oncogenic protein interfaces: small molecules, big challenges. Nat Rev Cancer 14(4):248–262
21. Ran X, Gestwicki JE (2018) Inhibitors of protein-protein interactions (PPIs): an analysis of scaffold choices and buried surface area. Curr Opin Chem Biol 44:75–86
22. Potts MB, Kim HS, Fisher KW, Hu Y, Carrasco YP, Bulut GB et al (2013) Using functional signature ontology (FUSION) to identify mechanisms of action for natural products. Sci Signal 6(297):ra90
23. Wang K, Saito M, Bisikirska BC, Alvarez MJ, Lim WK, Rajbhandari P et al (2009) Genome-wide identification of post-translational modulators of transcription factor activity in human B cells. Nat Biotechnol 27(9):829–839
24. Babur O, Gonen M, Aksoy BA, Schultz N, Ciriello G, Sander C et al (2015) Systematic identification of cancer driving signaling pathways based on mutual exclusivity of genomic alterations. Genome Biol 16:45

25. Ciriello G, Cerami E, Sander C, Schultz N (2012) Mutual exclusivity analysis identifies oncogenic network modules. Genome Res 22 (2):398–406
26. Rosenbluh J, Nijhawan D, Cox AG, Li X, Neal JT, Schafer EJ et al (2012) Beta-catenin-driven cancers require a YAP1 transcriptional complex for survival and tumorigenesis. Cell 151 (7):1457–1473
27. Cowley GS, Weir BA, Vazquez F, Tamayo P, Scott JA, Rusin S et al (2014) Parallel genome-scale loss of function screens in 216 cancer cell lines for the identification of context-specific genetic dependencies. Sci Data 1:140035
28. Tsherniak A, Vazquez F, Montgomery PG, Weir BA, Kryukov G, Cowley GS et al (2017) Defining a cancer dependency map. Cell 170 (3):564–576.e16
29. Alanis-Lobato G, Andrade-Navarro MA, Schaefer MH (2017) HIPPIE v2.0: enhancing meaningfulness and reliability of protein-protein interaction networks. Nucleic Acids Res 45(D1):D408–DD14
30. Meng X, Wang J, Yuan C, Li X, Zhou Y, Hofestadt R et al (2015) CancerNet: a database for decoding multilevel molecular interactions across diverse cancer types. Oncogene 4:e177
31. Van Coillie S, Liang L, Zhang Y, Wang H, Fang JY, Xu J (2016) OncoBinder facilitates interpretation of proteomic interaction data by capturing coactivation pairs in cancer. Oncotarget 7(14):17608–17615
32. Li Z, Ivanov AA, Su R, Gonzalez-Pecchi V, Qi Q, Liu S et al (2017) The OncoPPi network of cancer-focused protein-protein interactions to inform biological insights and therapeutic strategies. Nat Commun 8:14356. https://doi.org/10.1038/ncomms
33. Ivanov AA, Gonzalez-Pecchi V, Khuri L, Niu T, Wang Y, Xu R et al (2017) OncoPPi-informed discovery of mitogen-activated protein kinase kinase 3 as a novel binding partner of c-Myc. Oncogene 36(42):5852–5860
34. Xiong J, Pecchi VG, Qui M, Ivanov AA, Mo X, Niu Q et al (2018) Development of a time-resolved fluorescence resonance energy transfer ultrahigh-throughput screening assay for targeting the NSD3 and MYC interaction. Assay Drug Dev Technol 16(2):96–106
35. Ivanov AA, Revennaugh B, Rusnak L, Gonzalez-Pecchi V, Mo X, Johns MA et al (2018) The OncoPPi Portal: an integrative resource to explore and prioritize protein-protein interactions for cancer target discovery. Bioinformatics 34(7):1183–1191
36. Mo XL, Qi Q, Ivanov AA, Niu Q, Luo Y, Havel J et al (2017) AKT1, LKB1, and YAP1 revealed as MYC interactors with NanoLuc-based protein-fragment complementation assay. Mol Pharmacol 91(4):339–347
37. Harati S, Cooper LA, Moran JD, Giuste FO, Du Y, Ivanov AA et al (2017) MEDICI: mining essentiality data to identify critical interactions for cancer drug target discovery and development. PLoS One 12(1):e0170339
38. Gray KA, Yates B, Seal RL, Wright MW, Bruford EA (2015) Genenames.org: the HGNC resources in 2015. Nucleic Acids Res 43(Database issue):D1079–D1085
39. Aken BL, Ayling S, Barrell D, Clarke L, Curwen V, Fairley S et al (2016) The Ensembl gene annotation system. Database (Oxford) 2016
40. UniProt: the universal protein knowledgebase (2017) Nucleic Acids Res 45(D1):D158–DD69
41. NCBI Resource Coordinators (2017) Database resources of the National Center for Biotechnology Information. Nucleic Acids Res 45 (D1):D12–DD7
42. Aksoy BA, Dancik V, Smith K, Mazerik JN, Ji Z, Gross B et al (2017) CTD2 dashboard: a searchable web interface to connect validated results from the cancer target discovery and development network. Database (Oxford) 2017
43. Lawrence MS, Stojanov P, Mermel CH, Robinson JT, Garraway LA, Golub TR et al (2014) Discovery and saturation analysis of cancer genes across 21 tumour types. Nature 505 (7484):495–501
44. Cerami E, Gao J, Dogrusoz U, Gross BE, Sumer SO, Aksoy BA et al (2012) The cBio cancer genomics portal: an open platform for exploring multidimensional cancer genomics data. Cancer Discov 2(5):401–404
45. Gao J, Aksoy BA, Dogrusoz U, Dresdner G, Gross B, Sumer SO et al (2013) Integrative analysis of complex cancer genomics and clinical profiles using the cBioPortal. Sci Signal 6 (269):pl1
46. Rees MG, Seashore-Ludlow B, Cheah JH, Adams DJ, Price EV, Gill S et al (2016) Correlating chemical sensitivity and basal gene expression reveals mechanism of action. Nat Chem Biol 12(2):109–116
47. Kim S, Thiessen PA, Bolton EE, Chen J, Fu G, Gindulyte A et al (2016) PubChem substance and compound databases. Nucleic Acids Res 44(D1):D1202–D1213
48. Wishart DS, Knox C, Guo AC, Shrivastava S, Hassanali M, Stothard P et al (2006) DrugBank: a comprehensive resource for in silico drug discovery and exploration. Nucleic Acids Res 34(Database issue):D668–D672

49. Yang W, Soares J, Greninger P, Edelman EJ, Lightfoot H, Forbes S et al (2013) Genomics of drug sensitivity in cancer (GDSC): a resource for therapeutic biomarker discovery in cancer cells. Nucleic Acids Res 41(Database issue):D955–D961
50. He X, Zhang J (2006) Why do hubs tend to be essential in protein networks? PLoS Genet 2 (6):e88
51. Jeong H, Mason SP, Barabasi AL, Oltvai ZN (2001) Lethality and centrality in protein networks. Nature 411(6833):41–42
52. Tomczak K, Czerwinska P, Wiznerowicz M (2015) The Cancer Genome Atlas (TCGA): an immeasurable source of knowledge. Contemp Oncol (Pozn) 19(1A):A68–A77
53. Ashburner M, Ball CA, Blake JA, Botstein D, Butler H, Cherry JM et al (2000) Gene ontology: tool for the unification of biology. The Gene Ontology Consortium. Nat Genet 25 (1):25–29
54. The Gene Ontology Consortium (2017) Expansion of the gene ontology knowledgebase and resources. Nucleic Acids Res 45 (D1):D331–D3D8
55. Thul PJ, Akesson L, Wiking M, Mahdessian D, Geladaki A, Ait Blal H et al (2017) A subcellular map of the human proteome. Science 356 (6340)
56. Chakrabarti P, Janin J (2002) Dissecting protein-protein recognition sites. Proteins 47 (3):334–343
57. Mosca R, Ceol A, Stein A, Olivella R, Aloy P (2014) 3did: a catalog of domain-based interactions of known three-dimensional structure. Nucleic Acids Res 42(Database issue): D374–D379
58. Sillitoe I, Cuff A, Dessailly BH, Dawson NL, Furnham N, Lee D, Lees JG et al (2013) New functional families (FunFams) in CATH to improve the mapping of conserved functional sites to 3D structures. Nucleic Acids Res 41 (Database issue):D490–D498
59. Finn RD, Coggill P, Eberhardt RY, Eddy SR, Mistry J, Mitchell AL et al (2016) The Pfam protein families database: towards a more sustainable future. Nucleic Acids Res 44(D1): D279–D285
60. Stamenkovic I, Yu Q (2010) Merlin, a "magic" linker between extracellular cues and intracellular signaling pathways that regulate cell motility, proliferation, and survival. Curr Protein Pept Sci 11(6):471–484
61. Guerrero PA, Yin W, Camacho L, Marchetti D (2015) Oncogenic role of Merlin/NF2 in glioblastoma. Oncogene 34(20):2621–2630
62. Chang CI, Xu BE, Akella R, Cobb MH, Goldsmith EJ (2002) Crystal structures of MAP kinase p38 complexed to the docking sites on its nuclear substrate MEF2A and activator MKK3b. Mol Cell 9(6):1241–1249
63. Lawrence MS, Stojanov P, Polak P, Kryukov GV, Cibulskis K, Sivachenko A et al (2013) Mutational heterogeneity in cancer and the search for new cancer-associated genes. Nature 499(7457):214–218
64. Beroukhim R, Mermel CH, Porter D, Wei G, Raychaudhuri S, Donovan J et al (2010) The landscape of somatic copy-number alteration across human cancers. Nature 463 (7283):899–905

Chapter 13

Perform Pathway Enrichment Analysis Using ReactomeFIViz

Robin Haw, Fred Loney, Edison Ong, Yongqun He, and Guanming Wu

Abstract

Modern large-scale biological data analysis often generates a set of significant genes, frequently associated with scores. Pathway-based approaches are routinely performed to understand the functional contexts of these genes. Reactome is the most comprehensive open-access biological pathway knowledge base, widely used in the research community, providing a solid foundation for pathway-based data analysis. ReactomeFIViz is a Cytoscape app built upon Reactome pathways to help users perform pathway- and network-based data analysis and visualization. In this chapter we describe procedures on how to perform pathway enrichment analysis using ReactomeFIViz for a gene score file. We describe two types of analysis: pathway enrichment based on a set of significant genes and GSEA analysis using gene scores without cutoff. We also describe a feature to overlay gene scores onto pathway diagrams, enabling users to understand the underlying mechanisms for up- or down- regulated pathways collected from pathway analysis.

Key words Biological pathway, Pathway enrichment analysis, GSEA, Reactome, ReactomeFIViz, Cytoscape, Gene score

1 Introduction

High-throughput technologies are routinely employed in modern biological studies. For example, RNA-seq is widely used for differential gene expression studies between groups of samples having different phenotypes. Typically, the outputs of these studies are lists of genes that are significantly related to the understudied biological phenotypes or associated with scores based upon some types of statistical analysis. Understanding biological mechanisms associated with these lists of genes is a significant challenge in the data analysis of high-throughput technologies.

Biological pathways summarize existing knowledge generated from experiments into a graphical framework illustrating cellular activities within and between cells, enabling biologists to understand the underlying mechanisms and disease etiologies in a systematic way. Projecting a list of significant genes onto pathways places seemingly unconnected genes into the same pathway

Stefan Canzar and Francisca Rojas Ringeling (eds.), *Protein-Protein Interaction Networks: Methods and Protocols*,
Methods in Molecular Biology, vol. 2074, https://doi.org/10.1007/978-1-4939-9873-9_13,

context, assisting researchers to uncover the common functional mechanisms related to these genes. Pathway-based analytical methods are powerful and popular approaches, widely used in large-scale "-omics" data analysis and visualization. Pathway enrichment analysis is designed to find significant pathways related to a list of genes.

Almost all pathway-based approaches are composed of two components: (a) a statistical test and (b) a pathway database. The pathway database provides gene sets that can be tested for a list of genes, while the statistical test provides the actual algorithm to perform such a test. Reactome [1] is arguably the most comprehensive open-source biological database, widely used in the community for researchers to explore biological pathways and perform pathway-based data analysis and visualization. The contents in Reactome are routinely released every quarter. The latest release is 65 (released in June 2018), which covers 10,763 human protein-coding genes, annotated across 11,674 complexes, 11,896 reactions, and 2,222 pathways and supported by over 28,000 literature references. The highly reliable Reactome content provides a solid foundation for pathway-based data analysis.

Cytoscape [2] is a Java-based standalone software application, widely used in the research community for biological network analysis and visualization. Cytoscape provides a framework for third-party developers to develop applications (called apps) to extend the visualization and analysis functions [3]. Currently over 200 apps can be downloaded from the Cytoscape app store [4], assisting users to perform a variety of pathway- and network-based data analysis and visualization.

Based on high-quality Reactome pathways, we have developed an app, called "ReactomeFIViz" [5], to help users to perform pathway- and network-based data analysis and visualization. ReactomeFIViz is very popular, downloaded around 34,000 times since it was released to the Cytoscape app store in August 2013. Many features have been provided in ReactomeFIViz, including features for network-based data analysis and visualization and features for pathway mathematical modeling based on probabilistic graphical models [6] and Boolean networks [7]. In this chapter, we focus on features allowing users to perform pathway enrichment analysis for a set of significant genes or based upon a gene score file.

2 Materials

To use the ReactomeFIViz app, the reader should first install the latest version of Cytoscape from http://www.cytoscape.org. The following hardware and software requirements are suggested to use Cytoscape:

Hardware requirements: For small network visualization, 1 Ghz processor, 512 MB or more physical memory, integrated video memory, and a display that supports XGA (1024×768) or higher resolution. With large network analysis and visualization, as fast as possible processor (with dual or multiple cores), 2 GB or more physical memory, high-end graphics cards, and a wide or dual monitor. Furthermore, a fast Internet connection is necessary.

Software requirements: To use the latest version of Cytoscape (version 3.6.1 in July 2018), the reader is required to install Java 8 first. Java 8 is supported under Windows, Mac OS, and Linux and can be downloaded from http://www.java.com. Note: Java 9 or higher has not been supported by Cytoscape as of this writing (July 2018).

To describe the procedures for pathway enrichment analysis, we use an example gene score file generated from differential gene expression analysis using the data file downloaded from NCBI GEO with accession number GSE13486: https://www.ncbi.nlm.nih.gov/geo/query/acc.cgi?acc=GSE13486. The original data was generated to study differential gene expression caused by yellow fever YF-17D vaccine [8]. The example file was regenerated for comparison between two expression profiles for Day 0 and Day 7, respectively. Empty rows and genes that could not be mapped to HUGO symbols (https://www.genenames.org) were removed. The example data file can be downloaded via this URL: https://github.com/reactome-fi/CytoscapePlugIn/blob/path-x/example_data/Querec_New_D7_D0_All.txt.

3 Methods

3.1 Software Installation

3.1.1 Cytoscape Installation

1. Install Cytoscape: Use your preferred web browser and load the Cytoscape website using URL http://www.cytoscape.org.
2. Select the "Download #.#.#" button (#.#.# should be the latest version number), and follow the instruction for user registration and installing the software. For notes related to Cytoscape installation (*see* **Note 1**).
3. Double-click or use a platform-specific way (e.g., a shell script under Linux) to launch Cytoscape. For more information about the Java version (*see* **Note 2**). For information on how to launch Cytoscape (*see* **Note 3**).

3.1.2 ReactomeFIViz App Installation

1. Launch the Cytoscape software.
2. Go to the "Apps" drop-down menu and then click App Manager to bring up the App Manager dialog.
3. In the dialog's search box (Fig. 1), type "ReactomeFI" (remove the quotations marks) to find the app hosted in the Cytoscape app store.

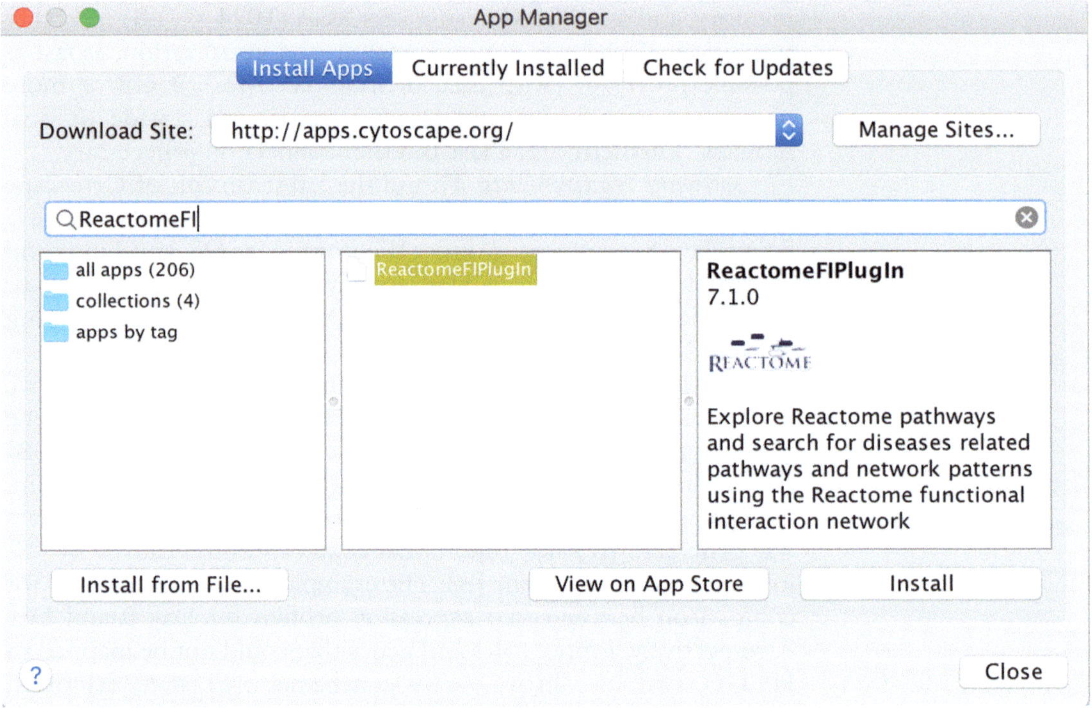

Fig. 1 Install ReactomeFIViz via the Cytoscape app store by using App Manager

4. Select "ReactomeFIPlugIn" app (the old name of ReactomeFIViz), and then click the "Install" button. For more information relating to other ways to install the ReactomeFIViz app (*see* **Note 4**).

3.2 Perform Pathway Enrichment Analysis Based on a Set of Significant Genes

Pathway enrichment analysis is used to generate a list of pathways that significantly enrich genes from the user's gene list based on a binomial test. A gene list can be generated from differential gene expression study, mutation frequency analysis, genome-wide association studies (GWAS), or some other high-throughput data analysis approaches. In the following example, we select genes from the gene score file with FDR (false discovery rate) < 0.05, and then subject those genes for pathway enrichment analysis.

1. Open the example file in Microsoft Excel or other similar software and then sort rows based on FDR.
2. Select the first 139 genes having FDR <0.05 and copy these cells into an empty Excel sheet.
3. Save the Excel sheet containing these 139 genes into a tab-delimited text file as GeneList.txt.
4. If you have not already launched Cytoscape, start Cytoscape and then select menu Apps/Reactome FI/Reactome Pathways to open the pathway hierarchical tree.

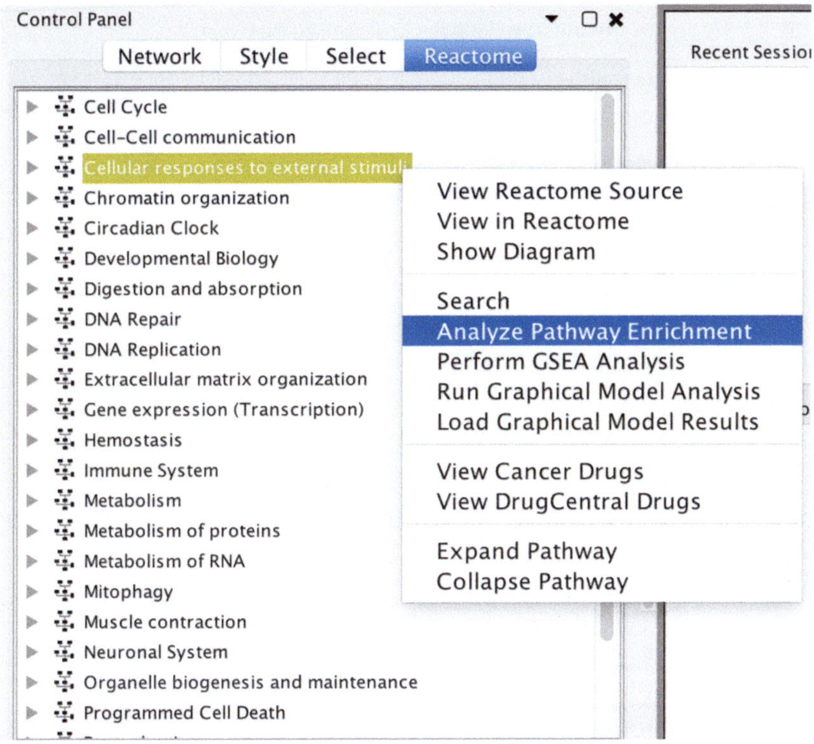

Fig. 2 Use menu "Analyze Pathway Enrichment" for performing pathway enrichment analysis for a list of genes in a file

5. Select any pathway in the tree and then right-click to bring up the popup menu and select "Analyze Pathway Enrichment" (Fig. 2). The dialog for configuring the gene list data source should be opened.
6. Choose the saved gene list file, GeneList.txt, and then click "OK" button (Fig. 3). *See* **Note 5** for other file formats supported by ReactomeFIViz for pathway enrichment analysis.
7. After the analysis is complete, the results are displayed in a table labeled as "Reactome Pathway Enrichment" (Fig. 4) in Cytoscape's Table Panel at the bottom of the window. Pathways in the Reactome hierarchical tree are highlighted based on FDRs.
8. To investigate where hit genes are in a pathway, select the pathway in the table, right-click to bring out the popup menu, and then choose "View in Diagram" (*see* **Note 6** for some extra information about this feature). The diagram for the selected pathway will be open in the Cytoscape main panel. Entities containing genes from your list will be highlighted in the purple background.
9. Entities in a pathway diagram may be a complex or a set of entities (called EntitySet in Reactome). To see what genes are

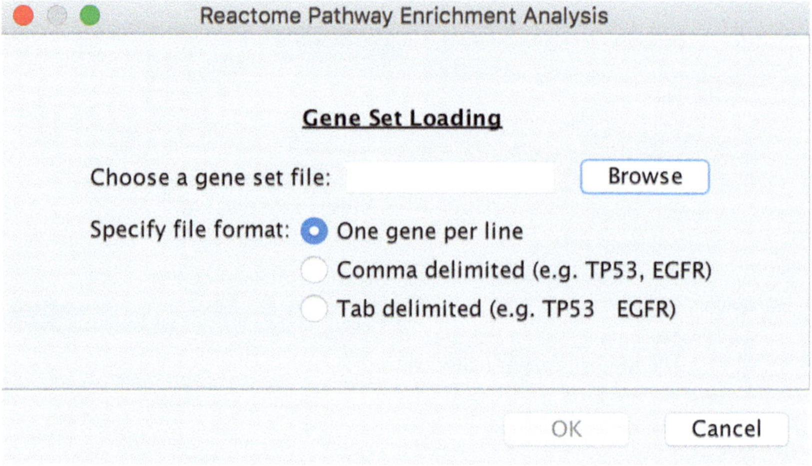

Fig. 3 The configuration dialog for choosing the gene list file for pathway enrichment analysis

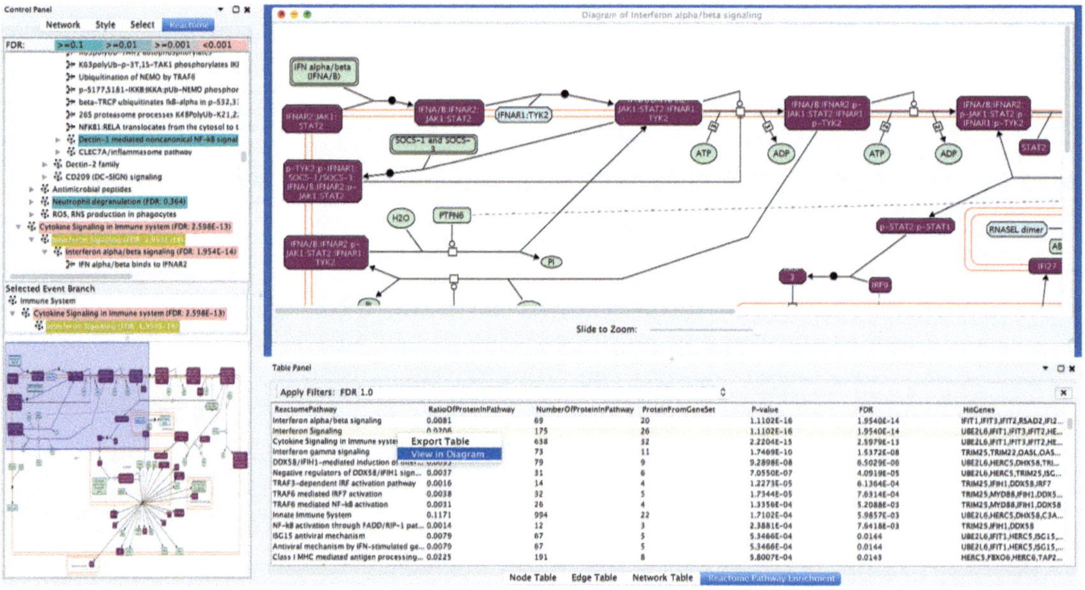

Fig. 4 Pathway enrichment analysis results

from your gene list, select the highlighted entity, right-click to get the popup menu for the selected entity, and then choose "List Genes" (Fig. 5). Hit genes are highlighted in purple in the "Gene List" dialog (Fig. 6).

10. You can click a gene to view its page at the GeneCards website (http://www.genecard.org). For example, clicking STAT2 in Fig. 6 takes you to https://www.genecards.org/cgi-bin/carddisp.pl?gene=STAT2 for detailed annotation for gene STAT2.

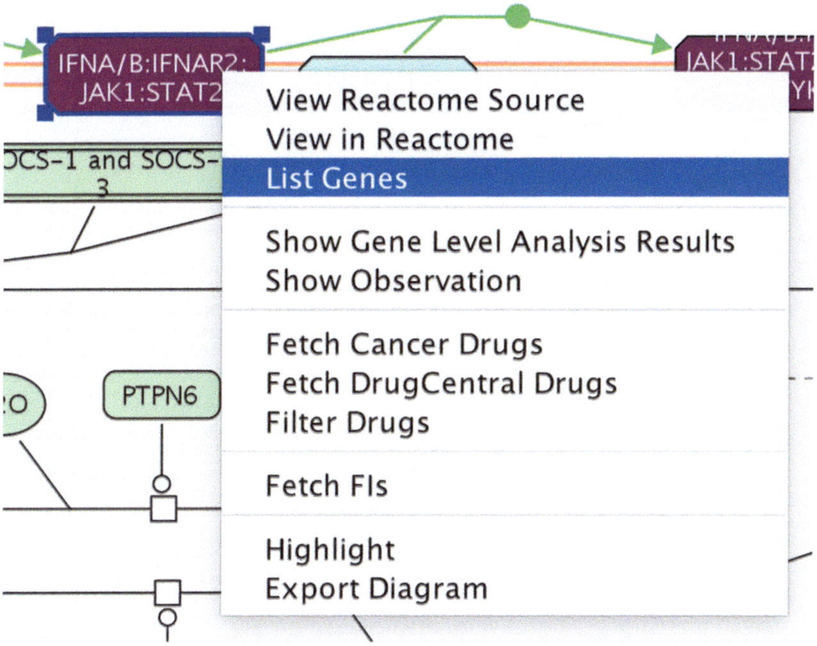

Fig. 5 Use popup "List Genes" to see what genes are from the user's gene list

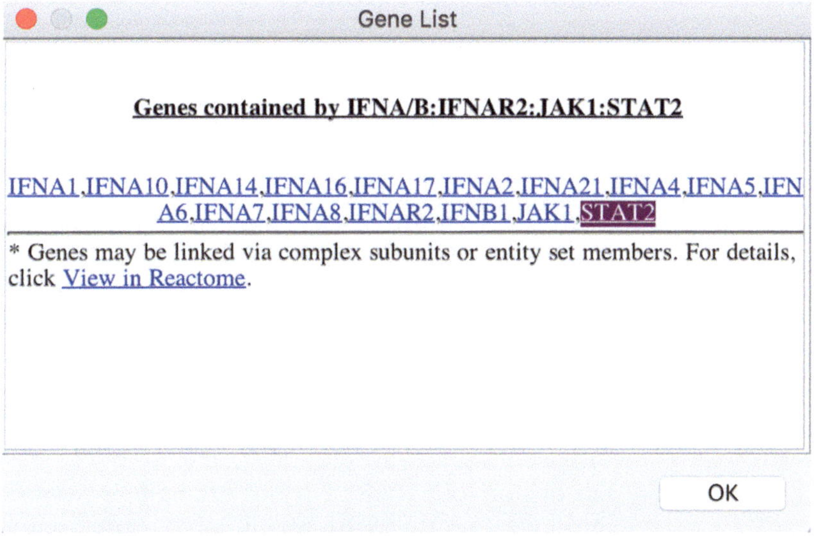

Fig. 6 Genes listed for a selected entity in the pathway diagram. Genes can be clicked to open the GeneCards website for detailed annotation. Genes from the user's list are highlighted in purple

11. To save the results from your pathway enrichment analysis, use popup menu, "Export Table" (Fig. 4), to export results into a tab-delimited text file. You may also choose an FDR cutoff for the table before export. For our example, we choose FDR

cutoff = 0.05, and export the results into a text file, "Pathways-FromEnrichment.txt."

12. To clear up the display of enrichment results in pathway diagrams and delete the results table, choose popup menu "Remove Enrichment Results" in the pathway tree (*see* **Note 7**).

3.3 Perform Pathway Enrichment Analysis Based on a Gene Score File

Pathway enrichment analysis based on a simple gene list discards the scores associated with each gene by using a threshold value to filter to significant genes. This type of analysis has been defined as the first-generation pathway analysis by Khatri et al. [9]. Gene set enrichment analysis (GSEA) was originally published by Subramanian et al. in 2005 [10]. Since then it has become one of the most popular approaches for pathway analysis. Contrary to gene list-based pathway enrichment, GSEA considers all genes with their scores based on a weighted Kolmogorov–Smirnov-like test, which is a representative approach of the second-generation pathway analysis. ReactomeFIViz implements features to perform GSEA analysis using Reactome pathways for a gene score file. We will still use the same example file to describe how to perform GSEA analysis using ReactomeFIViz.

1. Launch Cytoscape and then open Reactome pathway tree as described in Subheading 3.1. Choose popup menu "Perform GSEA Analysis" in the pathway tree (Fig. 2) to bring out the GSEA configuration dialog.
2. Browse the gene score file by clicking the "Browse" button, and then enter the numbers for the minimum size and the maximum size of pathways subject to analysis and the number of permutations (Fig. 7). *See* **Note 8** for detailed information about the GSEA configuration. Click "OK" to start the analysis (*see* **Note 9**).
3. Similar to pathway enrichment results as described above, GSEA analysis results are displayed in a table labeled as "Reactome GSEA Analysis" in Cytoscape's Table Panel (Fig. 8) (*see* **Note 10**). Pathways subject to the analysis are colored based on FDRs in the left-hand pathway tree.
4. To export the GSEA analysis results, choose popup menu "Export Table" in the table and save the results into a tab-delimited text file. For our example, we export the results into file "PathwaysFromGSEA.txt" after choosing FDR cutoff = 0.05.
5. Results from pathway enrichment analysis using a set of significant genes (*see* Subheading 3.2) and GSEA using all genes with scores may be different. To compare these two results, open two exported tab-delimited text files in Excel to find significant pathways shared and not shared from these two approaches. *See* **Note 11** for results using the example file.

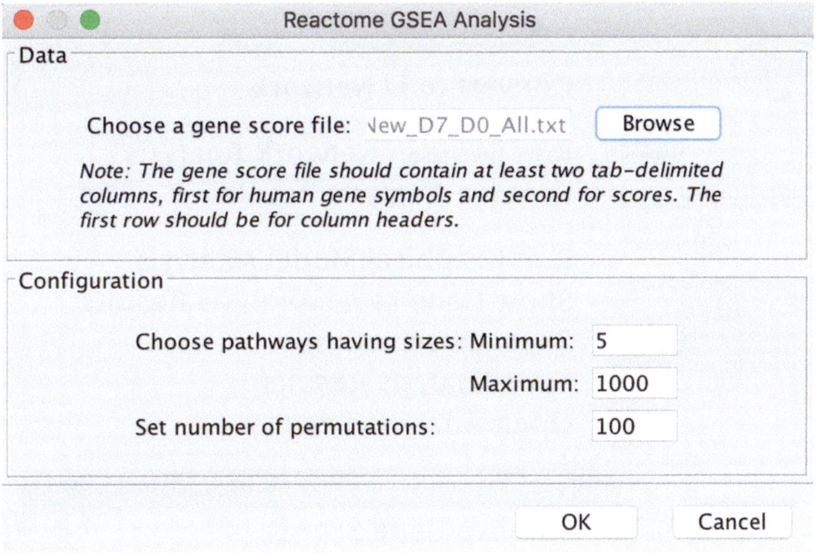

Fig. 7 GSEA analysis configuration dialog

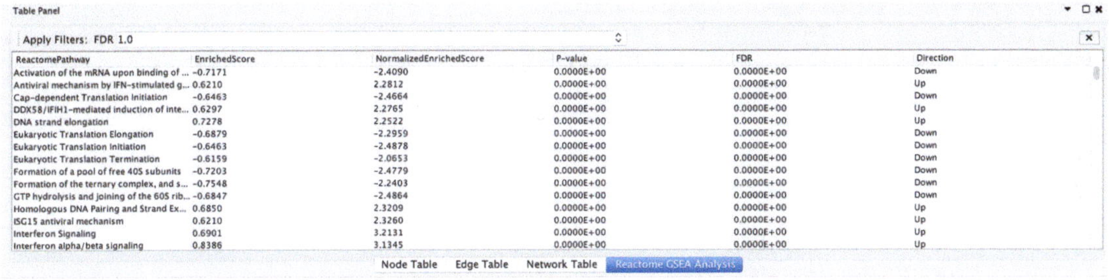

ReactomePathway	EnrichedScore	NormalizedEnrichedScore	P-value	FDR	Direction
Activation of the mRNA upon binding of ...	-0.7171	-2.4090	0.0000E+00	0.0000E+00	Down
Antiviral mechanism by IFN-stimulated g...	0.6210	2.2812	0.0000E+00	0.0000E+00	Up
Cap-dependent Translation Initiation	-0.6463	-2.4664	0.0000E+00	0.0000E+00	Down
DDX58/IFIH1-mediated induction of inte...	0.6297	2.2765	0.0000E+00	0.0000E+00	Up
DNA strand elongation	0.7278	2.2522	0.0000E+00	0.0000E+00	Up
Eukaryotic Translation Elongation	-0.6879	-2.2959	0.0000E+00	0.0000E+00	Down
Eukaryotic Translation Initiation	-0.6463	-2.4878	0.0000E+00	0.0000E+00	Down
Eukaryotic Translation Termination	-0.6159	-2.0653	0.0000E+00	0.0000E+00	Down
Formation of a pool of free 40S subunits	-0.7203	-2.4779	0.0000E+00	0.0000E+00	Down
Formation of the ternary complex, and s...	-0.7548	-2.2403	0.0000E+00	0.0000E+00	Down
GTP hydrolysis and joining of the 60S rib...	-0.6847	-2.4864	0.0000E+00	0.0000E+00	Down
Homologous DNA Pairing and Strand Ex...	0.6850	2.3209	0.0000E+00	0.0000E+00	Up
ISG15 antiviral mechanism	0.6210	2.3260	0.0000E+00	0.0000E+00	Up
Interferon Signaling	0.6901	3.2131	0.0000E+00	0.0000E+00	Up
Interferon alpha/beta signaling	0.8386	3.1345	0.0000E+00	0.0000E+00	Up

Fig. 8 GSEA analysis results displayed in the "Reactome GSEA Analysis" table

3.4 Overlay Gene Scores onto Pathway Diagrams

GSEA analysis produces a list of significant pathways for a provided gene score file based on a statistical test. However, we still do not know how the activity of a pathway is changed according to gene scores. To help users to answer this question, ReactomeFIViz provides a feature for users to overlay gene scores onto pathway diagrams, therefore, generating testable hypothesis about the underlying mechanisms of pathway activity changes.

1. Choose a pathway in the "Reactome GSEA Analysis" table (Fig. 8), and then use popup menu "View Diagram" (Fig. 4) to open the diagram for the selected pathway. The diagram should be opened in the central panel of Cytoscape. For the example output, we choose "interferon gamma signaling."
2. Click any empty space in the pathway diagram and then right-click to bring up the popup menu for the pathway view. Choose the "Overlay Gene Scores" menu (Fig. 9) to open the dialog to

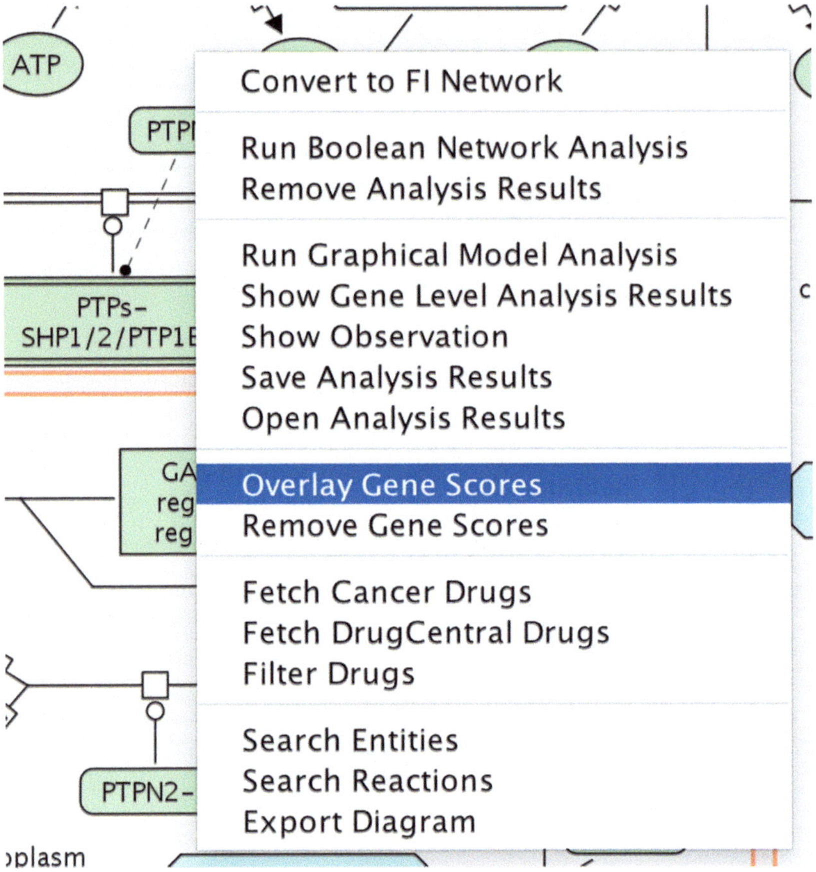

Fig. 9 Choose "Overlay Gene Scores" menu for overlaying gene scores onto a pathway diagram

choose a file containing gene scores, which should be the same file used for the GSEA analysis (Fig. 10).

3. Click the "OK" button in the dialog to let ReactomeFIViz perform gene score overlay. After the operation is complete, entities in the pathway diagram are colored based on average scores calculated according to annotated genes for them (*see* **Note 12** on how to adjust colors). A new tab labeled as "Gene Scores" is added in Cytoscape's Results Panel at the right side of the window. This tab contains two views for scores of genes annotated for the pathway: Table View and Plot View (Fig. 11).
4. Table view has three columns: column "Gene" for all genes having scores annotated for the diagram, column "Score" for scores loaded from the file, and column "Rank" for the rankings of genes across all genes from the file. You may click each column to sort rows. Selections are synchronized in the pathway diagram and this table. Figure 11 shows this synchronization of selection by choosing genes ranked at top

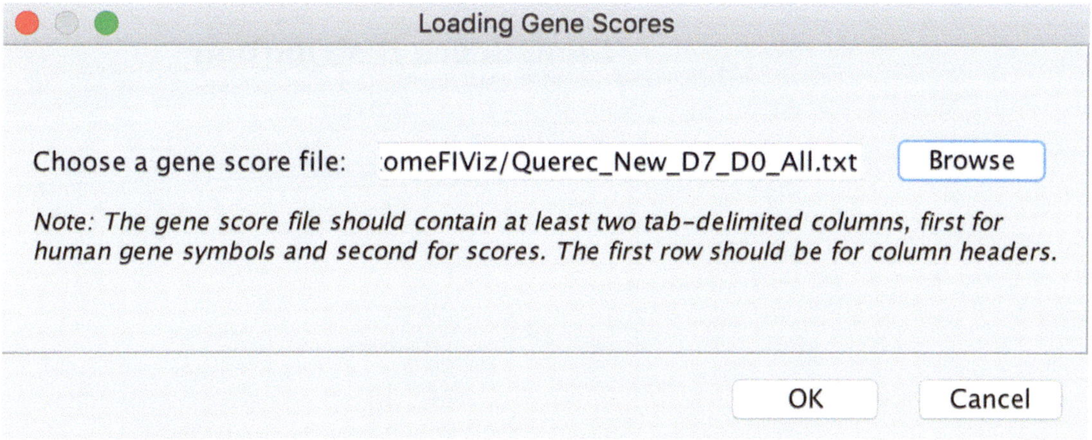

Fig. 10 Browse the gene score file for overlaying scores onto a pathway diagram

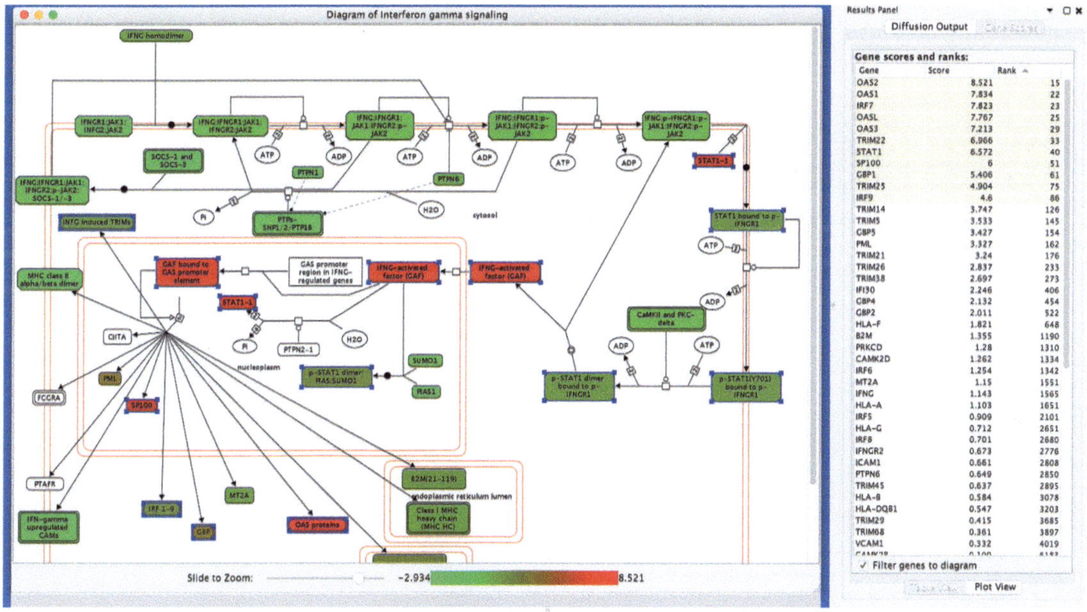

Fig. 11 Results of overlaying gene scores loaded from a local file

100, suggesting a large upregulation of outputs of this pathway (e.g., expression of GAS proteins and SP100).

5. Plot View places scores for pathway genes in the context of scores of all genes. Figure 12 shows the distribution of scores for genes annotated for pathway "interferon gamma signaling," explaining why this pathway is significantly upregulated based on the GSEA analysis.

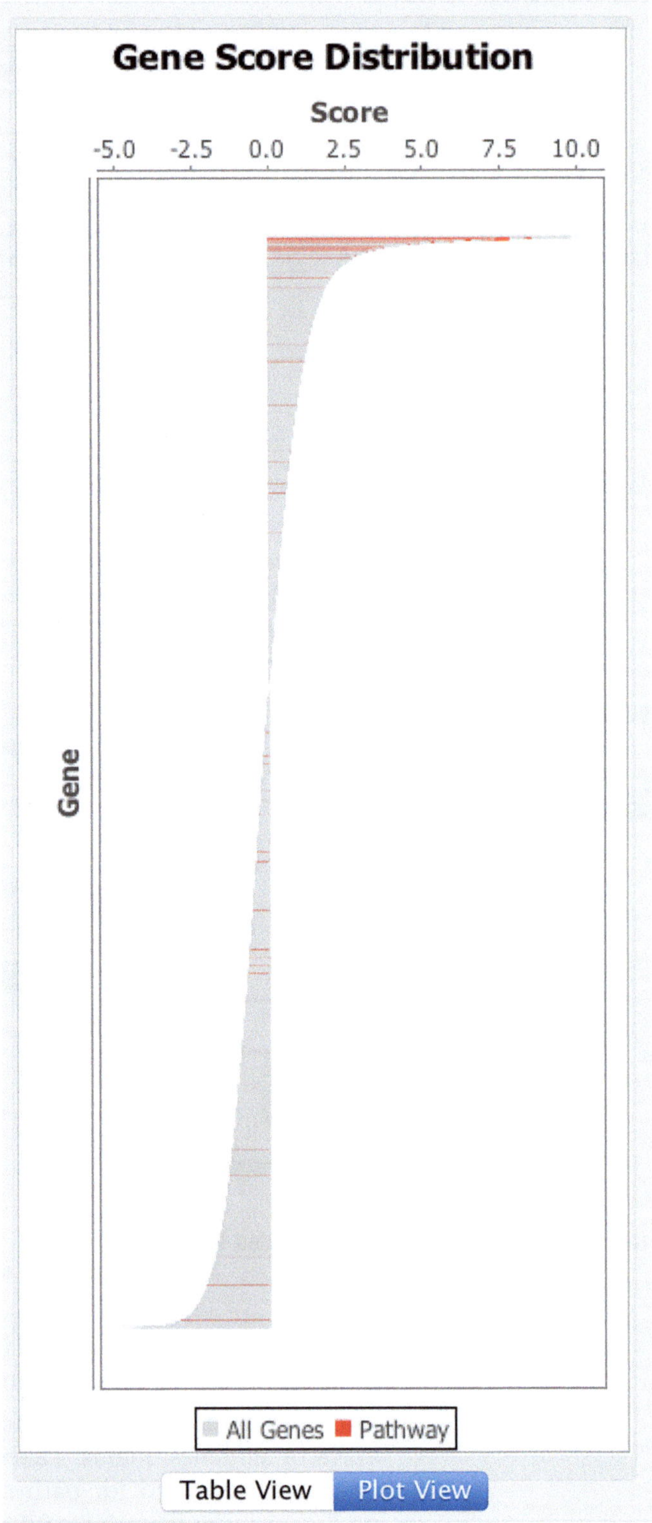

Fig. 12 Plot view of scores for genes annotated for a pathway, explaining the GSEA analysis results

4 Notes

1. *Cytoscape installation*. Cytoscape can also be installed from a compressed archive distribution, from source, or by using one of the automated installation packages that exist for Windows, Mac OS X, and Linux platforms.
2. *Java version*. Make sure you have the correct Java version installed by following the instruction in the Cytoscape download page. For example, Java 9 or higher is not supported for Cytoscape 3.6.0 or 3.6.1, which supports Java 8 only. You can download Java from http://www.java.com.
3. *Launching Cytoscape*. You may find a more detailed document by reading the user guide provided by the Cytoscape website: http://manual.cytoscape.org/en/stable/Launching_Cytoscape.html.
4. *ReactomeFIViz installation*. The reader can also directly install the ReactomeFIViz app from the Cytoscape app store website:
   (a) Launch Cytoscape software.
   (b) Use a web browser to load the Reactome web within the Cytoscape app store page: http://apps.cytoscape.org/apps/reactomefiplugin.
   (c) Click on the "Install" button. A dialog window should pop out showing the progress.
   (d) When installed, the button on the ReactomeFI plugin web page will change to "Installed."

   We frequently release new features for ReactomeFIViz when they are mature. The current version of ReactomeFIViz is 7.1.0 (July 2018), which should be compatible to Cytoscape 3.6.0 and 3.6.1. In rare cases, some of our features may not be compatible to the latest release of Cytoscape. We will update ReactomeFIViz as soon as we can fix the incompatibility. To uninstall or temporarily disable ReactomeFIViz:

   (a) Select the "Currently Installed" tab in the App Manager dialog (Fig. 1).
   (b) Choose ReactomeFIPlugIn.
   (c) Click the "Disable" or "Uninstall" button.
5. *File formats for pathway enrichment analysis*. Three text file formats are supported for pathway enrichment analysis: (a) one line per gene, (b) all genes in one line delimited by comma (,), and (c) all genes in one line delimited by tab.
6. *View in diagram*. Reactome organizes pathways in a hierarchical tree. Some pathways have their own hand-drawn diagrams, while others may be drawn inside these diagrams as sub-pathways. "View in Diagram" may open a diagram for

the pathway containing the selected one in the table. You can select a sub-pathway in the left-hand pathway tree to select all reactions annotated for a sub-pathway, focusing on your attention to a specific region in the diagram.

7. *Remove enrichment results.* You cannot see this popup menu in Fig. 2 since there are no results displayed. The popup menu is shown only after an analysis is performed for either "Analyze Pathway Enrichment" or "Perform GSEA Analysis."
8. *GSEA configuration.* Two columns are required for the gene score file to be used by GSEA in ReactomeFIViz: the first for human gene symbols and the second for gene scores. Also, the first row is reserved for the headers and will not be processed by ReactomeFIViz. The default number of permutations is 100 for a quick run. Though you may get similar results with 1000 permutations, it is recommended to use 1000 or higher permutations for final results.
9. *GSEA analysis time.* Depending on the configuration, the analysis may take different times. With the configuration in Fig. 7, it takes about 20 s for the GSEA analysis.
10. *GSEA analysis results.* See the original GSEA document for meanings of columns listed in the GSEA analysis results table: https://software.broadinstitute.org/gsea/doc/GSEAUserGuideFrame.html. We have not listed all columns described in the GSEA user guide. We have also added one extra column, Direction, showing the possible direction of pathway activities (down for downregulated or up for upregulated).
11. *Comparison of results from two pathway analysis approaches.* Using the FDR cutoff of 0.05, 14 pathways (Set 1) are significant based on the set of significant genes using pathway enrichment analysis and 84 pathways (Set 2) significant from GSEA. Overlap study shows that nine pathways in Set 1 are shared in Set 2, implying that these two approaches may yield overlapped but different results. Therefore, it is suggested to perform these two approaches together.
12. *Pathway diagram highlighting.* The default mapping from gene scores to colors is calculated based on the minimum and maximum values of gene scores. Users may adjust the maximum and minimum values used for this mapping by double-clicking the color bar at the bottom of the pathway diagram view.

References

1. Fabregat A, Jupe S, Matthews L et al (2018) The Reactome pathway knowledgebase. Nucleic Acids Res 46(D1):D649–D655. https://doi.org/10.1093/nar/gkx1132
2. Shannon P, Markiel A, Ozier O et al (2003) Cytoscape: a software environment for integrated models of biomolecular interaction networks. Genome Res 13(11):2498–2504

3. Pico AR, Bader GD, Demchak B et al (2014) The Cytoscape app article collection. F1000Res 3:138. https://doi.org/10.12688/f1000research.4642.1
4. Lotia S, Montojo J, Dong Y et al (2013) Cytoscape app store. Bioinformatics 29 (10):1350–1351. https://doi.org/10.1093/bioinformatics/btt138
5. Wu G, Dawson E, Duong A et al (2014) ReactomeFIViz: a Cytoscape app for pathway- and network-based data analysis. F1000Res 3:146. https://doi.org/10.12688/f1000research.4431.2
6. Vaske CJ, Benz SC, Sanborn JZ et al (2010) Inference of patient-specific pathway activities from multi-dimensional cancer genomics data using PARADIGM. Bioinformatics 26(12): i237–i245. https://doi.org/10.1093/bioinformatics/btq182
7. Morris MK, Saez-Rodriguez J, Clarke DC et al (2011) Training signaling pathway maps to biochemical data with constrained fuzzy logic: quantitative analysis of liver cell responses to inflammatory stimuli. PLoS Comput Biol 7 (3):e1001099. https://doi.org/10.1371/journal.pcbi.1001099
8. Querec TD, Akondy RS, Lee EK et al (2009) Systems biology approach predicts immunogenicity of the yellow fever vaccine in humans. Nat Immunol 10(1):116–125. https://doi.org/10.1038/ni.1688
9. Khatri P, Sirota M, Butte AJ (2012) Ten years of pathway analysis: current approaches and outstanding challenges. PLoS Comput Biol 8 (2):e1002375. https://doi.org/10.1371/journal.pcbi.1002375
10. Subramanian A, Tamayo P, Mootha VK et al (2005) Gene set enrichment analysis: a knowledge-based approach for interpreting genome-wide expression profiles. Proc Natl Acad Sci U S A 102(43):15545–15550

Chapter 14

De Novo Pathway Enrichment with KeyPathwayMiner

Nicolas Alcaraz, Anne Hartebrodt, and Markus List

Abstract

Biomolecular networks such as protein–protein interaction networks provide a static picture of the interplay of genes and their products, and, consequently, they fail to capture dynamic changes taking place during the development of complex diseases. KeyPathwayMiner is a software platform designed to fill this gap by integrating previous knowledge captured in molecular interaction networks with OMICS datasets (DNA microarrays, RNA sequencing, genome-wide methylation studies, etc.) to extract connected subnetworks with a high number of deregulated genes. This protocol describes how to use KeyPathwayMiner for integrated analysis of multi-omics datasets in the network analysis tool Cytoscape and in a stand-alone web application available at https://keypathwayminer.compbio.sdu.dk.

Key words De novo pathway enrichment, Molecular interaction networks, Cytoscape, Multi-omics

1 Introduction

Cell differentiation and cell activity in health and disease are directed by the complex interplay of genes and their products [1]. A growing understanding of this interplay has led to the development of pathway databases (such as KEGG [2] or Reactome [3]) that describe how functionally associated genes interact to fulfill a particular role in the cellular machinery. Gene set overrepresentation [4] and gene set enrichment analysis [5] can be used to measure the activity or relevance of these pathways in specific experimental conditions. However, high-throughput molecular profiling methodologies enable us to draw a more comprehensive picture of gene interactions (gene regulatory networks) or of the transcribed proteins (protein–protein interaction networks). Typical examples for such molecular interaction networks are listed in Table 1. These molecular interaction networks offer the opportunity to detect functional modules de novo by searching for subnetworks that are active under particular

Stefan Canzar and Francisca Rojas Ringeling (eds.), *Protein-Protein Interaction Networks: Methods and Protocols*, Methods in Molecular Biology, vol. 2074, https://doi.org/10.1007/978-1-4939-9873-9_14,

Table 1
A non-exhaustive list of commonly used molecular interaction networks

Network	Reference	Description	Species
BioGrid	[12]	Curated resource integrating protein, genetic, and phosphorylation interaction data from publications	13 major model organisms and many more
GeneMania	[13]	Uses an algorithm to determine association strength based on publicly available data for protein and genetic interactions, pathways, co-expression, co-localization, and protein domain similarity	9
HTRIdb	[14]	Network of interactions between transcription factors and target genes	Human
HPRD	[15]	The first comprehensive human protein–protein interaction network	Human
IntAct	[16]	Curated resource for all types of molecular interactions	Many
I2D	[17]	Integration of known and predicted protein–protein interactions	6
STRING	[18]	Protein–protein interactions, experimental and predicted	>2000

Countless more molecular interaction networks exist most of which can be browsed and accessed via the NDEX web service (http://www.ndexbio.org)

experimental conditions or in a specific phenotype. These subnetworks are expected to partially represent known pathways but also to reveal new pathways with unknown functional role. The activity of a candidate subnetwork is here typically measured by the enrichment of active, e.g., differentially expressed or methylated, genes. We have developed KeyPathwayMiner (KPM) [6] as a user-friendly software tool that enables de novo pathway enrichment either through a web interface [7] or as app in the network analysis tool Cytoscape [8–10]. As input, KPM expects a molecular interaction network and one or several datasets that describe if a gene is active in a given sample. Furthermore, KPM offers the choice between two models for extracting subnetworks with a local and a global view on exceptions, e.g., cases in which genes are not active in one or several samples. Finally, KPM supports three inference methods, i.e., a fast greedy approach, an exact fixed parameter tractable approach, and an ant colony optimization heuristic approach. In this protocol, we will briefly describe the principles of the models and methods used in KPM, explain how the input data needs to be formatted, and guide interested users through the steps and parameter choices for performing de novo pathway enrichment.

2 Models

Extracting de novo pathways from a large biological network, where genes/proteins in the pathways are active in one or more OMICs datasets, can be modeled as a graph-theoretical problem.

Let $G(V,E)$ be an undirected graph representing a biological network (e.g., protein–protein interaction network); $A^{n,m}$ an indicator matrix for n molecules (e.g., proteins) vs. m samples, obtained after applying a cut-off from the measurements (e.g., log fold change or p-value); and a mapping $Z: V \rightarrow \{0, 1\}^m$ from each node $v \in V$ to its corresponding vector row i in A. The main objective is to extract subnetworks $G'(V',E') \subset G$ such that:

(a) G' is connected (any pair of nodes in V' is reachable by a path traversing edges in E').

(b) The set of nodes V' satisfy a certain criteria based on their values in A.

Here we propose two models that can be applied to one or multiple OMICS datasets:

2.1 Individual Node Exceptions (INEs)

Given a series of indicator matrices $A_1, \ldots, A_p$, parameters K, $L_1, \ldots, L_p$ and a proposition logical formula F with Boolean variables $x_1, \ldots, x_p$. The Boolean variables are assigned a value by each node r in the following way:

$$x_i{}^r := \left(\sum_{s=1}^{m} A^{r,s}\right) > m - L_i.$$

In other words, a node r will be considered "active" in dataset i if all but at most L_i samples contain a 1 in their corresponding entry in the indicator matrix. For a node to be considered valid (non-exception node), it must satisfy the formula F_i. The formula is satisfied if it evaluates to *True*, after assigning values to all the Boolean variables. As an example, let us say we have two indicator matrices: A_1 corresponds to gene expression measurements, while A_2 to gene promoter methylation. If one would like to search for pathways containing only genes that are silenced (not expressed) due to promoter methylation, the formula: $\neg x_1 \land x_2$ (read as NOT x_1 AND x_2) can be provided.

The objective of INEs is to extract all maximal connected subnetworks $G'(V',E')$ such that all, for the exception of at most K nodes in V', satisfy F. The parameters $L_1, \ldots, L_p$ can be interpreted as the amount of noise in the datasets we tolerate, while parameter K controls the number of gene "outliers" in our resulting subnetworks.

2.2 Global Node Exceptions (GloNE)

Though parameters in INEs are straightforward to interpret, results can be biased toward highly connected nodes, since the search will be optimized to link as many nodes as possible to the solution resulting in the preferential inclusion of the so-called hub-nodes that have a high node degree (many connections) as part of the solution, even though these nodes are not "active" in the experiments. Since this is often undesired, we developed the GloNE model. Here we omit the K parameter and redefine $L_1, \ldots, L_p$.

Now instead of each node assigning values to the variable x_i in the propositional formula, the whole subnetwork G' will assign the Boolean values in the following way:

$$x_i := \left(\sum_{r \in V(G')} \sum_{s=1}^{m} A^{r,s}\right) > m - L_i.$$

Hence, if the total number of inactive cases in the subnetwork does not exceed L_I, the network will be considered active in the dataset i. The subnetwork will be considered a valid solution if after assigning values to $x_1, \ldots, x_p$ it will satisfy the given formula F. The objective of GloNE is to extract maximal subnetworks that satisfy formula F.

3 Methods

Both INEs and GloNE are computationally hard problems. While INEs is exponential [6] in parameter K, GloNE is proven to be NP-hard (by reduction to the maximal weighted connected subnetwork problem [11]). To tackle these problems, we implemented the following algorithms.

3.1 Fixed Parameter Tractable

The INEs problem can be solved exactly for low values of K (typically up to $K = 4$ exceptions); in other words, it is possible to implement an exact fixed parameter tractability (FPT) approach to tackle this problem. We developed an exact branch and bound algorithm method where the current best solution is used as a lower bound. If for some y, the algorithm has a partial solution which already has K - y exception nodes, then it proceeds to compute an upper bound as follows: (1) determines all possible new exception nodes it can add to the solution; (2) takes the n nodes with the highest weights and adds them together, which is the upper bound for the partial solution; and (3) performs exhaustive search using the two bounds to find the optimal solution.

3.2 Greedy

We developed a fast greedy approach that can provide good solutions for prohibitively large problems (e.g., dense networks). The algorithm starts from a seed node v which is added to the partial solution S_v and proceeds by iteratively adding an adjacent node u to the solution that maximizes the score:

$$S = \sum\nolimits_{r\in(V(S)\cap\{u\})} \sum\nolimits_{i=1}^{p} \sum\nolimits_{s=1}^{m} A_i^{r,s}.$$

This process continues until there are no new nodes to add that would keep the solution S_u to be valid.

3.3 Ant Colony Optimization

Additionally, we implemented an heuristic method based on ant colony optimization (ACO), called max-min ant system (MMAS) [19]. The MMAS algorithm is able to outperform the greedy algorithm if given enough time to search for the solutions. Similar to the greedy algorithm, we start constructing a solution S_v from a seed node v. Then, we imagine an ant running through the network and iteratively adding valid adjacent nodes to the solution by selecting them on a probability proportional to $\eta(u)^{\alpha} * \tau(u)^{\beta}$. In this case $\eta(u)$ is the heuristic value of the node (which is proportional to the number of active cases), $\tau(u)$ is the pheromone value representing the memory of previous ants that have used this to construct good solutions, and α, β are parameters to control their importance. After the ant cannot find any new valid nodes to add to the solution, the fitness (in this case the size of the solution) is evaluated, and the best ant is allowed to update the pheromone values for each edge. In the max-min ant system, pheromone values must have a minimum and maximum value (in this case the value is bound by the interval (0,1). Additionally, after each iteration all edge pheromone values are decreased by a percentage ρ, in order to avoid bad solutions to keep appearing in future searches due to early pheromone drops.

4 Parameters

4.1 Datasets (Single and Multi-Omics)

KPM expects as input indicator matrices containing either 1 or 0 values (active or inactive), one for each case study or measurement. Each matrix must be in a tab-delimited text file where rows correspond to molecules (genes, transcripts, proteins, etc.) and columns to cases/samples. It is important that the first column contains the molecule ids and that those are the same with that used in the input network. After the datasets have been provided and if there is more than one, the user must provide a logical formula connecting the datasets.

4.2 Input Network

The biological network is expected in a tab-delimited format, where each row corresponds to an interaction with at least two columns that represent the interaction partners. The node ids must correspond to molecule ids of the input dataset(s). Self-loops and duplicated edges will be removed before any computations are made, since these are not required for the algorithms.

4.3 Previous Knowledge

Previous knowledge can be provided in the form of positive/negative node lists. These can only be considered if they are present in the network. Providing these lists will set all cases for those particular nodes to active (positive list) or inactive (negative list), biasing the search to include or exclude them in the solutions.

4.4 Run Parameters

1. Search strategy: Which model will be used, either INEs or GloNE.
2. $L_1, \ldots, L_p$ parameters: One for each input matrix, where they represent either the maximum number of allowed exceptions (inactive cases) in each node in the solution (INEs) or the total in the whole solution (GloNE).
3. K (only for INEs): The maximum number of allowed nodes in the solution not satisfying the provided logical formula connecting the datasets.
4. Algorithm: One of the three algorithms to extract the pathways—greedy, ACO, or exact FPT.
5. Border exception nodes: Often solutions overlap with respect to active genes and differ only in different exception genes added on the periphery of the solution. We coin these border exception nodes which can be removed in a post-processing step to avoid this issue.

The default values parameters should be adequate for most cases, and changing them is only suggested for advanced users familiar to the algorithm.

5 KPM in Cytoscape

5.1 Installation

1. Download Cytoscape 3 from http://www.cytoscape.org/ and install it (consider **Note 1**).
2. To install the KPM app in Cytoscape 3 and above, go to the Apps -> App Manager and search for "KeyPathwayMiner."
3. Click on the "KeyPathwayMiner" entry in the search result and select "install." After closing the App Manager, the KPM panel should now appear as a tab in the Cytoscape Control Panel on the left-hand side (if not follow **Note 2**).

5.2 Network Selection

Before you can use KPM, you need to make sure a suitable molecular interaction network is loaded into Cytoscape (*see* **Notes 3** and **4**). Networks can be imported via File -> Import -> Network with three options:

1. From File: Here you can import a local network from your hard disk. Make sure you use a network file format that is supported by Cytoscape (*see* http://manual.cytoscape.org/en/stable/Supported_Network_File_Formats.html for more details). A list of commonly used molecular interaction networks is found in Table 1.
2. From URL: Instead of loading a supported network file type from disk, you can provide a URL to a website where this file is available.
3. From Public Databases: Here, a curated list of publicly available databases is available (*see* http://manual.cytoscape.org/en/stable/Creating_Networks.html#import-networks-from-public-databases) for further details (also consider **Note 5**).

5.3 Dataset Selection, Linking, and Previous Knowledge

1. You can add a dataset from your local hard disk by clicking on "Load" in the "Data" tab (*also see* **Notes 6–8** as well as Fig. 1).
2. You can load additional datasets by clicking on the green button and subsequently on the new "Load" button.
3. Proceed to the "Links" tab if you have loaded more than one dataset; otherwise continue to the next section.

Fig. 1 The "data" tab of the KPM Cytoscape app where one or multiple datasets can be added by selecting a corresponding file from the hard drive (Load button). Additional datasets can be added and already selected datasets can be removed

4. You can control how KPM considers information from multiple datasets by linking them in various ways as described in Subheading 4.1. In the Cytoscape app, you can choose between three modes, namely:
   (a) OR—Example: A node/gene is considered active if it is active in dataset A OR B OR C.
   (b) AND—Example: A node/gene is considered active when it is active in datasets A AND B AND C.
   (c) Custom—Using custom you can create complex links. Example: A node/gene is considered active when it is active in datasets (A AND (B OR C)). For the last case, you can use a built-in formula editor as described in the following and shown in Fig. 2.
5. Datasets (yellow) can be linked through operator nodes (blue). Add an operator node by clicking anywhere on the empty space.
6. Change the type of operator (OR, AND, XOR) in a blue node by left clicking on it and selecting the option from the pop-up menu.
7. Connect two nodes by right clicking on the source node and dragging it to the destination node (*see* **Note 9** for details).
8. Continue to the tab "Pos/Neg" if you have previous knowledge as described in Subheading 4.3. You can enter positive and negative nodes by entering one node identifier (such as a gene symbol or accession number) per line or by clicking on "Load positive nodes from file" or "Load negative nodes from file." Make sure that the file lists one node identifier per line (*see* **Note 10**).

5.4 Setting Up a KPM Analysis

1. Continue to the "Run" tab to configure KPM (Fig. 3).
2. Select a search algorithm (one of greedy, ACO, or exact/FPT) as described in Subheading 3 (*see* **Note 11** for advanced options of the ACO algorithm).
3. Select a search graph or network to use (*see* **Note 12**).
4. Decide whether you want to remove border exception nodes as described in Subheading 4.4.
5. Choose the number of CPU cores used for the analysis (*see* **Note 13**).
6. Decide how to treat unmapped nodes, i.e., add them to the negative or positive list as described in Subheading 4.4.
7. Select the number of pathways to be computed (*see* **Note 14**).
8. Select the number of node exceptions K (if INEs is selected). Check "use range" if several options for K should be explored (*see* **Note 15**).

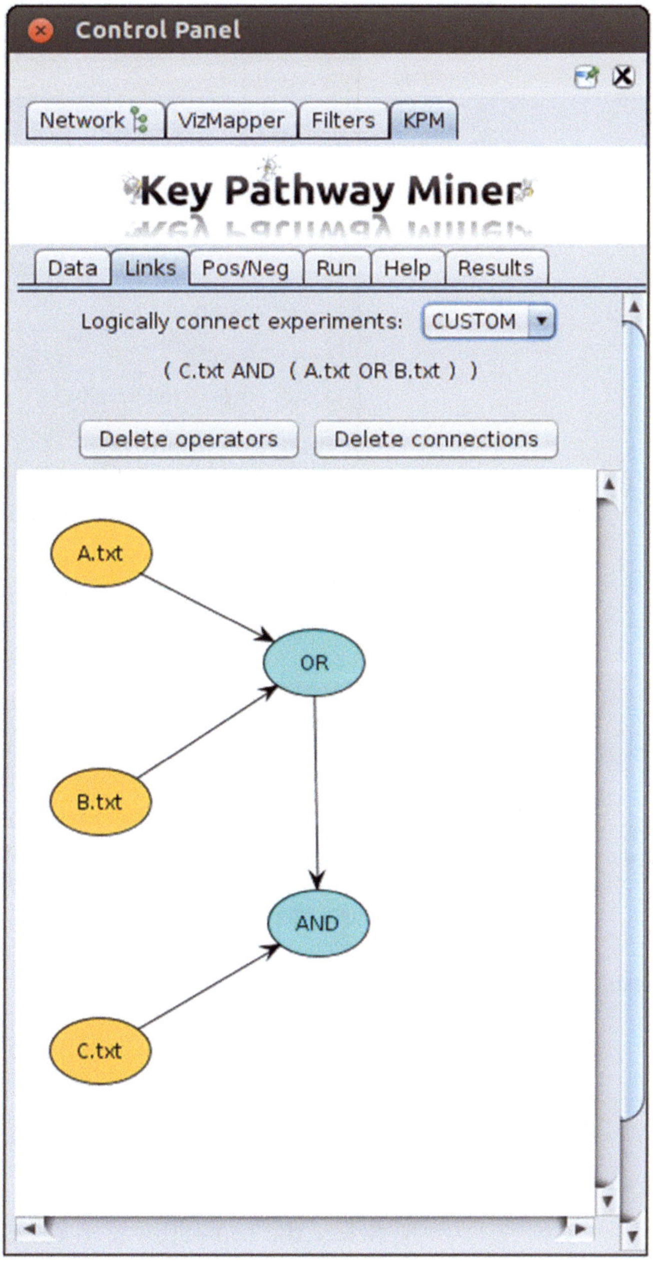

Fig. 2 The "links" tab of the KPM Cytoscape app. Here, users can interactively combine different datasets logically using AND and OR operators

9. Similarly, select the number of case exceptions *L*. This can either be done globally or for each dataset individually. *L* can be expressed as a fixed value or interpreted as percentage (check box). Moreover, a range of parameters can be defined by checking "use range" (*see* **Notes 15** and **16**).
10. Press "Search key pathways" to start the KPM analysis (*see* **Notes 17** and **18**).

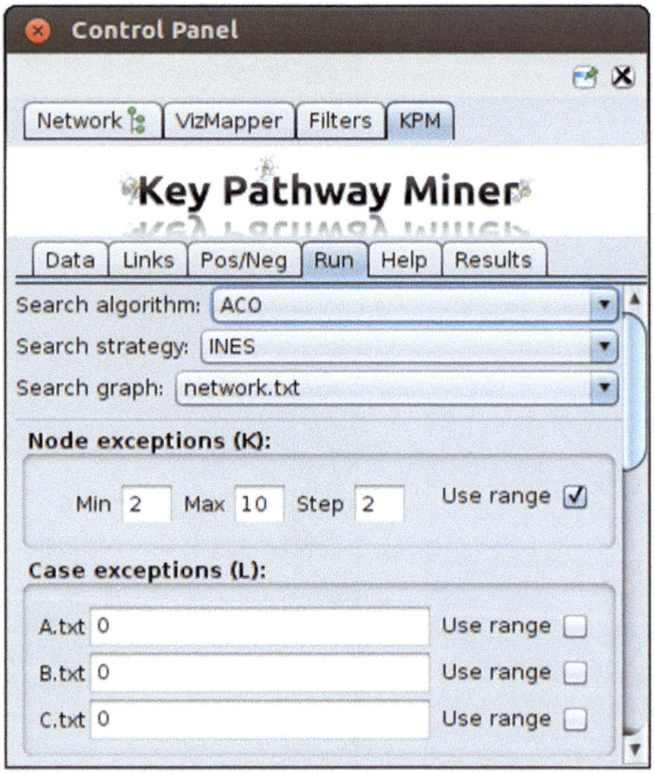

Fig. 3 The "run" tab of the KPM Cytoscape app. Here, users can select a search algorithm, a search strategy, the input network, as well as parameters for node and case exceptions (*K* and *L*). Users can also define a range of parameters with a particular step size

5.5 Results

1. Select results to view: In case a range of parameter values (K and/or L) was selected, a slider will appear on the results for each parameter. Move the slider to the corresponding value, and the results table will update with the results for that specific parameter combination.
2. View the results table and sort it by column value by clicking the corresponding field.
3. Select the top pathway in the table and view it by clicking on "View selected." Multiple rows can be selected by clicking on them while holding down the "Control" key to visualize multiple solutions.
4. Inspect the pathway in the Cytoscape viewer. There are three different shapes:
   - Circle: the node was fully mapped to all measurements.
   - Hexagon: the node was partially mapped to some, but not all measurements.
   - Triangle: the node was not mapped to any measurement.

5. Update the current pathway coloring in the results table and select one of the following:
   - Active cases: nodes will be colored based on the number of active cases.
   - Mapped: nodes will be colored based on the number of mapped datasets.
   - Pathway hits: nodes will be colored based on the number of times they appear on the full list of pathways.
   - Exceptions (only INEs): nodes will be colored differently if they are "exception" or non-exception nodes.
6. View node stats in the bottom table where you can sort the nodes by their ID, number of pathways containing the node, number of expressed cases, and degree by clicking on the corresponding header column. Click on one node to select it in the current viewed network. You can also select multiple genes from the list.
7. Pathways can be saved into different formats by selecting the pathways in the results table (*see* **step 3**) and clicking on "Save Pathways."

6 KPM in the Web Browser

The web version of KPM can be accessed at https://keypathwayminer.compbio.sdu.dk.

6.1 Network Selection

In order to perform your analysis, you will need to provide a suitable molecular interaction network to KPM.

1. Default networks: You can use one of the networks provided on the website. In order to view the available networks, click on the "networks" tab in the toolbar (Fig. 4). You will see a list of networks with a short description (*see* **Note 19**).
2. Custom network: By clicking on the "Upload network" button in the "networks" tab, you will access a page that allows you to upload a network from your local drive. You can select the file and provide a name, the species, and an optional description. Submit your network via the "Upload" button. The network will now be visible in the networks tab, where you can also delete it, if you wish (*see* **Note 20**).

6.2 Dataset Selection

1. Preselected datasets: We provide example data on the website which you can access by clicking on the "datasets" tab in the toolbar.
2. Custom dataset: Via the "Upload dataset" button in the "datasets" tab, you can upload your own dataset. Currently two types of matrices are supported, an indicator matrix containing

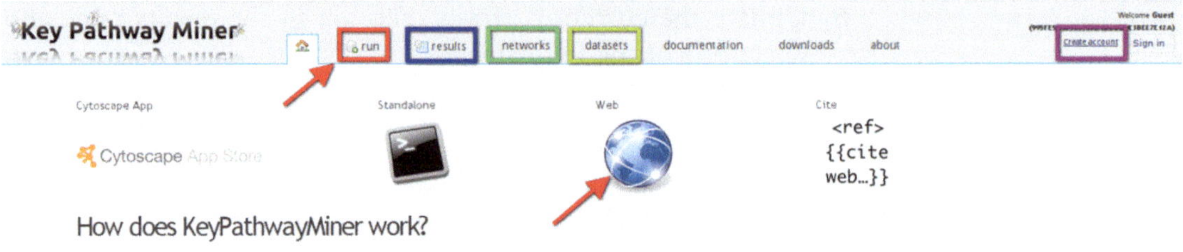

Fig. 4 An overview of the menu of KPM web. Run (red): Click on this tab to start a new KeyPathwayMiner analysis. The same can be achieved by clicking on Web (also indicated by a red arrow). Results (blue): If a run was previously started, its progress can be monitored in this tab, and the selected settings can be reviewed. For runs that have already been completed, the results can be opened. Networks (green): KeyPathwayMiner Web provides selected interaction networks by default. However, should you wish to study other organisms than human or should your dataset contain other IDs than Entrez gene IDs, you should go to this tab to upload your own network file. Note: Only SIF files can be accepted at the moment. Datasets (yellow): KeyPathway-Miner comes with a small selection of demo datasets. Before you start your analysis, you probably want to upload your own experimental data here. Note that uploaded datasets are only visible within the current guest session or to the logged in user. Create account (purple): Optionally, it is also possible to sign up for a user account, which has the advantage that results can later be opened without the need to bookmark the session link. This is done via the "create account" link in the right side of the header

0 or 1 or a numerical matrix. For the numerical matrix, KPM creates an indicator matrix internally. Simply select your file, the type, and the species, give a name and a short description, and click on "Upload." You can then see your dataset in the "data-sets" tab (*see* **Notes 20** and **21**).

6.3 Setting Up a KPM Analysis

Once you have uploaded your network and your datasets, you can proceed to set up your analysis. To do so, click on the "run" tab on the website.

1. First, select one or more datasets and decide on a threshold in the case of a numerical matrix (*see* **Note 22**). Choose your dataset by clicking on the list entry. Added datasets will appear in the right handed list (Fig. 5).
2. If you want to use several datasets, you need to connect them with a logical operator, for instance, we provide logical "AND" and "OR" which you can select from the dropdown menu (*see* **Note 23**).
3. Previous knowledge can be included via the positive list for genes that are preferentially included in the result and negative list for genes that will preferentially be excluded in the result. These genes can either be provided in the text input or by selecting a file (*see* **Note 24**).
4. Click on "Next" to continue your setup.
5. Make sure to select the correct network from the dropdown menu (Fig. 6).

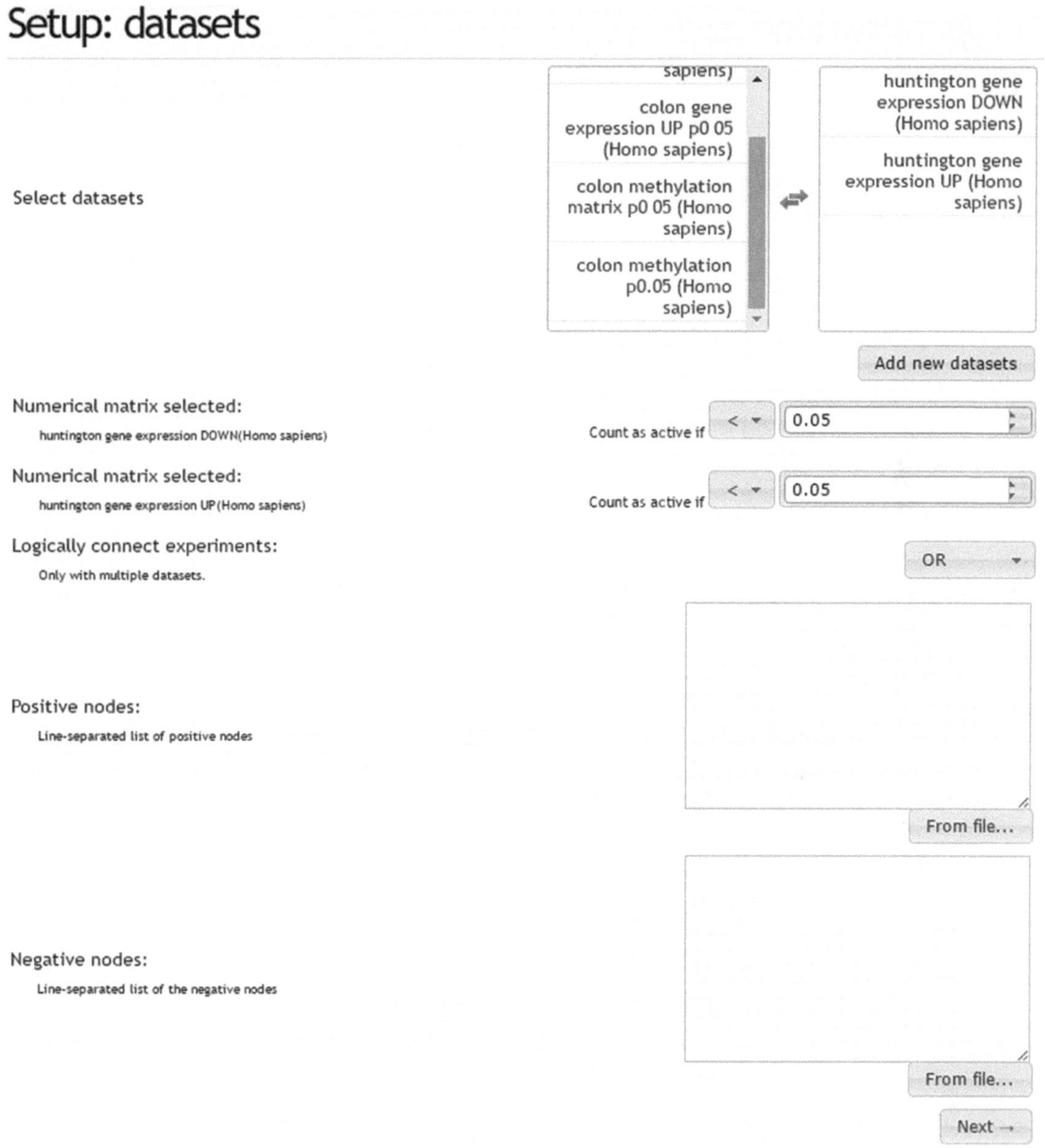

Fig. 5 Screenshot of the dataset setup of KPM web. Here, users can select one or several existing datasets or add new ones. Next, users need to determine how active genes/nodes are identified and how multiple datasets are connected. Positive and negative nodes can also be defined here

6. Select your search algorithm and search strategy and optionally give a name to your run (*see* **Note 25**).
7. Decide whether you want to remove border exception nodes as described in Subheading 4.4.
8. Select the number of node exceptions K (if INEs is selected). Check "use range" if several options for K should be explored.

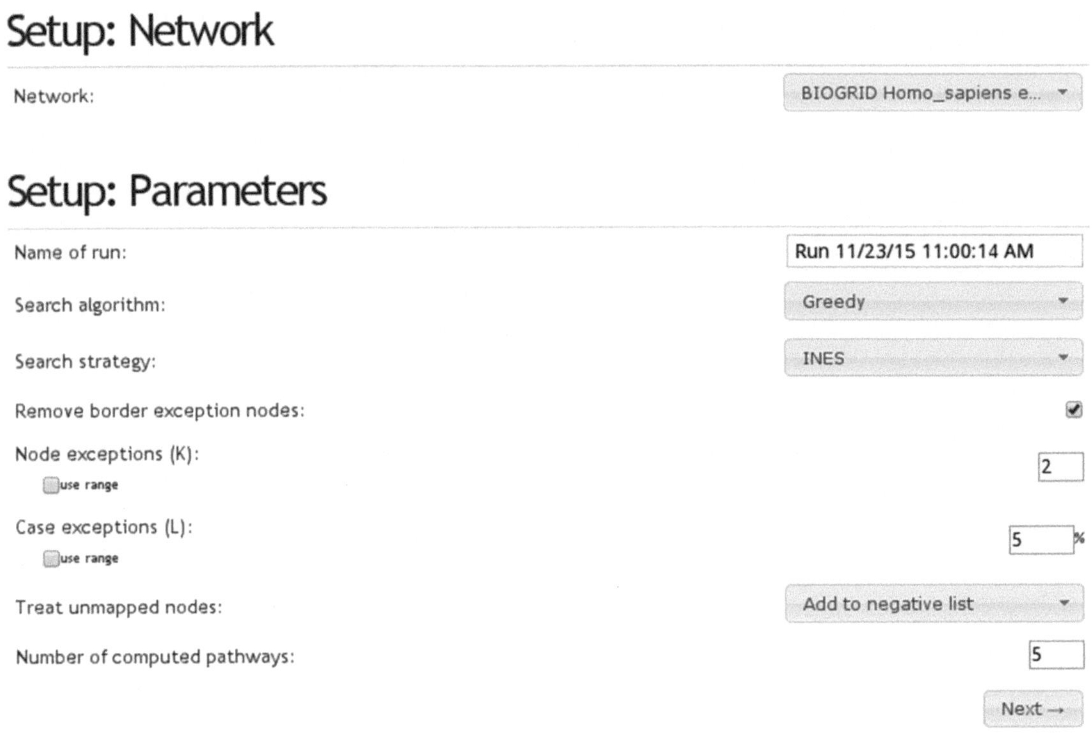

Fig. 6 Screenshot of the network and parameter setup of KPM web. Here, users can select a search algorithm, a search strategy, the input network, as well as parameters for node and case exceptions (*K*, *L* in %). Users can also define a range of parameters with a particular step size

9. Similarly, select the allowed percentage of case exceptions *L*. Again, a range of parameters can be defined by checking "use range."
10. Decide whether you want to add the unmapped nodes to the positive or to the negative list.
11. Choose a number of pathways you want to compute and click on "next."
12. Check whether all the settings are correct and click on "START." You will be redirected to the "results" tab, where you can view the results, as well as all other results of your current session (*see* **Note 26**).

6.4 Results

1. Select the result you want to see and click on "See results" (Fig. 7).
2. The chosen *K* and *L* parameters are summarized at the top of the page. A slider can be used, to select parameter combinations.
3. Users can replace node identifiers by clicking the "Substitute node labels" checkbox and selecting source and target identifiers (*see* **Note 27**).

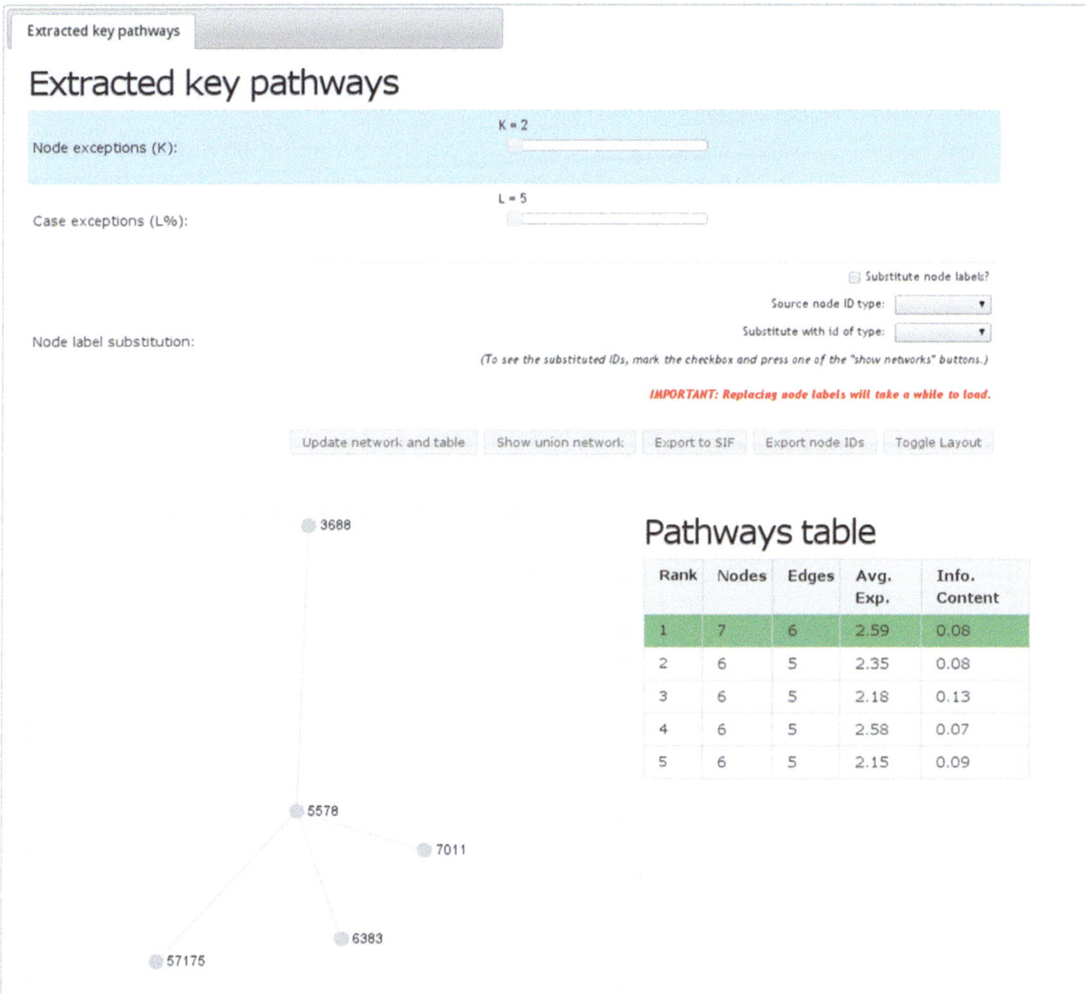

Fig. 7 Exemplary screenshot of the results of KPM web. Users can browse through solutions for different parameter combinations and select individual pathways from a results table. Alternatively, a union of the computed pathways can be shown. In addition to export functionality, users can also replace node ids using an external service

4. "Update network and table" will refresh the graph.
5. "Show union network" will create a network containing all nodes from all the results, and the number of times a node appears is color coded, to indicate the evidence level.
6. A convenient way to access information on the genes is to click on the nodes which allows you to directly view the NCBI gene record (https://www.ncbi.nlm.nih.gov/gene/).
7. Export of the network to .sif format and the corresponding list of genes is possible via the "Export to SIF" and "Export node IDs" buttons.
8. The layout of the network can be optimized using the "Toggle Layout" button or by dragging nodes with the mouse.

7 Notes

Cytoscape

1. Cytoscape expects a specific Java version to function correctly. Make sure that you have the correct Java version installed. For instance, Cytoscape v. 3.6.1 requires Java 8 which can be downloaded from http://www.oracle.com/technetwork/java/javase/overview/java8-2100321.html.
2. Once the KPM Cytoscape app is installed, it should immediately appear in the Apps menu as "KeyPathwayMiner." In rare cases where this is not the case, it helps to simply restart Cytoscape to fix the issue.
3. KPM will always use the currently selected network when you start an analysis. Be aware to select the network you intend to use in the "Network" tab of the Control Panel.
4. If you delete an existing network, Cytoscape may report an error (null pointer exception) and then hang. You will have to restart Cytoscape to fix this issue.
5. Since Cytoscape 3.6, the curated list of publicly available networks can also be searched conveniently from within the "Network" tab in the Control Panel.
6. If you accidentally loaded the wrong dataset, you can remove it by clicking on the red cross.
7. Make sure your input file is in the right format that KPM expects. The indicator matrix format is described in Subheading 4.1.
8. After loading the dataset, the number of entities (e.g., number of genes) and number of samples will be shown below the file name. These numbers should be checked carefully since they indicate if the dataset was loaded correctly or not. For instance, if the wrong column delimiter was used, columns cannot be separated and only 1 sample might be shown.
9. You can connect dataset nodes (yellow) and/or operator nodes (blue) to build more complex relationships (links). Considering an example of three datasets A, B, and C, you can first connect A and B by adding and linking an AND operator. You can subsequently incorporate a third dataset by introducing a second OR operator that is linked to dataset C as well as the first AND operator, leading to the operation ((A and B) OR C).
10. Below the positive and negative node lists, you find a button "Check" which allows you to test if the selected node identifiers are found in the dataset. If node identifiers are not mapped successfully, they are highlighted in red font, allowing you to

check if the spelling or identifier type is correct. Note that nodes not mapped to the network will be ignored.

11. In case ACO is chosen, click on the arrow next to the "ACO Advanced Parameters" option to see and edit the following parameters:
    - Importance of pheromone level (alpha)—Relevance to the amount of pheromone when constructing solutions.
    - Importance of edge heuristic value (beta)—Relevance given to the heuristic value when constructing the solutions.
    - Pheromone decay rate (rho)—This is the parameter for pheromone decay rate in order to avoid fast convergence to suboptimal solutions.
    - Minimal pheromone value—This is the parameter for the minimum level of pheromone a node can have.
    - Number of solutions per iteration—The number of solutions computed per iteration (this can affect the runtime).
    - Number of Startnodes—used by the GloNE approach, affects runtime.
    - Version to use—used by the GloNE approach, iteration based or global search.
    - Maximum number of iterations—This maximum number of iterations the algorithm will be allowed to run.
    - Maximum iterations without change—The number of iterations the algorithm is allowed to run with no improvement to find solutions.

    The default parameters should be sufficient in most cases; tweaking the values from these parameters is only recommended for advanced users.
12. If none can be selected, make sure you imported a network correctly as shown in Subheading 5.2.
13. Selecting a higher number of CPU cores (processors) can significantly speed up the computations performed by KPM. The highest number that can be selected corresponds to the number of cores reported by your operating system. Note that exhausting your machine's resources by selecting the maximal available number of cores may block other processes, and as a result your machine may be unresponsive until the analysis is completed.
14. Choosing a large number of pathways to report may dramatically increase the runtime of the method.
15. A range is defined by a lower and upper boundary as well as a step size, e.g., min = 4, max = 12, and step size = 4 results in parameter values 4, 8, and 12.

16. If "Set same percentage for all datasets" is selected, it is no longer possible to set a specific L value since the number of samples may strongly vary between individual datasets.
17. An advanced feature of KPM that was introduced in version 5 is robustness and validation analysis. Here, the network can be perturbed to study how robust the solution is using one of several techniques. Further details can be found in [9].
18. If the algorithm cannot detect a pathway, the message "No pathways could be detected with these parameters. Please try again with different parameters" is shown, but the program does not terminate as expected. This can be fixed by manually canceling the search.

Web Interface

19. The default networks all use Entrez Gene IDs.
20. The correct data format for the files is described in Subheading 4.
21. Without prior login, the datasets you upload and the results of your analyses will be stored anonymously and tied to a session id. You can bookmark a session if you wish to retain access to the results. The session bookmark URL is shown in the results tab. Alternatively, you can sign up for a user account and gain access to your results independent of the session id. Note that anonymous user data and results are deleted frequently. Note that we do not track the IP address of our users but use cookies for the purpose of keeping a session alive.
22. The threshold value determines how the matrix is discretized. A sensible threshold depends on your dataset. When using p-values, all values smaller than your significance level should be set to 1 (=active). Other data types (e.g., read counts) might require other thresholds.
23. Selecting "AND" means that, in order to be counted as active, the gene needs to be active (=1) in both datasets. "OR" means that it is sufficient for a gene to be active in one dataset in order to be counted as active.
24. In the text input as well as the file upload, provide one gene per line.
25. Currently, the web version of KPM does not provide the complete set of functionalities implemented in the Cytoscape version. The algorithm available on the web is "Greedy" with the search strategies GloNE and INEs.
26. The results tab provides an overview of all KPM runs executed by the user (if logged in) or during the anonymous session (if not logged in). Progress bars indicate the current status of the run. Moreover, if demand is highly queried, runs are not

executed immediately but are queued. In the latter case, the position in the queue will be shown. Previously queried runs can also be cancelled here.

27. Note that exchanging the identifiers might take some time, especially if the number of nodes is large, because it makes use of another web service. Note that even when logging out or closing the browser window, the scheduled runs will still be executed (consider bookmarking the session id, **Note 21**).

References

1. Cordell HJ (2009) Detecting gene–gene interactions that underlie human diseases. Nat Rev Genet 10:392
2. Kanehisa M, Goto S (2000) KEGG: Kyoto encyclopedia of genes and genomes. Nucleic Acids Res 28:27–30
3. Joshi-Tope G, Gillespie M, Vastrik I et al (2005) Reactome: a knowledgebase of biological pathways. Nucleic Acids Res 33: D428–D432
4. Falcon S, Gentleman R (2008) Hypergeometric testing used for gene set enrichment analysis. In: Hahne F, Huber W, Gentleman R, Falcon S (eds) Bioconductor case studies. Springer New York, New York, NY, pp 207–220
5. Subramanian A, Tamayo P, Mootha VK et al (2005) Gene set enrichment analysis: a knowledge-based approach for interpreting genome-wide expression profiles. Proc Natl Acad Sci U S A 102:15545–15550
6. Alcaraz N, Kücük H, Weile J et al (2011) KeyPathwayMiner: detecting case-specific biological pathways using expression data. Internet Math 7:299–313
7. List M, Alcaraz N, Dissing-Hansen M et al (2016) KeyPathwayMinerWeb: online multi-omics network enrichment. Nucleic Acids Res 44(W1):W98–W104
8. Alcaraz N, Pauling J, Batra R et al (2014) KeyPathwayMiner 4.0: condition-specific pathway analysis by combining multiple omics studies and networks with Cytoscape. BMC Syst Biol 8:99
9. Alcaraz N, List M, Dissing-Hansen M et al (2016) Robust de novo pathway enrichment with KeyPathwayMiner 5. F1000Res 5:1531
10. Shannon P, Markiel A, Ozier O et al (2003) Cytoscape: a software environment for integrated models of biomolecular interaction networks. Genome Res 13:2498–2504
11. Ideker T, Ozier O, Schwikowski B, Siegel AF (2002) Discovering regulatory and signalling circuits in molecular interaction networks. Bioinformatics 18(Suppl 1):S233–S240
12. Chatr-aryamontri A, Breitkreutz B-J, Oughtred R et al (2015) The BioGRID interaction database: 2015 update. Nucleic Acids Res 43:D470–D478
13. Zuberi K, Franz M, Rodriguez H et al (2013) GeneMANIA prediction server 2013 update. Nucleic Acids Res 41:W115–W122
14. Bovolenta LA, Acencio ML, Lemke N (2012) HTRIdb: an open-access database for experimentally verified human transcriptional regulation interactions. BMC genomics, 13(1):405
15. Keshava Prasad TS, Goel R, Kandasamy K et al (2009) Human protein reference database—2009 update. Nucleic Acids Res 37:D767–D772
16. Orchard S, Ammari M, Aranda B et al (2014) The MIntAct project—IntAct as a common curation platform for 11 molecular interaction databases. Nucleic Acids Res 42:358–363
17. Kotlyar M, Pastrello C, Sheahan N, Jurisica I (2016) Integrated interactions database: tissue-specific view of the human and model organism interactomes. Nucleic Acids Res 44: D536–D541
18. Franceschini A, Szklarczyk D, Frankild S et al (2013) STRING v9.1: protein-protein interaction networks, with increased coverage and integration. Nucleic Acids Res 41: D808–D815. https://doi.org/10.1093/nar/gks1094
19. Dorigo M, Stützle T (2004) Ant colony optimization. MIT Press, ISBN: 9780262042192

Chapter 15

De Novo Pathway-Based Classification of Breast Cancer Subtypes

Markus List, Nicolas Alcaraz, and Richa Batra

Abstract

Breast cancer is a heterogeneous disease for which various clinically relevant subtypes have been reported. These subtypes are characterized by molecular differences which direct treatment selection. The state of the art for breast cancer subtyping utilizes histochemistry or gene expression to measure a few selected markers. However, classification based on molecular pathways (rather than individual markers) is a more robust way to classify breast cancer samples into known subtypes.

Here, we present PathClass, a web application that allows its users to predict breast cancer subtypes using various traditional as well as advanced methods. This includes methods based on classical gene expression panels as well as de novo pathway-based predictors. Users can predict labels for datasets in the Gene Expression Omnibus or upload their own expression profiling data.

Availability: https://pathclass.compbio.sdu.dk/.

Key words De novo pathways, Disease subtyping, Breast cancer, Classification

1 Introduction

1.1 Clinical Relevance of Breast Cancer Subtypes

Prior to the availability of gene expression profiling data, histopathological methods were used to separate clinically relevant subtypes, namely, the luminal, basal, and HER2 subtypes. Subsequently, clustering of breast cancer patients by their gene expression profiles has revealed that two groups of luminal breast cancer patients can be distinguished with distinct clinical outcome, namely, lumA and lumB [1]. The properties of cancer cells in these different subtypes closely resemble different differentiation stages of cells in the breast tissue (Fig. 1). It is noteworthy that subtypes that are more similar to the undifferentiated and more potent stem cells are more aggressive and treatment resistant as postulated in the cancer stem cell hypothesis [2]. The largest fraction of breast cancer cases is of luminal subtypes, which resemble the fully differentiated luminal cells that are most abundant in the

Stefan Canzar and Francisca Rojas Ringeling (eds.), *Protein-Protein Interaction Networks: Methods and Protocols*, Methods in Molecular Biology, vol. 2074, https://doi.org/10.1007/978-1-4939-9873-9_15,

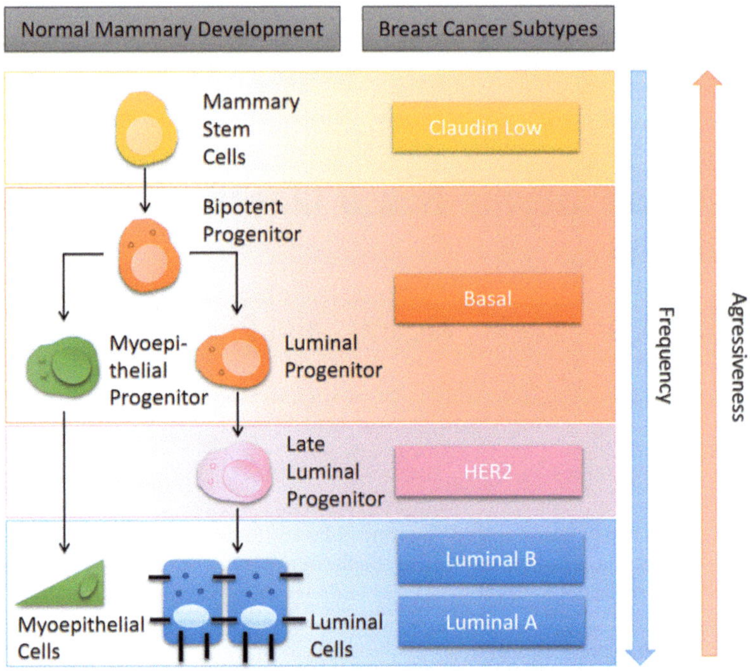

Fig. 1 Overview of breast cancer subtypes which vary in frequency and aggressiveness and closely resemble different stages in mammary tissue development. More frequent cells correspond to more differentiated cells, whereas the molecular profile of more aggressive and rare subtypes corresponds to undifferentiated stem cells or progenitor cells

breast tissue. Due to the differences in the molecular profile and the aggressiveness, the treatment choice critically depends on the correct subtype identification. For instance, administering tamoxifen, which is further metabolized in the liver to target the estrogen receptor, is only sensible for estrogen-receptor positive luminal breast cancer.

1.2 Gene Expression Panels

Histopathological methods for breast cancer subtyping are subjective, which motivated the search for gene expression signatures that could be utilized for a clinical test (known as gene expression panels). The idea is to measure the expression of a few genes to robustly infer the subtype, thus saving costs. The first gene expression signature that was developed for breast cancer is known as PAM50 [3]. While PAM50 is still considered the gold standard for breast cancer subtyping in biomedical research, various clinical gene panels have been developed based on different gene signatures, including MammaPrint [4] and Oncotype DX [5], which have found routine clinical application [6]. In spite of a similar prognostic quality, different gene expression signatures show very little overlap, indicating that various genes are suitable as reporters for specific functional programs in the cell [7]. However, the expected penetration of next-generation sequencing technologies in the clinical sector suggests that it may no longer be necessary to restrict subtype classification to a few genes but to incorporate genome-wide transcriptomic profiles.

1.3 Pathway-Based Classification

The lack of feature stability in and agreement between gene signatures advocates the use of composite features for breast cancer subtyping. While such composite features do not necessarily outperform classifiers based on single genes [8, 9], they can achieve similar performance with the added value that features represent genes associated in functional modules and thus increase interpretability. Composite features are built based on prior knowledge of gene associations that are captured in pathway definitions (such as KEGG [10] or Reactome [11]). In addition, functional modules can be extracted de novo from molecular interaction networks (such as BioGrid [12] or I2D [13]). In [14], we could show that features extracted from known or de novo discovered pathways are indeed significantly more stable than single gene classifiers in breast cancer subtyping and less susceptible to batch effects. To this end, we developed a pathway-based classification pipeline that, given either a large human interaction network, or a set of pathways (composite features) and a set of labelled training samples (genome-wide transcriptome profiles), outputs a classifier after four steps (Fig. 2):

1. Feature extraction: In case a biological network is provided as input, our pipeline first extracts subtype-specific de novo pathways using KeyPathwayMiner [15]. The training datasets are used to extract a list of pathways which will be used in the next steps.
2. Feature scoring: The activity of each pathway in each sample is scored through single sample gene set enrichment analysis [16].
3. Feature reduction: Since pathways may be overlapping, we reduce the number of features by clustering them based on their pairwise Spearman correlation using transitivity clustering [17]. For each cluster, we retain the most representative pathway as feature.
4. Feature selection: We employ random forest [18] to build a classifier based on the most informative features.

2 PathClass Web Application

Following the strategy above, we trained classifiers on breast cancer data from The Cancer Genome Atlas [19] and evaluated the performance in repeated cross-validation as well as on independent test data from the Amsterdam Classification Evaluation Suite [8]. We

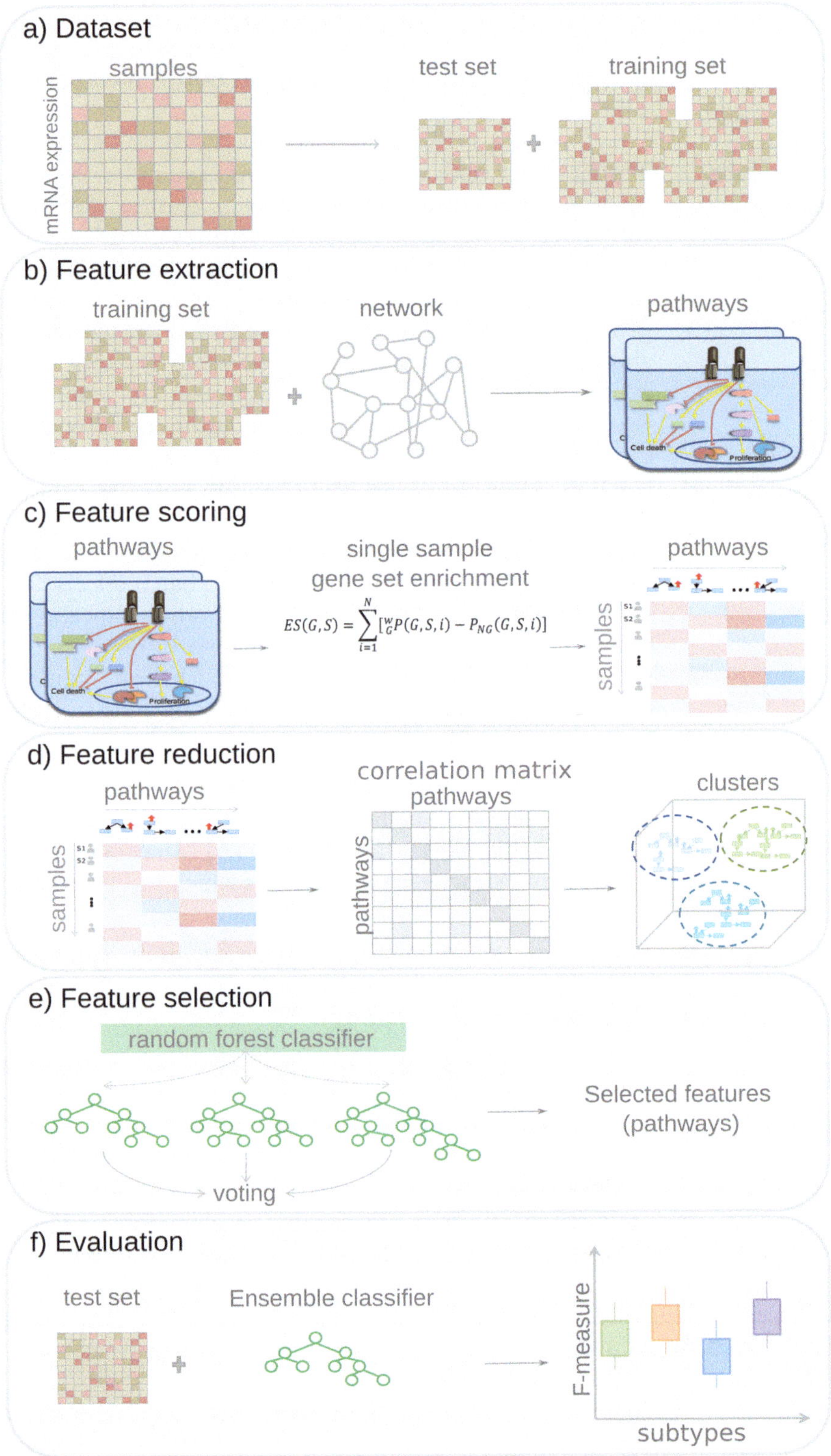

Fig. 2 Workflow for de novo pathway-based classification of breast cancer subtypes. Datasets (**a**): dataset is split into fivefold: four for training and one for testing. Feature extraction (**b**): subtype-specific pathways are

considered various approaches for obtaining the initial set of candidate pathways, giving rise to a number of pathway-based predictors which can be applied to predict the breast cancer subtype of unseen samples. To make this functionality available to a wide range of users, we developed PathClass, a web application that allows users to upload transcriptomic profiling data for one or several samples and to predict the breast cancer subtype using a selection of trained pathway-based and classical gene signature-based classifiers. The web application is available at https://pathclass.compbio.sdu.dk/.

3 Available Pathway Predictors

3.1 De Novo Pathway-Based Predictors

We provide four predictors constructed by training classifiers on de novo pathways as features. These pathways were extracted by applying KeyPathwayMiner [15] to the following human interaction networks:

1. *The Human Protein Reference Database* (HPRD) [20]: a protein–protein interaction network with 39,277 interactions found in the literature, covering 9520 proteins. We used the latest update: version 9 (released April 2010).
2. *The Interologous Interaction Database* (I2D) [21]: a large protein–protein interaction network curated from different sources such as HPRD, BIND [22], and MINT [23], among others. This is the version downloaded in February 2015, containing 209,203 interactions and 15,379 proteins.
3. *The Human Transcriptional Regulation Interaction Database (HTRIdb)* [24]: a gene regulatory network containing validated transcription factor that targets gene interactions in human, with 51,833 interactions and 18,310 genes, downloaded in September 2015.
4. *HumanNet* [25]: a functional network constructed with a Bayesian probabilistic method integrating 21 different types of OMICs data. Each interaction represents the probability of two genes being functionally related. We used version 1.0, containing 13,022 genes and 123,052 edges having a log-likelihood score of at least 1.5.

Fig. 2 (continued) extracted by projecting the training sets on the input network and running KeyPathway-Miner (KPM). Feature scoring (**c**): the extracted pathways are scored using single sample gene set enrichment analysis (ssGSEA) to produce a matrix of samples versus pathways. Feature reduction (**d**): pathways are clustered based on their Spearman's correlation coefficient across all training samples, and the most representative feature for each cluster is selected for the next step. Feature selection (**e**): the best features are selected using random forests (RFs) with recursive feature elimination, and a final RF model is built with the selected features. Evaluation (**f**): the final model is used to predict the subtype labels of the test set. The process is repeated for all fold splits and 10 repeats, recording the performance for each run. (Taken from [14])

3.2 Predefined Gene Set-Based Predictors

1. *The Molecular Signature Database* (*MSigDB*, v5.1) [26]: containing several cancer-related and other biological relevant gene sets constructed from gene expression knockout studies. The complete set contains 4726 gene sets.
2. *The Consensus Pathway Database* (*CPDB*) [27]: a meta-database containing human pathways from 32 different sources such as protein interactions, signaling reactions, and metabolic reactions, among many others. We downloaded release 31, containing 4593 human gene sets.

3.3 Classification Single Gene-Based Predictors

In addition, it is also possible to use a variety of classical gene signature-based predictors that were described in the literature and implemented as part of the genefu R package [28]. In addition to the well-known PAM50 gene signature [3], this includes SSP2003 with 500 genes [29], SSP2006 with 306 genes [30], SCMOD1 with 726 genes [31], SCMOD2 with 663 genes [7], and SCMGENE with only three genes [28]. These predictors are included per default for comparison.

3.4 Selecting Pathway-Based Predictors

1. Click on "Selected Pathway Predictors" in the navigation panel.
2. Check boxes for all predictors you want to use (*see* **Note 1**).

4 Classification of Samples in the Gene Expression Omnibus

The Gene Expression Omnibus (GEO) [32] is the largest database for gene expression data to date. It also hosts a fair amount of studies related to breast cancer patients, including information pertaining to the different subtypes. In this chapter, we enable users to predict breast cancer subtypes for GEO samples.

1. Click on "Gene Expression Omnibus" in the navigation panel.
2. A GEO dataset with several samples is characterized by a GSE identifier. A GSE identifier has to be entered to download the corresponding expression data into PathClass (*see* **Notes 2** and **3**). An example identifier (GSE45827) is already provided for testing.
3. After the data has been successfully downloaded, users may select a reference/gold standard for comparison. The labels of this reference are used as a baseline to assess the accuracy of the individual predictors. The reference can either be a prediction from one of the methods (continue to **step 5**) or a set of labels obtained from the GEO metadata.
4. To use breast cancer subtype labels from GEO, first select a metadata column in the "BRCA class labels" section.

Afterward, you need to map an entry of this column to each of the subtypes LumA, LumB, Her2, and Basal (*see* **Notes 4** and **5**).

5. Once the setup is complete, press the "Start" button in the red panel to trigger the prediction of breast cancer subtype labels using the predictors selected in Subheading 2.
6. Continue with Subheading 6 for interpreting the results.

5 Classification of User-Submitted Samples

Users can also directly upload gene expression data to PathClass in order to determine breast cancer subtypes.

1. Click on "Upload Data" in the navigation panel.
2. Click on "Browse" for selecting a correctly formatted gene expression file on your hard drive (*see* **Notes 6–8**). Subsequently, a progress bar will indicate the progress of the upload.
3. Various types of gene identifiers are commonly used in the scientific community. Since there is no de facto standard we can rely on, PathClass has a built-in support for mapping gene identifiers. To this end, it is essential that you select the correct gene identifier type in the "ID mapping" panel (*see* **Note 9**).
4. Next, users may select a reference/gold standard for comparison. The labels of this reference are used as a baseline to assess the accuracy of the individual predictors. The reference can either be a prediction from one of the methods (continue to **step 6**) or an uploaded list of labels (*see* **Note 10**).
5. Upload a custom class label file. The labels in the file need to be mapped to the breast cancer subtypes known to PathClass (*see* **Note 11**).
6. Once the setup is complete, press the "Start" button in the red panel to trigger the prediction of breast cancer subtype labels using the predictors selected in Subheading 2.
7. Continue with Subheading 6 for interpreting the results.

6 Interpretation of Results

Once you have followed the steps mentioned in Subheading 4 or 5 to classify your samples into breast cancer subtypes, you will be automatically taken to the result screen of the application once the processing has finished. Various tabs will appear presenting the classification results from various perspectives as described below.

Data Table: Here, we display the user-uploaded dataset. Each row corresponds to a gene, and each column corresponds to a

Table 1
Example of the prediction table output with predictors in columns and samples in rows

Sample	Predictors			
	MSIG	KPM-HPRD	KPM-I2D	PAM50
GSM177885	Her2	Basal	Basal	Basal
GSM177886	LumB	LumB	LumA	LumB
GSM177887	LumA	LumA	LumB	LumB
...	...	...	...	...
...	...	...	...	...

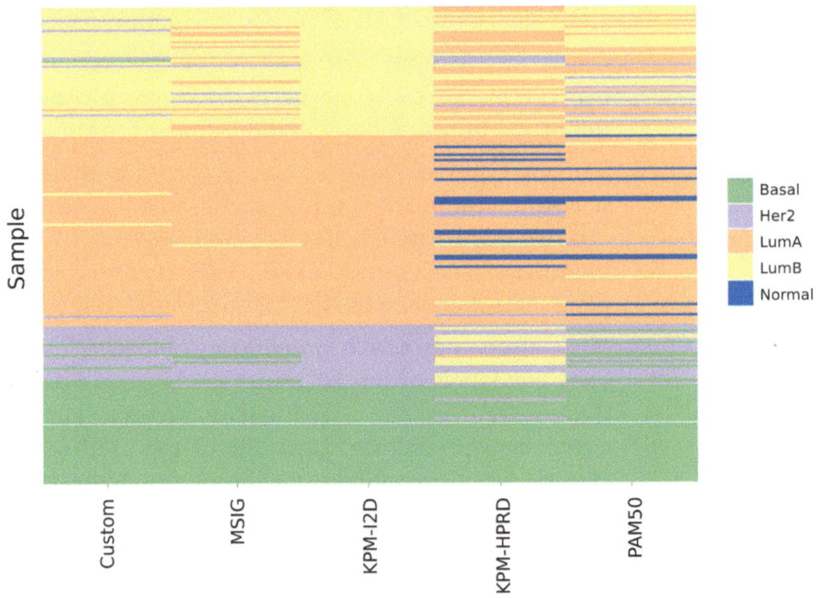

Fig. 3 Consensus of classifiers illustrated as heatmap. Different colors correspond to different breast cancer subtypes. For each sample and predictor, the heatmap shows the predicted subtype

sample. Gene ids and sample ids are visible allowing users to manually inspect a gene/sample of interest.

Gene ID Mapping: If the user chose to perform gene identifier conversion, this table displays the results of mapping the provided identifiers to Entrez IDs (*see* **Note 9**).

Predictions Table: Classification results are displayed here for each sample and predictor (*see* Table 1 for an example). Results can be sorted by sample identifier or by a specific predictor. Results can also be searched and filtered (full text search or search by column). Users can download the results in a tab-separated format and perform post hoc analysis.

Predictions Plot: The heatmap in this tab is a visual representation of the predictions table shown in the previous tab (Fig. 3).

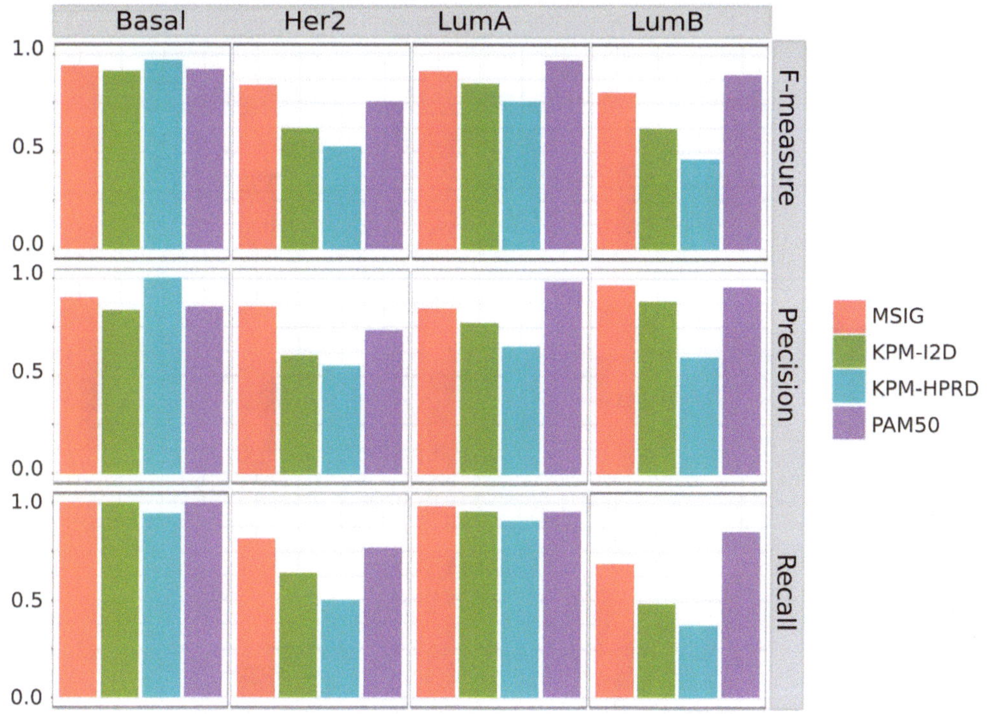

Fig. 4 Visualization of prediction performance with a focus on subtypes. For each subtype, three bar plots show the achieved performance in *F*-measure, precision, and recall, respectively, for the different predictors

Each row corresponds to a sample, and each column corresponds to a predictor. One of the predictors can be used as reference (defaults to PAM50). This plot helps to visualize the consensus among the predictors and to identify the classes with frequent disagreements among the predictors.

Performance (*Subtypes*): Here, users can compare the performance of each predictor across breast cancer subtypes (Fig. 4). We provide three performance measures, namely, precision, recall, and the *F*-measure (harmonic mean of precision and recall). The *F*-measure provides one reference value for the performance of the classifier, i.e., a compromise between positive predictive value and true positive rate. Similar to *P*-values threshold, the acceptable value of *F*-measure is case-dependent.

Performance Classifiers: This tab is similar to the previous tab but focuses on the comparison of different predictors (Fig. 5).

Confusion Matrices: The last tab provides the confusion matrices for each classifier (Table 2). Here, rows correspond to true labels and columns correspond to predicted labels. Thus, such matrices help to understand which subtypes are confused with each other. For instance, in our breast cancer cohort, predictors often confuse samples from class LumA and LumB with each other.

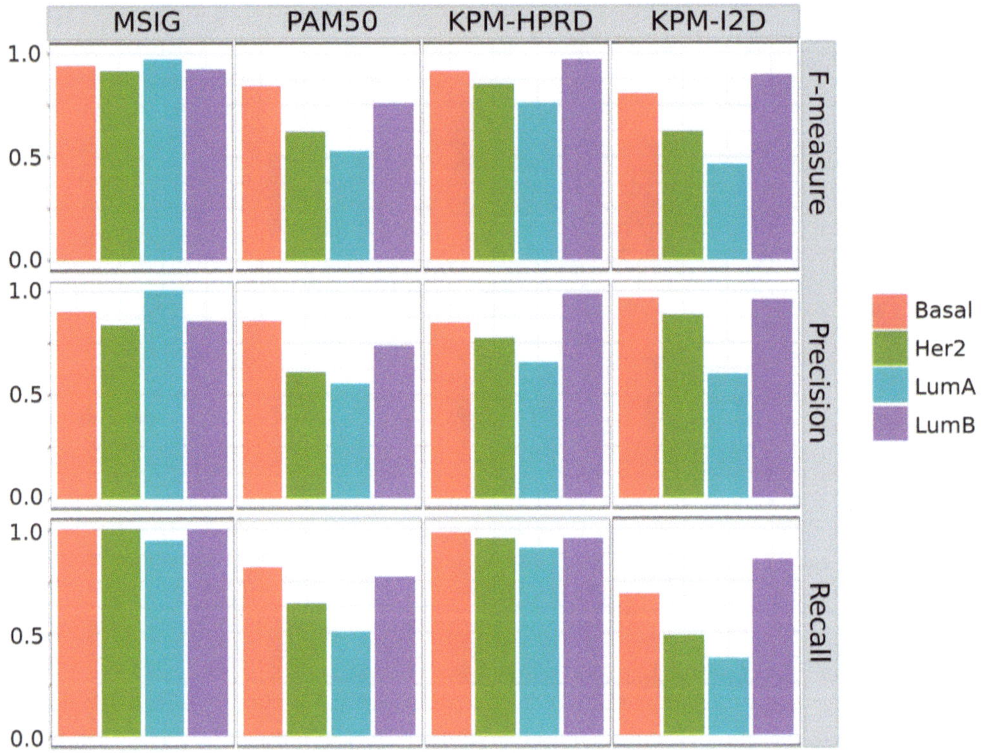

Fig. 5 Visualization of prediction performance with a focus on predictors. For each predictor, three bar plots show the achieved performance in *F*-measure, precision, and recall, respectively, for the different subtypes

Table 2
Example for a confusion matrix that compares for a given predictor the true class (based on the user-selected reference) against the predicted class

		True class			
Predicted class		LumA	LumB	Basal	Her2
	LumA	69	1	0	0
	LumB	12	33	0	3
	Basal	0	0	36	0
	Her2	0	0	4	18

7 Notes

1. For every predictor selected, the runtime of the method will increase considerably. Consider limiting the selection to a few predictors of interest to save some time.
2. There are different types of gene expression omnibus identifiers. PathClass expects dataset identifiers starting with GSE,

Table 3
Example for the input file with samples in columns and genes in rows

	GSM177885	GSM177886	GSM177887	...
780	-0.445880156708764	0.0835223251198922	-0.261878535304206	...
5982	0.0483928122144004	0.601880034706467	0.0681142566168065	...
7318	-0.633116560416459	-0.0789442588323661	-0.428397244018818	...
2049	-0.320556324126809	-0.859121339455566	0.961707402983496	...
4323	0.118571981099018	-0.33219320782046	0.061128902229127	...
...	...	...	...	...
...	...	...	...	...

e.g., GSE45827. Identifiers starting with GSM characterize individual samples and are part of a dataset.

3. Downloading data from the gene expression omnibus takes quite a while. Please be patient and do not reload the page.
4. The gene expression omnibus datasets typically come with a rich sample annotation. However, since this is not standardized, users need to go through the individual columns to see if the breast cancer subtype is annotated. Typical columns that hold this information start with characteristic, e.g., characteristics_ch1.1. However, for some datasets, this information is not present. PathClass can then still be used to predict breast cancer subtypes, but users need to select a different gold standard/reference.
5. PathClass has no way to check if the provided GSE identifier is even related to breast cancer. It is the responsibility of the user to check if the selected dataset is meaningful.
6. The gene expression file is expected to have samples in columns and genes in rows (*see* Table 3 for an example). Columns can be separated by tab, comma, or semicolon. Make sure to select the correct column separator *before* uploading the file. Note that Excel files (such as .xlsx or .xls) are *not supported*. Excel files need to be converted to a suitable text file before upload. This can be achieved in Excel by clicking on File -> save as where a text-based format such as (text (tab-delimited)) needs to be selected as file type.
7. In the above example, you may note that the first column has no name, i.e., the first entry in the header is empty. Should your input file have an entry here (for example "gene"), you have to uncheck the box "Gene identifiers are row names (alternatively it is the first column)."

8. In the panel "Demo Data," you can download example input files that may be used to get to know the file upload feature of PathClass and to see examples of accepted input formats.
9. After selecting the "Gene ID mapping" tab in the "Upload Data" section of PathClass, you can investigate if the ID mapping was successful. If the ID mapping was not successful, double check if you selected the correct gene identifier type. Alternatively, you can use an external service for converting the gene identifiers in your dataset to the Entrez identifiers (sometimes called GeneID) used internally in PathClass. An example for such an external web service is http://biodb.jp/idc.cgi. Note that in case several Entrez IDs map to a single identifier in the uploaded dataset, all possible Entrez identifiers will be considered in the subsequent analysis.
10. The "Upload custom class labels" panel allows users to upload their own breast cancer subtype labels for use as a reference. In this file, each line is expected to contain a single label, where the order of the labels and the number of rows of the file correspond to the number of samples in the previously uploaded expression data. An example for such a file can be retrieved from the "Demo Data" panel.
11. If, for example, your custom class labels refer to the luminal A subtype as "Luminal A," or "LumA," you simply select this label in the "LumA" checkbox. Similarly, you need to assign labels for LumB, Her2, and Basal. Other labels than the ones mapped here are ignored.

Acknowledgments

R.B. would like to thank 676858-IMCIS for funding.

References

1. Perou CM, Sørlie T, Eisen MB et al (2000) Molecular portraits of human breast tumours. Nature 406:747–752
2. Donnenberg VS, Donnenberg AD (2005) Multiple drug resistance in cancer revisited: the cancer stem cell hypothesis. J Clin Pharmacol 45:872–877
3. Parker JS, Mullins M, Cheang MCU et al (2009) Supervised risk predictor of breast cancer based on intrinsic subtypes. J Clin Oncol 27:1160–1167
4. Slodkowska EA, Ross JS (2009) MammaPrint™ 70-gene signature: another milestone in personalized medical care for breast cancer patients. Expert Rev Mol Diagn 9:417–422
5. Cronin M, Sangli C, Liu M-L et al (2007) Analytical validation of the Oncotype DX genomic diagnostic test for recurrence prognosis and therapeutic response prediction in node-negative, estrogen receptor-positive breast cancer. Clin Chem 53:1084–1091
6. Thakur S, Das AM, Das BC (2016) Utility of gene expression signature in treatment decision of breast cancer. Transl Cancer Res 5: S1469–S1472
7. Wirapati P, Sotiriou C, Kunkel S et al (2008) Meta-analysis of gene expression profiles in breast cancer: toward a unified understanding of breast cancer subtyping and prognosis signatures. Breast Cancer Res 10:R65

8. Staiger C, Cadot S, Kooter R et al (2012) A critical evaluation of network and pathway-based classifiers for outcome prediction in breast cancer. PLoS One 7:e34796
9. Allahyar A, de Ridder J (2015) FERAL: network-based classifier with application to breast cancer outcome prediction. Bioinformatics 31:i311–i319
10. Kanehisa M, Goto S (2000) KEGG: Kyoto encyclopedia of genes and genomes. Nucleic Acids Res 28:27–30
11. Joshi-Tope G, Gillespie M, Vastrik I et al (2005) Reactome: a knowledgebase of biological pathways. Nucleic Acids Res 33: D428–D432
12. Chatr-aryamontri A, Breitkreutz B-J, Oughtred R et al (2015) The BioGRID interaction database: 2015 update. Nucleic Acids Res 43:D470–D478
13. Kotlyar M, Pastrello C, Sheahan N, Jurisica I (2016) Integrated interactions database: tissue-specific view of the human and model organism interactomes. Nucleic Acids Res 44: D536–D541
14. Alcaraz N, List M, Batra R et al (2017) De novo pathway-based biomarker identification. Nucleic Acids Res 45:e151–e151
15. Alcaraz N, List M, Dissing-Hansen M et al (2016) Robust de novo pathway enrichment with KeyPathwayMiner 5. F1000Res 5:1531
16. Barbie DA, Tamayo P, Boehm JS et al (2009) Systematic RNA interference reveals that oncogenic KRAS-driven cancers require TBK1. Nature 462:108–112
17. Wittkop T, Emig D, Lange S et al (2010) Partitioning biological data with transitivity clustering. Nat Methods 7:419–420
18. Diaz-Uriarte R (2009) varSelRF: Variable selection using random forests. R package version 0 7-1. http://ligarto.org/rdiaz/Software/Software.html
19. Atlas TCG (2012) Comprehensive molecular portraits of human breast tumours. Nature 490:61–70
20. Keshava Prasad TS, Goel R, Kandasamy K et al (2009) Human protein reference database—2009 update. Nucleic Acids Res 37: D767–D772
21. Brown KR, Jurisica I (2007) Unequal evolutionary conservation of human protein interactions in interologous networks. Genome Biol 8:R95
22. Isserlin R, El-Badrawi RA, Bader GD (2011) The biomolecular interaction network database in PSI-MI 2.5. Database 2011:baq037
23. Licata L, Briganti L, Peluso D et al (2012) MINT, the molecular interaction database: 2012 update. Nucleic Acids Res 40:D857–D861
24. Bovolenta LA, Acencio ML, Lemke N (2012) HTRIdb: an open-access database for experimentally verified human transcriptional regulation interactions. BMC Genomics 13:405
25. Lee I, Blom UM, Wang PI et al (2011) Prioritizing candidate disease genes by network-based boosting of genome-wide association data. Genome Res 21:1109–1121
26. Subramanian A, Tamayo P, Mootha VK et al (2005) Gene set enrichment analysis: a knowledge-based approach for interpreting genome-wide expression profiles. Proc Natl Acad Sci U S A 102:15545–15550
27. Kamburov A, Stelzl U, Lehrach H, Herwig R (2013) The ConsensusPathDB interaction database: 2013 update. Nucleic Acids Res 41: D793–D800
28. Haibe-Kains B, Desmedt C, Loi S et al (2012) A three-gene model to robustly identify breast cancer molecular subtypes. J Natl Cancer Inst 104:311–325
29. Sørlie T, Tibshirani R, Parker J et al (2003) Repeated observation of breast tumor subtypes in independent gene expression datasets. Proc Natl Acad Sci U S A 100:8418–8423
30. Hu Z, Fan C, Oh DS et al (2006) The molecular portraits of breast tumors are conserved across microarray platforms. BMC Genomics 7:96
31. Desmedt C, Haibe-Kains B, Wirapati P et al (2008) Biological processes associated with breast cancer clinical outcome depend on the molecular subtypes. Clin Cancer Res 14:5158–5165
32. Edgar R, Domrachev M, Lash AE (2002) Gene Expression Omnibus: NCBI gene expression and hybridization array data repository. Nucleic Acids Res 30:207–210

Chapter 16

Vienna Graph Clustering

Sonja Biedermann, Monika Henzinger, Christian Schulz, and Bernhard Schuster

Abstract

This paper serves as a user guide to the Vienna graph clustering framework. We review our general memetic algorithm, VieClus, to tackle the graph clustering problem. A key component of our contribution are natural recombine operators that employ ensemble clusterings as well as multi-level techniques. Lastly, we combine these techniques with a scalable communication protocol, producing a system that is able to compute high-quality solutions in a short amount of time. After giving a description of the algorithms employed, we establish the connection of the graph clustering problem to protein–protein interaction networks and moreover give a description on how the software can be used, what file formats are expected, and how this can be used to find functional groups in protein–protein interaction networks.

Key words Graph clustering, Evolutionary algorithms, Protein–Protein interaction

1 Introduction

Graph clustering is the problem of detecting tightly connected regions of a graph. Depending on the task, knowledge about the structure of the graph can reveal information such as voter behavior, the formation of new trends, existing terrorist groups and recruitment [36], or a natural partitioning of data records onto pages [14]. Further application areas include the study of protein interaction [30], gene expression networks [42], fraud detection [1], program optimization [12, 24], and the spread of epidemics [27]—possible applications are plentiful, as almost all systems containing interacting or coexisting entities can be modeled as a graph.

It is common knowledge that there is no single best strategy for graph clustering, which justifies a plethora of existing approaches. Moreover, most quality indices for graph clusterings have turned out to be NP-hard to optimize and are rather resilient to effective approximations, see, e.g., [3, 11, 40], allowing only heuristic approaches for optimization. The majority of algorithms for

Stefan Canzar and Francisca Rojas Ringeling (eds.), *Protein-Protein Interaction Networks: Methods and Protocols*, Methods in Molecular Biology, vol. 2074, https://doi.org/10.1007/978-1-4939-9873-9_16,

graph clustering are based on the paradigm of intra-cluster density versus inter-cluster sparsity. One successful heuristic to cluster large graphs is the *multi-level* approach [9], e.g., the Louvain method for the optimization of modularity [8]. Here, the graph is recursively *contracted* to obtain smaller graphs which should reflect the same general structure as the input. After applying an *initial clustering* algorithm to the smallest graph, the contraction steps are undone and, at each level, a *local search* method is used to improve the clustering induced by the coarser level w.r.t some objective function measuring the quality of the clustering. The intuition behind this approach is that a good clustering at one level of the hierarchy will also be a good clustering on the next finer level. Hence, depending on the definition of the neighborhood, local search algorithms are able to explore local solution spaces very effectively in this setting. However, these methods are also prone to get trapped in local optima. The multi-level scheme can help to some extent since local search has a more global view on the problem on the coarse levels and a very fine-grained view on the fine levels of the multi-level hierarchy. In addition, several repeated runs can be made in order to improve the final result at the expense of running time.

Still, even a large number of repeated executions can only scratch the surface of the huge search space of possible clusterings. In order to explore the global solution space extensively, we need more sophisticated meta-heuristics. This is where memetic algorithms (MAs), i.e., genetic algorithms combined with local search [22], come into play. Memetic algorithms allow for effective exploration (global search) and exploitation (local search) of the solution space. The general idea behind genetic algorithms is to use mechanisms inspired by biological evolution such as selection, mutation, recombination, and survival of the fittest. A genetic algorithm (GA) starts with a population of individuals (in our case clusterings of the graph) and evolves the population over several generational cycles (rounds). In each round, the GA uses a selection rule based on the fitness of the individuals of the population to select good individuals and combines them to obtain improved offspring [17]. When an offspring is generated, an eviction rule is used to select a member of the population to be replaced by the new offspring. For an evolutionary algorithm, it is of major importance to preserve diversity in the population [4], i.e., the individuals should not become too similar in order to avoid a premature convergence of the algorithm. This is usually achieved using mutation operations and using eviction rules that take similarity of individuals into account.

We present a memetic algorithm, VieClus (Vienna Graph Clustering), for the graph clustering problem. A key component of our contribution are natural recombine operators that employ ensemble clusterings as well as multi-level techniques. In machine learning, ensemble methods combine multiple weak classification (or clustering) algorithms to obtain a strong algorithm for classification (or clustering). More precisely, given a number of clusterings, the *overlay/ensemble clustering* is a clustering in which two vertices belong to the same cluster if and only if they belong to the same cluster in each of the input clusterings. Our recombination operators use the overlay of two clusterings from the population to decide whether pairs of vertices should belong to the same cluster [29, 37]. This is combined with a local search algorithm to find further improvements and also embedded into a multi-level algorithm to find even better clusterings. Our general principle is to randomize tie-breaking whenever possible. This diversifies the search and also improves solutions. Lastly, we combine these techniques with a scalable communication protocol, producing a system that is able to compute high-quality solutions in a short amount of time. The algorithm is able to compute a result that is better than the currently reported modularity value in literature on *every* instance under consideration. More precisely, in 98 out of 115 runs of VieClus, the previous benchmark result of the 10th DIMACS implementation challenge is outperformed. In further 15 out of 115 runs, our algorithm reproduces the results. Moreover, all recently published solvers have been outperformed by the VieClus algorithm. From another point of view, while previous results have been computed by a variety of solvers, our algorithm can now be used as a single tool to compute the result. We refer the reader to [7] for all experimental details. After giving a description of the algorithms employed, we establish the connection of the graph clustering problem to protein–protein interaction networks and moreover give a description on how the software can be used, what file formats are expected, and how this can be used to cluster protein–protein interaction networks.

2 Preliminaries

2.1 Basic Concepts

Let $G = (V = \{0, \ldots, n-1\}, E)$ be an undirected graph and $N(v) := \{u : \{v, u\} \in E\}$ denote the neighbors of v. The degree of a vertex v is $d(v) := |N(v)|$. The problem that we tackle in this paper is the *graph clustering* problem. A clustering $\mathcal{C}$ is a partition of the set of vertices, i.e., a set of disjoint *blocks/clusters* of vertices $V_1, \ldots, V_k$ such that $V_1 \cup \cdots \cup V_k = V$. However, k is usually not given in advance. A size-constraint clustering constrains the size of the blocks of a clustering by a given upper bound U. A clustering is *trivial* if there is only one block, or all clusters/blocks contain only one element, i.e., are singletons. We identify a cluster V_i with its node-induced subgraph of G. The set $E(\mathcal{C}) := E \cap (\cup_i V_i \times V_i)$ is the set of *intra-cluster edges*, and $E \backslash E(\mathcal{C})$ is the set of *inter-cluster edges*. We set $|E(\mathcal{C})| =: m(\mathcal{C})$ and $|E \backslash E(\mathcal{C})| =: \overline{m}(\mathcal{C})$. An edge running between two blocks is called *cut edge*. There are different objective functions that are optimized in the literature. We review some of them in Subheading 2.3. Our main focus in this work is on *modularity*. However, our algorithm can be generalized to optimize other objective functions. The *graph partitioning problem* is also looking for a partition of the vertices. Here, a *balancing constraint* demands that all blocks have weight $|V_i| \leq (1+\varepsilon)\lceil \frac{|V|}{k} \rceil =: L_{\max}$ for some imbalance parameter ε. A vertex is a *boundary vertex* if it is incident to a vertex in a different block. The objective is to minimize the total *cut* $\omega(E \cap \cup_{i<j} V_i \times V_j)$.

In general, our input graphs G have unit edge weights and vertex weights. However, even those will be translated into weighted problems in the course of a multi-level algorithm. In order to avoid tedious notation, G will denote the current state of the graph before and after a (un)contraction in the multi-level scheme throughout this paper.

2.2 Ensemble/Overlay Clusterings

In machine learning, ensemble methods combine multiple weak classification algorithms to obtain a strong classifier. These base clusterings are used to decide whether pairs of vertices should belong to the same cluster [29, 38]. Given two clusterings, the *overlay clustering* is a clustering in which two vertices belong to the same cluster if and only if they belong to the same cluster in each of the input clusterings. More precisely, given two clusterings C_1 and C_2 the *overlay clustering* is the clustering where each block corresponds to a connected component of the graph $G_{\mathcal{E}} = (V, E \backslash \mathcal{E})$ where $\mathcal{E}$ is the union of the cut edges of C_1 and C_2, i.e., all edges that run between blocks in either C_1 or C_2. Intuitively, if the input clusters agree that two vertices belong to the same cluster, the two vertices belong together with high confidence. It is easy to see that the number of clusters in the overlay clustering cannot be smaller than the number of clusters in each of the input clusterings.

An overlay clustering can be computed in expected linear-time. More precisely, given two clusterings $\{C_1, C_2\}$, we use the following approach to compute the overlay clustering. Initially, the overlay clustering O is set to the clustering C_1. We then iterate through the remaining clusterings and incrementally update the current solution O. To this end, we use pairs of cluster IDs (i, j) as a key in a hash map $\mathcal{H}$, where i is a cluster ID of O and j is a cluster ID of the current clustering C. We initialize $\mathcal{H}$ to be the empty hash map and set a counter c to zero. Then we iterate through the nodes. Let v be the current node. If the pair $(O[v], C[v])$ is not contained in $\mathcal{H}$, we set $\mathcal{H}(O[v], C[v])$ to c and increment c by one. Afterwards, we update the cluster ID of v in O to be $\mathcal{H}(O[v], C[v])$. At the end of the algorithm, c is equal to the number of clusters contained in the overlay clustering and each vertex is labeled with its cluster ID in O.

2.3 Objective Functions

There are a variety of measures used to assess the quality of a clustering, such as *coverage* [9], *performance* [39], *inter-cluster conductance* [20], *surprise* [2], *map equation* [33], and *modularity* [28]. The most simple index realizing a traditional measure of clustering quality is coverage. The coverage of a graph clustering C is defined as the fraction of intra-cluster edges within the complete set of edges $\text{cov}(C) := \frac{m(C)}{m}$. Intuitively, large values of coverage correspond to a good quality of a clustering. However, one principal drawback of coverage is that the converse is not necessarily true: coverage takes its largest value of 1 in the trivial case, where there is only one cluster. *Modularity* fixes this issue by comparing the coverage of a clustering to the same value after rearranging edges randomly. *Performance* is the fraction of correctly classified vertex pairs, w.r.t the set of edges. *Inter-cluster conductance* returns the worst (i.e., the thickest) bottleneck created by separating a cluster from the rest of the graph. *Surprise* measures the probability that a random graph $\mathscr{R}$ has more intra-cluster edges. *Map equation* is a flow-based and information-theoretic method to assess clustering quality. Our focus in this work is on modularity as it is a widely accepted quality function and has been the main objective function in the 10th DIMACS implementation challenge [5]. For further discussions of these indices, we refer the reader to the given references and simply state the formal definition of modularity here:

$$\mathcal{Q}(C) := \text{cov}(C) - \mathbb{E}[\text{cov}(C)] = \frac{m(C)}{m} - \frac{1}{4m^2} \sum_{V_i \in C} \left(\sum_{v \in V_i} d(v) \right)^2$$

2.4 Related Work

This paper is a summary and extension of the bachelor's thesis by Sonja Biedermann [6] that has later been published [7]. There has been a *significant* amount of research on graph clustering. We refer the reader to [16, 18] for thorough reviews of the results in this area. Here, we focus on results closely related to our main contributions. It is common knowledge that there is no single best strategy for graph clustering. Moreover, most quality indices for graph clusterings have turned out to be NP-hard to optimize and rather resilient to effective approximations, see, e.g., [3, 11, 40], allowing only heuristic approaches for optimization. Other approaches often rely on specific strategies with high running times, e.g., the iterative removal of central edges [28], or the direct identification of dense subgraphs [13]. Provably, good methods with a decent running time include algorithms that have a spectral embedding of the vertices as the basis for a geometric clustering [10], min-cut tree clustering [15], a technique which guarantees certain bounds on bottlenecks, and an approach which relies on random walks in graphs staying inside dense regions with high probabilities [39]. The Louvain method is a multi-level clustering algorithm introduced by Blondel et al. [8] that optimizes modularity as an objective function. As we use this method in our algorithm to create the initial population as well as to improve individuals after recombination, we go into more detail for that method in Subheading 2.6. For more detailed related work on evolutionary and high-quality graph clustering, we refer the reader to [7].

2.5 Karlsruhe High-Quality Partitioning

Within this work, we use the open source multi-level graph partitioning framework KaHIP [35] (Karlsruhe High-Quality Partitioning). More precisely, we employ partitioning tools contained therein to create high-quality partitions of the graphs. We shortly outline its main components. KaHIP implements many different algorithms, for example, flow-based methods and more-localized local searches within a multi-level framework, as well as several coarse-grained parallel and sequential meta-heuristics. Recently, specialized methods to partition social networks and web graphs have been included in the framework [25].

2.6 Multi-Level Louvain Method

The Louvain method is a multi-level clustering algorithm introduced by Blondel et al. [8]. It is an approach to graph clustering that optimizes modularity as an objective function. Since we instantiate our memetic algorithm to optimize for modularity, we give more detail in order to be self-contained. The algorithm is organized in two phases: a local movement phase and contraction/uncontraction phase. The first phase, *local movement*, works in rounds and is done as follows: In the beginning, each vertex is a singleton cluster. Vertices are then traversed in random order and always moved to the neighboring cluster yielding the highest modularity increase. Computing the best move can be done in time proportional to the degree of a vertex by storing cumulative vertex degrees of clusters. More precisely, the gain in modularity by removing a vertex from a cluster can be computed by

$$\Delta Q(u) = \frac{s}{m} + \frac{d(u)}{4m^2} + \frac{(\sum_{v \in C} d(v))^2}{4m^2} - \frac{(d(u) + \sum_{v \in C} d(v))^2}{4m^2}.$$

A similar formula holds when adding a singleton vertex to a cluster. Once a vertex is moved, the cumulated degrees of the affected clusters are updated. Hence, the best move can be found in time proportional to the degree of a vertex. The local movement algorithm stops when a local maximum of the modularity is attained, that is, when no vertex move yields a modularity gain. The *second phase* of the algorithm consists in contracting the clustering. Contracting the clustering works as follows: each block of the clustering is contracted into a single node. There is an edge between two vertices u and v in the contracted graph if the two corresponding blocks in the clustering are adjacent to each other in G, i.e., block u and block v are connected by at least one edge. The weight of an edge (A, B) is set to the sum of the weight of the edges that run between block A and block B of the clustering. Moreover, a self-loop is inserted for each vertex in the contracted graph. The weight of this edge is set to be the cumulative weight of the edges in the respective cluster. Note that due to the way the contraction is defined, a clustering of the coarse graph corresponds to a clustering of the finer graph with the same objective. The algorithm then continues with the local movement phase on the contracted graph. In the end, the clustering contractions are undone and at each level local movement improves the current clustering w.r.t. modularity.

3 Memetic Graph Clustering

We now explain the components of our memetic graph clustering algorithm. Our algorithm starts with a population of individuals (in our case one individual is a clustering of the graph) and evolves the population into different populations over several rounds. In each round, the EA uses a selection rule based on the fitness of the individuals (in our case the objective function of the clustering problem under consideration) of the population to select good individuals and recombine them to obtain improved offspring. Our selection process is based on the tournament selection rule [26], i.e., C is the fittest out of two random individuals R_1, R_2 from the population. When an offspring is generated, an elimination rule is used to select a member of the population and replace it with the new offspring. In general, one has to take both into consideration, the fitness of an individual and the distance between individuals in the population [31] in order to avoid premature convergence of the algorithm. We evict the solution that is *most similar* with respect to the edges that run in between clusters with the offspring among those individuals in the population that have a worse or equal objective than the offspring itself. If there is no such individual, then the offspring is rejected and not inserted into the population. The difference between two individuals is defined as the size of the symmetric difference between their sets of cut edges. Our algorithm generates only one offspring per generation.

The core of our algorithm are our novel recombination and mutation operations. We provide two different kinds of operations, flat- and multi-level recombination operations. We also define a mutation operator that splits clusters by employing graph partitioning techniques. In any case, the offspring is typically improved using a local search algorithm that optimizes for the objective function of the graph clustering problem. We continue this section by explaining the details of our algorithm with a focus on modularity—the results are transferable to other clustering problems using local search algorithms that optimize the objective function of the problem under consideration.

3.1 Initialization/Creating Individuals

We initialize our population using the following modified Louvain algorithm. We also use the following algorithm to create an individual in recombine and mutation operations. Depending on the objective of the evolutionary algorithm, the choice of algorithms to initialize the population may be different. We describe our scheme for modularity.

We mainly use the Louvain method to create clusterings. Due to the non-deterministic nature of this algorithm—the order of nodes visited is randomized—we obtain different initial graph clusterings and later on individuals. In order to introduce more diversification, we modify the approach using a different coarsening strategy based on the size-constrained label propagation algorithm [25]. Label propagation works similar to the local movement phase of the Louvain method. However, the objective of the algorithm is different. Without size-constraints, it was proposed by Raghavan et al. [32]. Initially, each vertex is in its own cluster/block, i.e., the initial block ID of a vertex is set to its vertex ID. The algorithm then works in rounds. In each round, the vertices of the graph are traversed in a random order. When a vertex v is visited, it is *moved* to the block that has the strongest connection to v, i.e., it is moved to the cluster V_i that maximizes $\omega(\{(v, u)|u \in N(v) \cap V_i\})$ while not overloading the target cluster w.r.t to the size-constrained bound U. Ties are broken randomly. The process is repeated until the process has converged. Here, we perform at most ℓ rounds of the algorithm, where ℓ is a tuning parameter, and stop the algorithm if less than five percent of the vertices changed its cluster during one round. One LPA round can be implemented to run in $O(n + m)$ time.

We modify the coarsening stage of the Louvain method using size-constrained label propagation as follows: For the first $\lambda \in_{\text{rnd}} [0, 4]$ levels of the multi-level hierarchy, we use size-constrained label propagation to compute the clustering to be contracted instead of the local movement phase to compute a clustering. We choose $U \in_{\text{rnd}} [n/10, n]$ in the beginning of the multi-level algorithm. Afterwards, we switch to the local movement phase of the Louvain method to compute a clustering to be contracted. Note that for $\lambda = 0$ the method is the Louvain method. In any case, in the uncoarsening phase local movement improves the current clustering by optimizing modularity.

3.2 Recombination

We define flat- and multi-level recombination operations which we are going to explain now. *Both* operations use the notion of overlay clustering to pass on good parts of the solutions to the offspring. Some of our recombination operators ensure that the offspring has non-decreasing fitness. Moreover, our recombine operations can recombine a solution from the population with an arbitrary clustering of the graph. Due to the fact that our local search and multi-level algorithms are randomized, a recombine operation performed twice using the same parents can yield different offsprings.

3.2.1 Flat Recombination

The *basic flat recombination operation* starts by taking two clusterings C_1, C_2 from the population as input and computes its overlay. The overlay is then contracted such that vertices represent blocks in the overlay clustering and edges represent edges between vertices of two blocks. Edge weights are equal to the summed weight of the edges that run between the respective blocks, and self-loops are inserted with the edge weight being equal to the summed internal edge weight of that corresponding block. Note that due to the way contraction is defined, a clustering of the contracted graph can be transformed into a clustering of the input network having the same modularity score. Hence, we use the Louvain method to cluster the contracted graph, which then constitutes the offspring. Also, note that the contraction of the overlay ensures that vertices that are clustered together in both inputs will belong to the same cluster in the offspring. Only vertices which are split by one of the input clusterings will be affected by the Louvain method. What follows are multiple variations of the basic flat recombination method that ensure non-decreased fitness of the input individual, i.e., the offspring has fitness at least as good as the better input individual.

Apply Input Clustering: This operator uses the better of the two parent clusterings as a starting point for the Louvain method on the contracted graph. Due to the way contraction is defined, the objective function of the starting point on the contracted graph is the same as of the corresponding clustering on the input network. The resulting offspring can be expected to be less diverse than when using the plain recombination operator, but may significantly more improve upon the parent.

Cluster-/Partition Recombination: This operator differs from the previous two operators insofar that only one of the parents is selected from the population. The other parent is manufactured on the spot by running either the size-constrained label propagation (SCLP) or a graph partitioning algorithm on the input graph. The resulting clustering can be very different from the other parent. This introduces more diversification which helps local search to explore a larger search space. When using SCLP clustering for the recombination, we use the parameters as described in Subheading 3.1. When using a partition, we use the KaHIP graph partitioning framework with the fast social preconfiguration using $k \in_{\text{rnd}} [2, 64]$ and $\varepsilon \in_{\text{rnd}} [0.03, 0.5]$.

3.2.2 Multi-Level Recombination

We now explain our multi-level recombine operator. This recombine operator also ensures that the solution quality of the offspring is *at least as good as the best of both parents.* For our recombine operator, let C_1 and C_2 again be two individuals from the population. Both individuals are used as input for the multi-level Louvain method in the following sense. Let $\mathcal{E}$ be the set of edges that are cut edges, i.e., edges that run between two blocks, in either C_1 *or* C_2. All edges in $\mathcal{E}$ are blocked during the coarsening phase, i.e., they are *not* contracted during the coarsening phase. In other words, these edges cannot be contracted during the multi-level scheme. We ensure this by modifying the local movement phase of the Louvain method, i.e., clusters can only grow inside connected components of the overlay $(V, E\backslash\mathcal{E})$. We stop contracting clusterings when no contractable edge is left.

When coarsening is stopped, we then apply the better out of both input individuals w.r.t. the objective to the coarsest graph and use this as initial clustering instead of running the Louvain method on the coarsest graph. However, during uncoarsening local search still optimizes modularity on each level of the hierarchy. Note that this is possible since we did not contract any cut edge of both inputs. Also, note that this way we obtain a clustering of the coarse graph having a modularity score being equal to the better of both input individuals. Local movement guarantees no worsening of the clustering. Hence, the offspring is at least as good as the best input individual. Note that the coarsest graph is the same as the graph obtained in a flat recombination. However, the operation now is able to pass on good parts of the solution on multi-levels of the hierarchy during uncoarsening and hence has a more fine-grained view on both individuals.

3.3 Mutation

Our recombination operators can only decrease the number of clusters present in the solution. To counteract this behavior, we define a mutation operator that selects a subset of clusters and splits each of them in half. Splitting is done using the KaHIP graph partitioning framework. The number of clusters to be split is set to $p_s|C|$, where p_s is the splitting probability and $|C|$ is the number of clusters that the clustering selected from the population has. That means we split a cluster into two balanced blocks such that there are only a small amount of edges running between them. We do this operation with two individuals from the population that are found by tournament selection. This results in two output individuals which are used as input to a multi-level recombination operation. The computed offspring of that operation is then inserted into the population as described above.

3.4 Parallelization

We now explain the island-based parallelization that we use. We use a parallelization scheme that has been successfully used in graph partitioning [34]. Each processing element (PE) basically performs the same operations using different random seeds. First, we estimate the population size S: each PE creates an individual and measures the time $\bar{t}$ spent. We then choose S such that the time for creating S clusterings is approximately t_{total}/c, where the fraction c is a tuning parameter and t_{total} is the total running time that the algorithm is given to produce a clustering of the graph. The minimum amount of individuals in the population is set to 3, the maximum amount of the individuals in the population is set to 100. The lower bound on the population size is chosen to ensure a certain minimum of diversity, while the upper bound is used to ensure convergence.

Each PE then builds its own population. Afterwards, the algorithm proceeds in rounds as long as time is left. Either a mutation or recombination operation is performed. Our communication protocol is similar to *randomized rumor spreading* which has shown to be scalable in previous work [34]. Let p denote the number of PEs used. A communication step is organized in rounds. In each round, a PE chooses a communication partner and sends her the currently best individual C of the local population. The selection of the communication partner is done uniformly at random among those PEs to which C not already has been sent to. Afterwards, a PE checks if there are incoming individuals and if so inserts them into the local population using the elimination strategy described above. If the best local individual has changed, all PEs are again eligible. This is repeated $\log p$ times. The algorithm is implemented *completely asynchronously*, i.e., there is no need for a global synchronization.

4 Application to Protein–Protein Interaction Networks

Complex networks are prevalent in every aspect of our lives: from technological networks to biological systems like the human brain. These networks are composed of billions of entities that give rise to emerging properties and structures. A prominent example are large-scale protein–protein interaction data which make it feasible to decode components and interactions in biological processes. The important observation here is that proteins rarely act as single isolated unit when performing their functions; Instead, very often different proteins that are involved in the same biological processes have tight interactions with each other. This can be used to predict the functions of uncharacterized proteins, i.e., by observing and comparing the interactions with similar known proteins. Many studies have shown that graph clustering protein–protein interaction network is a highly useful approach for identifying protein complexes or functional modules [23, 30, 41].

The general model is described very easily: the vertices of a protein–protein interaction network are proteins and edges are inserted between two vertices if the corresponding proteins interact with each other. In a more complex model, edges can have an assigned weight which reflects the probability that there is an interaction between the associated proteins. A clustering of the network then groups the proteins into clusters/blocks. The properties of the objective function of the graph clustering problem, in our case modularity, ensure that proteins inside a block have a high interaction rate with each other while proteins from different blocks do interact very little. Using this model, one can directly employ our graph clustering framework to cluster the protein–protein interaction network and then predict functions of previously uncharacterized proteins. If the desired application requires a different clustering objective, there are two ways to go: first, one can directly modify the objective function of the evolutionary algorithm and/or use a local search algorithm within our multilevel framework that optimizes for that objective in addition. The latter directly yields an evolutionary algorithm that optimizes for the "correct" objective function.

5 VieClus Software

VieClus is open source and available at http://vieclus.taa.univie.ac.at/ or https://github.com/schulzchristian/vieclus. It contains the following programs: vieclus, evaluator, and graph_checker. The first program in this list is the clustering program, the second program can be used to evaluate solutions and compute their modularity score, and the last program checks if the file provided as input to vieclus has the correct file format. To compile these programs, you need to have Argtable, g++, OpenMPI, and scons installed. Once you have that you can execute *compile.sh* in the main folder of the release. When the process is finished, the binaries can be found in the folder *deploy*. The remainder of this section is a description of the expected file format as well as the output format that we use. Lastly, we explain the parameters of each of the programs briefly.

5.1 Input Graph File Format

The graph format used by our clustering programs is the same as used by Metis [21], Chaco [19], and the graph format that has been used during the 10th DIMACS Implementation Challenge on Graph Clustering and Partitioning. The input graph has to be undirected and without parallel edges (Fig. 1).

To give a description of the graph format, we follow the description of the Metis 4.0 user guide very closely. A graph $G = (V, E)$ with n vertices and m edges is stored in a plain text file that contains $n + 1$ lines (excluding comment lines). The first line contains information about the size and the type of the graph, while the remaining n lines contain information for each vertex of G. Any line that starts with % is a comment line and is skipped.

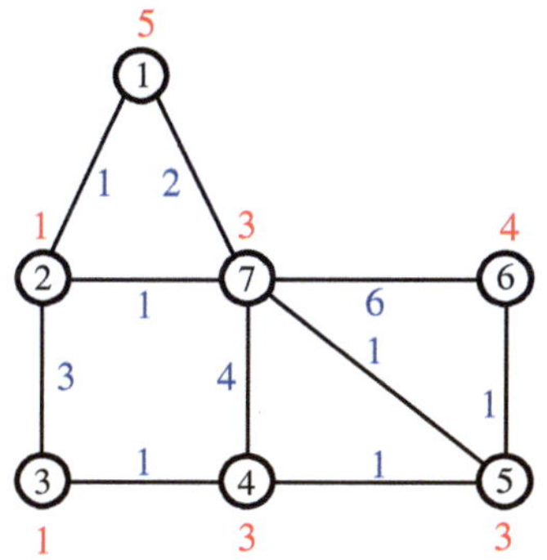


```
7 10 11
5 2 1 7 2
1 1 1 3 3 7 1
1 2 3 4 1
3 3 1 7 4 5 1
3 4 1 7 1 6 1
4 5 1 7 6
3 1 2 2 1 4 4 5 1 6 6
```

Fig. 1 An example graph and its representation in the graph format. The IDs of the vertices are drawn within the cycle, the vertex weight is shown next to the circle (red), and the edge weight is plotted next to the edge (blue)

The first line in the file contains either two integers, *nm*, or three integers, *nmf*. The first two integers are the number of vertices n and the number of undirected edges of the graph, respectively. Note that in determining the number of edges m, an edge between any pair of vertices v and u is counted *only once* and not twice, i.e., we do not count the edge (v, u) from (u, v) separately. The third integer f is used to specify whether or not the graph has weights associated with its vertices, its edges, or both. If the graph is unweighted, then this parameter can be omitted. It should be set to 1 if the graph has edge weights, 10 if the graph has node weights, and 11 if the graph has edge and node weights.

The remaining n lines of the file store information about the actual structure of the graph. In particular, the ith line (again excluding comment lines) contains information about the ith vertex. Depending on the value of f, the information stored in each line is somewhat different. In the most general form (when $f = 11$, i.e., we have node and edge weights) each line has the following structure:

$$c\ v_1\ w_1\ v_2\ w_2 \ldots v_k\ w_k$$

where c is the vertex weight associated with this vertex, $v_1, \ldots, v_k$ are the vertices adjacent to this vertex, and $w_1, \ldots, w_k$ are the weights of the edges. Note that the vertices are numbered starting from 1 (not from 0). Furthermore, the vertex weights must be integers greater or equal to 0, whereas the edge weights must be strictly greater than 0.

5.2 Output Clustering File Formats

The output format of a clustering is also similar to the output format of a clustering provided by Metis. It is basically a text file named *tmpclustering*. This file contains n lines. In each line, the block ID of the corresponding vertex is given, i.e., line i contains the block ID of the vertex i (here the vertices are numbered from 0 to $n - 1$).

5.3 User Interface

5.3.1 VieClus

Description: This is the distributed memetic algorithm to tackle the graph clustering problem. For information on troubleshooting *see* **Note 1**.

Usage: mpirun -n P ./vieclus file [--help] [--seed=<int>] [--time_limit=<double>] [--output_filename=<string>]

Options:

P	Number of processes to use.
file	Path to graph file that you want to cluster.
--help	Print help.
--seed=<int>	Seed to use for the random number generator.
--time_limit=<double>	Time limit in seconds s. The default value is set to 0s, i.e., one clustering call will be made by each PE. In order to use combine operations you *have to* set a time limit $t > 0$. VieClus will return the best solution after the time limit is reached. A time limit $t = 0$ means that the algorithm will only create the initial population.
--output_filename=<string>	Specify the output filename (default tmpclustering).

5.3.2 Graph Format Checker

Description: This program checks if the graph specified in a given file is valid.

Usage: graphchecker file

Options:

file Path to the graph file.

5.3.3 Evaluation

Description: This is the program to compute the modularity of a clustering.

Usage: ./evaluator file --input_partition=<string>

Options:

file	Path to graph file
--input_partition=file	Path to clustering file to evaluate.

6 Conclusion

We presented a parallel memetic algorithm, VieClus, that tackles the graph clustering problem. A key component of our contribution are natural recombine operators that employ ensemble clusterings as well as multi-level techniques. We combine these techniques with a scalable communication protocol, producing a system that is able to reproduce or improve previous all entries of the 10th DIMACS implementation challenge under consideration as well as results recently reported in the literature in a short amount of time. Moreover, while the previous best result for different instances has been computed by a variety of solvers, our algorithm can now be used as a single tool to compute the result. We also reviewed the connection of the graph clustering problem to protein–protein interaction networks and gave a description on how the software can be used, what file formats are expected, and how this can be used to find functional groups in protein–protein interaction networks. In addition, we shortly outline how our method can be modified for different objective functions.

7 Note

1. If VieClus crashes, it is mostly due to the following reasons: the provided graph contains parallel edges, there exists a forward edge but the backward edge is missing or the forward and backward edges have different weights, or the number of vertices or edges specified does not match the number of vertices or edges provided in the file. Please use the *graphcheck* tool provided in our framework to verify whether your graph has the right input format.

Acknowledgements

The research leading to these results has received funding from the European Research Council under the European Community's Seventh Framework Programme (FP7/2007–2013)/ERC grant agreement No. 340506. The authors acknowledge support by the state of Baden-Württemberg through bwHPC. Parts of this paper has appeared in the proceedings of the *17th Intl. Symp. on Exp. Algorithms* [7]; licensed under Creative Commons License CC-BY Leibniz International Proceedings in Informatics Schloss Dagstuhl – Leibniz-Zentrum für Informatik, Dagstuhl Publishing, Germany.

References

1. Akoglu L, Tong H, Koutra D (2015) Graph based anomaly detection and description: a survey. Data Min Knowl Disc 29(3):626–688. http://dx.doi.org/10.1007/s10618-014-0365-y
2. Arnau V, Mars S, Marín I (2004) Iterative cluster analysis of protein interaction data. Bioinformatics 21(3):364–378
3. Ausiello G, Crescenzi P, Gambosi G, Kann V, Marchetti-Spaccamela A, Protasi M (2012) Complexity and approximation: combinatorial optimization problems and their approximability properties. Springer Science & Business Media, Berlin
4. Bäck T (1996) Evolutionary algorithms in theory and practice: evolution strategies,

evolutionary programming, genetic algorithms. PhD thesis

5. Bader D, Meyerhenke H, Sanders P, Wagner D (eds) (2012) Proc. of the 10th DIMACS Impl. Challenge, Cont. Mathematics. AMS, Providence
6. Biedermann S (2017) Evolutionary graph clustering. Bachelor's Thesis, Universität Wien
7. Biedermann S, Henzinger M, Schulz C, Schuster B (2018) Memetic graph clustering. In: D'Angelo G (ed) 17th International symposium on experimental algorithms, SEA 2018, June 27–29, 2018, L'Aquila, Italy, volume 103 of LIPIcs, pp 3:1–3:15. Schloss Dagstuhl - Leibniz-Zentrum fuer Informatik. https://doi.org/10.4230/LIPIcs.SEA.2018.3
8. Blondel VD, Guillaume J-L, Lambiotte R, Lefebvre E (2008) Fast unfolding of communities in large networks. J Stat Mech: Theory Exp 2008(10):P10008. http://stacks.iop.org/1742-5468/2008/i=10/a=P10008
9. Brandes U (2005) Network analysis: methodological foundations, vol 3418. Springer Science & Business Media, Berlin
10. Brandes U, Gaertler M, Wagner D (2007) Engineering graph clustering: models and experimental evaluation. ACM J Exp Algorithmics 12(1.1):1–26
11. Brandes U, Delling D, Gaertler M, Gorke R, Hoefer M, Nikoloski Z, Wagner D (2008) On modularity clustering. IEEE Trans Knowl Data Eng 20(2):172–188
12. Demme J, Sethumadhavan S (2012) Approximate graph clustering for program characterization. ACM Trans Archit Code Optim 8 (4):21:1–21:21. http://doi.acm.org/10.1145/2086696.2086700
13. Derényi I, Palla G, Vicsek T (2005) Clique percolation in random networks. Phys Rev Lett 94(16):160202
14. Diwan AA, Rane S, Seshadri S, Sudarshan S (1996) Clustering techniques for minimizing external path length. In: Proceedings of the 22th international conference on very large data bases, VLDB '96. Morgan Kaufmann Publishers Inc, San Francisco, pp 342–353. http://dl.acm.org/citation.cfm?id=645922.673636
15. Flake GW, Tarjan RE, Tsioutsiouliklis K (2004) Graph clustering and minimum cut trees. Internet Math 1(4):385–408
16. Fortunato S (2010) Community detection in graphs. Phys Rep 486(3):75–174
17. Goldberg DE (1989) Genetic algorithms in search, optimization, and machine learning. Addison-Wesley, Boston
18. Hartmann T, Kappes A, Wagner D (2016) Clustering evolving networks. In: Algorithm engineering. Springer, Berlin, pp 280–329
19. Hendrickson B. Chaco: software for partitioning graphs. http://www.cs.sandia.gov/~bahendr/chaco.html
20. Kannan R, Vempala S, Vetta A (2004) On clusterings: good, bad and spectral. J ACM 51 (3):497–515
21. Karypis G, Kumar V (1998) A fast and high quality multilevel scheme for partitioning irregular graphs. SIAM J Sci Comput 20 (1):359–392
22. Kim J, Hwang I, Kim YH, Moon BR (2011) Genetic approaches for graph partitioning: a survey. In: Proceedings of the 13th annual genetic and evolutionary computation conference (GECCO'11). ACM, New York, pp 473–480
23. Lin C, Cho Y, Hwang W, Pei P, Zhang A (2007) Clustering methods in protein-protein interaction network. In: Knowledge discovery in bioinformatics: techniques, methods and application. Wiley, New York, pp 1–35
24. McFarling S (1989) Program optimization for instruction caches. SIGARCH Comput Archit News 17(2):183–191. http://doi.acm.org/10.1145/68182.68200
25. Meyerhenke H, Sanders P, Schulz C (2014) Partitioning complex networks via size-constrained clustering. In: SEA, volume 8504 of lecture notes in computer science. Springer, Berlin, pp 351–363
26. Miller BL, Goldberg DE (1996) Genetic algorithms, tournament selection, and the effects of noise. Evol Comput 4(2):113–131
27. Newman MEJ (2003) Properties of highly clustered networks. Phys Rev E 68(2):026121
28. Newman MEJ, Girvan M (2004) Finding and evaluating community structure in networks. Phys Rev E 69(2):026113
29. Ovelgönne M, Geyer-Schulz A (2013) An ensemble learning strategy for graph clustering. In: Graph partitioning and graph clustering, number 588 in contemporary mathematics. https://doi.org/10.1090/conm/588/11701
30. Pereira-Leal JB, Enright AJ, Ouzounis CA (2004) Detection of functional modules from protein interaction networks. Proteins Struct Funct Bioinformatics 54(1):49–57. http://dx.doi.org/10.1002/prot.10505
31. Porumbel DC, Hao J-K, Kuntz P (2011) Spacing memetic algorithms. In: 13th Annual genetic and evolutionary computation conference, GECCO 2011, Proceedings, Dublin, Ireland, July 12–16, 2011, pp 1061–1068

32. Raghavan UN, Albert R, Kumara S (2007) Near linear time algorithm to detect community structures in large-scale networks. Phys Rev E 76(3), 036106
33. Rosvall M, Axelsson D, Bergstrom CT (2009) The map equation. Eur Phys J Spec Top 178 (1):13–23
34. Sanders P, Schulz C (2012) Distributed evolutionary graph partitioning. In: Proc. of the 12th workshop on algorithm engineering and experimentation (ALENEX'12), pp 16–29
35. Sanders P, Schulz C (2013) Think locally, act globally: highly balanced graph partitioning. In: 12th International symposium on experimental algorithms (SEA'13). Springer, Berlin
36. Schaeffer SE (2007) Survey: graph clustering. Comput Sci Rev 1(1):27–64. http://dx.doi.org/10.1016/j.cosrev.2007.05.001
37. Staudt CL, Meyerhenke H (2013) Engineering high-performance community detection heuristics for massive graphs. In: Proceedings 42nd conference on parallel processing (ICPP'13)
38. Staudt CL, Meyerhenke H (2016) Engineering parallel algorithms for community detection in massive networks. IEEE Trans Parallel Distrib Syst 27(1):171–184. doi:10.1109/TPDS.2015.2390633
39. Van Dongen SM (2001) Graph clustering by flow simulation. PhD thesis
40. Wagner D, Wagner F (1993) Between min cut and graph bisection. In: Proceedings of the 18th international symposium on mathematical foundations of computer science. Springer, Berlin, pp 744–750
41. Wang J, Li M, Deng Y, Pan Y (2010) Recent advances in clustering methods for protein interaction networks. BMC Genomics 11(3): S10
42. Xu Y, Olman V, Xu D (2002) Clustering gene expression data using a graph-theoretic approach: an application of minimum spanning trees. Bioinformatics 18(4):536–545

Chapter 17

On TD-WGcluster: Theoretical Foundations and Guidelines for the User

Angela Re and Paola Lecca

Abstract

We review the TD-WGcluster (time delayed weighted edge clustering) software integrating static interaction networks with time series data in order to detect modules of nodes between which the information flows at similar time delays and intensities. The software has represented an advancement of the state of the art in the software for the identification of connected components due to its peculiarity of dealing with direct and weighted graphs, where the attributes of the physical entities represented by nodes vary over time. This chapter aims to deepen those theoretical aspects of the clustering model implemented by TD-WGcluster that may be of greater interest to the user. We show the instructions necessary to run the software through some exploratory cases and comment on the results obtained.

Key words Biological networks, Topological space, Graph clustering, Graph connected components, Graph entropy, Graph kernels, Geometric entropy, Eigenvector centrality, Time-lagged correlation

1 Introduction

TD-WGcluster is a tool implementing a procedure for the identification of connected components in a graph conceived as topological space. As a topological space, a graph X of nodes $v \in V$ connected by arches $e \in E$ is a simplicial 1-complex. In a topological space, the connectedness relation between two pairs of points is an equivalence relation, and the equivalence classes are the connected components. Using *pathwise-connectedness*, the pathwise-connected component containing $x \in X$ is the set of all y pathwise-connected to x. That is, it is the set of y such that there is a continuous path from x to y. In a topological space, *pathwise-connectedness* is generally different from *connectedness*. A subset S of X is connected if there is no way to write S as $S = U \cap V$ with $U \cap V = \emptyset$. Every topological space decomposes into a disjoint union $X = \Cup S_i$, where the S_i are connected. The S_i are called the connected components of X. The latter is the most general

Stefan Canzar and Francisca Rojas Ringeling (eds.), *Protein-Protein Interaction Networks: Methods and Protocols*, Methods in Molecular Biology, vol. 2074, https://doi.org/10.1007/978-1-4939-9873-9_17,

definition of connected component, and it is the one that allows to define the concept of belonging of a point (or node of the graph) to a connected component on the basis of not only spatial pathwise-connectedness but also other variables contributing to characterize the description of the physical process that the graph represents, such as the weight of the arch and the temporal dynamics of the node. Consequently, a graph can be generally conceived as a simplicial *k*-complex, where *k* accounts for the number of variables (and therefore dimensions) that describe the static and dynamic properties of the graph. For this reason, TD-WGcluster identifies a component connected in a multidimensional topological space, in which the dimensions are defined by the topological centrality of the graph and by the dynamics of its nodes. Unlike the results provided by the majority of the graph decomposition algorithms, the results provided by TD-WGcluster do not reflect the bare pairwise-connectedness between nodes, but rather a more general concept of node proximity. This concept includes the classical topological proximity measure, and also criteria of temporal correlation. In particular, the statistical significance of the topological structure and the nature of the dynamics (deterministic or stochastic) of the connected components are measured by its geometrical entropy.

Here, we present TD-WGcluster. We discuss the rationale behind the algorithm and showcase its application through the in silico generation of different network settings.

2 The Algorithm

TD-WGcluster has been extensively described by the authors in [25]. Here, we summarize the fundamental steps, giving particular importance to the purposes of each of them. TD-WGcluster implements four sequential computational steps.

Step 1 The calculation of the time-lagged correlations (TLC) between any couple of interacting nodes in the network. The TLC is the correlation between the nodes' time series shifted in time relative to one another. Then it analyzes the TLC curves to estimate the features describing their shape, i.e., the lag corresponding to the maximum of the curve, the trend index, the seasonality index, the autocorrelation test statistics, the non-linearity test statistics, the skewness index, the kurtosis index, and the Lyapunov coefficient.

Step 2 An unsupervised version of K-means algorithm for the detection of sub-graphs including nodes with similar TLC.

The TLC curves are then clustered by shape with a K-means algorithm taking as input the vectors of the shape features for each couple of linked nodes. The K-means clustering detects sub-graphs with edges for which the TLC between the two time series related to directly connected nodes have similar shape. Clustering the input graph by the shape of TLC between its nodes allows to obtain groups of nodes between which the information propagates from the source node to the target node at similar time lags. The similarity of the shapes of the TLC curves does not imply that those nodes have similar dynamics (i.e., similar time series), but only that the synchronization between the activities of directly linked nodes occurs at similar time lags. Therefore, each sub-graph can be characterized by its own time lag τ_r, representative of the time delay at which the correlation between the dynamics of the directly linked nodes reaches its maximum value.

Step 3 An optimized fast-greedy algorithm for the identification of connected components by sub-graph. It is a bottom-up hierarchical approach that optimizes the modularity in a greedy manner. Initially, every vertex belongs to a separate community, and communities are merged iteratively such that each merge is locally optimal (i.e., yields the largest increase in the current value of modularity). The algorithm stops when it is not possible to increase the modularity anymore, so it gives a grouping as well as a dendrogram. The method is fast, and it has no parameters to tune.

Step 4 The calculation of a geometric entropy measure for each connected component. The geometric entropy E_r of each connected component is calculated as the negative natural logarithm of the determinant of the covariance matrix of the time series of the nodes in the component at the representative τ_r. Geometric entropy is used as a measure of the complexity of the component since it encodes the information regarding the size of the component (i.e., number of nodes and number of edges) and the volume of space occupied by the time series data points of the nodes at the time shift of the maximum correlation [4, 11]. Large volumes may suggest the potential existence of significant differences in the levels of abundance of the variable defining the nodes' dynamics (e.g., concentration or number of molecules in the case in which nodes represent chemical species or expression levels if nodes represent genes). In these, the geometric entropy, as a global measure of the D-dimensional variance of the dynamics of the D nodes, is indicative of the nature of the dynamics of connected components. The most interesting cases are identified by extreme low and extreme high values of geometric entropy, as the first could indicate a purely stochastic or a purely deterministic nature of the interactions, whereas the second a hybrid deterministic/stochastic dynamic, as large volumes occupied by data points could mean high difference in the level of abundance of the species represented by the nodes. Furthermore, the distribution of the geometric entropy on the graph also summarizes the complexity of the global network. In fact, a network whose connected components differentiate significantly by entropy exhibits a more varied dynamic and thus a greater degree of complexity.

2.1 Time-Lagged Correlation

Time-lagged correlation (TLC) refers to the correlation between two time series shifted in time relative to one another. The lagged correlation is estimated by the cross-correlation function (CCF). The CCF of two time series is the product-moment correlation as a function of lag between the series. It is helpful to begin defining the CCF by the cross-covariance function (CCFV). Consider N pairs of observations on two time series, $x(t)$ and $y(t)$, the sample CCFV is

$$c_{xy} = \frac{1}{N}\sum_{t=1}^{N-\tau}(x(t) - \overline{x(t)})(y(t+k) - \overline{y(t)}), \quad k = 0, 1, \ldots, (N-1) \quad (1)$$

and, similarly,

$$c_{xy} = \frac{1}{N}\sum_{t=1-\tau}^{N}(x(t) - \overline{x(t)})(y(t+k) - \overline{y(t)}), \quad k = -1, -2, \ldots, -(N-1) \quad (2)$$

where $\overline{x(t)}$ and $\overline{y(t)}$ are the sample means, and τ is the lag. The sample CCF is the CCFV scaled by the variances $c_{xx}(0)$ and $c_{yy}(0)$ of the two series:

$$r_{xy}(\tau) = \frac{c_{xy}(\tau)}{\sqrt{c_{xx}(0)c_{yy}(0)}}. \quad (3)$$

The CCFV and CCF are asymmetrical functions. The asymmetry is specified in the definition given in Eq. 1. In this equation, the cross-correlation function is described in terms of *lead* and *lag* relationships. The first part of the equation applies to $y(t)$ shifted forward relative to $y(t)$. With this direction of shift, $x(t)$ is said to *lead* $y(t)$. This is equivalent to saying that $y(t)$ *lags* $x(t)$. The second part of Eq. 1 describes the reverse situation and summarizes lagged correlations when $y(t)$ *leads* $x(t)$ ($x(t)$ *lags* $y(t)$).

The analysis of the CCF permits to detect the lag at which the two time series are maximally correlated and to determine if this correlation is significant. The correlation is significant if its values do not belong to the confidence interval of the cross-correlation. This confidence interval relies on several simplifying assumptions and can be computed from the sample size alone. For a two-tailed test, the approximate γ confidence interval is $CI = 0 \pm z_\gamma \frac{1}{\sqrt{N}}$. The value z_γ is the γ probability point of the cumulative distribution function of the normal distribution.

2.2 Detection of Sub-graphs

The number of optimal sub-graphs which partition the input graph is estimated by minimizing the total within-clusters sum of squares (WCSS) obtained with a K-means procedure. K-means clustering is applied to the set of the feature vectors $\mathbf{tlc} = (tlc_1, tlc_2, \ldots, tlc_E)$, describing the shape of the time-lagged correlation between the time series of each couple of linked nodes. E is indeed the number of edges in the input graph. The components of the feature vector for the i-th TLC curve $C_i(\tau)$ $(i = 1, 2, \ldots, E)$ are

$$\mathbf{tlc}_i = \{ \\ (\text{lag of maximal correlation})_\text{i}, (\text{seasonality index})_\text{i}, \\ (\text{trend index})_\text{i}, (\text{non-linearity index})_\text{i}, \\ (\text{kurtosis index})_\text{i}, (\text{skewness index})_\text{i}, \\ (\text{Lyapunov coefficient})_\text{i}, (\text{autocorrelation statistics})_\text{i} \\ \} \quad (4)$$

The elements of $\mathbf{tlc}_i$ are the indices typically used in the analysis of the time series. However, in this study, they are not used to analyze the shape of the time series but to analyze the shape of the TLC curve. These measure report on the main behavioral characteristics of the TCL curve as follows.

- The lag of maximal correlation is the lag at which the cross-correlation (i.e., the TLC curve) is maximum. Hereafter, this lag will be denoted by τ_r and we will refer to it as the *representative* of a set of interactions in a sub-graph.
- Seasonality will be identified by regularly spaced peaks and troughs which have a consistent direction and approximately the same magnitude every fixed time period. In our case, seasonality is synonymous of periodicity.

  The seasonality index SI_i for the TCL $C_i(\tau)$ is calculated by a simple moving average-based approach on a detrended curve [27]:

$$SI_i = \frac{1}{N_\tau} \sum_{h=1}^{N_\tau} \left[\frac{C_i(\tau_h - p)}{MA(\tau_h - 1p)} + \frac{C_i(\tau_h - 2p)}{MA(\tau_h - 2)} + \frac{C_i(\tau_h - 3p)}{MA(\tau_h - 3p)} + \cdots \right] \quad (5)$$

where p is the period and N_τ is the total number of time lags, i.e., for a lags series $-\tau_m, -\tau_{m-1}, \ldots, 0, \ldots \tau_{m-1}, \tau_m$ $N_\tau = 2m$; MA indicates moving average.

- The trend is the long-term movement in the curve without periodic related and irregular effects, and is estimated by the moving average [27].
- Kurtosis index KI_i of $C_i(\tau)$ measures the frequency of extreme values and, for a TCL curve, $C_i(\tau)$ has been estimated by the following formula:

$$KI_i = N\frac{\sum_{h=1}^{N}(C_i(\tau_h) - \overline{C_i(\tau)})}{[\sum = h = 1^N (C_i(\tau_h) - \overline{C_i(\tau)})^{2}]^{2}} \tag{6}$$

- Skewness index measures the asymmetry of the curve, i.e., how much the curve appears distorted or skewed either to the left or to the right. Skewness quantifies the extent to which the curve differs from a Gaussian-shaped curve [14]. It is defined as

$$\Gamma_i = \mathbf{E}\left[\left(\frac{(C_i(\tau) - \overline{C_i(\tau)})}{\sigma_i}\right)^3\right] \tag{7}$$

where $\mathbf{E}$ is the expectation operator, and $\overline{C_i(\tau)}$ is the mean of $C_i(\tau)$, and σ_i its variance.

- Non-linearity index is the statistics of Teraesvirta's non-linearity test [34].
- The Lyapunov coefficient measures the level of deterministic chaos in the dynamics, i.e., the sensitivity of the dynamics to initial conditions [2, 24].
- The autocorrelation (also known as serial correlation) AC_i of the TLC curve $C_i(\tau)$ refers to the correlation of the curve with its own "past" and "future" values, i.e., this index measures how sequential observations in the curve affect each other. The autocorrelation statistics is the statistics of Box–Pierce test [13] that tests whether the autocorrelations in the data are different from zero.

Applying K-means algorithm to the feature vector set $\{\mathbf{tlc}_i\}$, $(i = 1, 2, \ldots, E)$, with an increasing putative number of clusters (sub-graphs) at each run, we obtained the values of WCSS. An elbow in the curve interpolating the WCSS points ($n_{\text{sub-graphs}}$, WCSS), where $n_{\text{sub-graphs}}$ is the increasing putative number of sub-graphs, suggests the appropriate number of sub-graphs n_{optimal}. n_{optimal} is estimated as the minimum value of n_{clusters} at which the first derivative of WCSS w.r.t. $n_{\text{sub-graphs}}$ is null within a tolerance $0 < \varepsilon \ll 1$, i.e.,

$$\left|\frac{d \quad WCSS}{dn_{\text{sub-graphs}}}\right| \leq \varepsilon. \tag{8}$$

The first derivative of the curve ($n_{\text{sub-graphs}}$, WCSS) is calculated by the Stineman algorithm [21]. The problem of WCSS minimization is known to be NP-hard. Furthermore, if the input data do not have a strong clustering structure, the procedure may not converge. For this reason, WGcluster adopts the Lloyd's algorithm whose complexity is linear in the number of edges and number of sub-graphs, and is recommended in case of data poorly clustered [9].

2.3 Cross-Correlation Depends on Kinetic Rate Constant

As an example, consider a reaction

$$X_1 \xrightarrow{k} X_2$$

The rate is

$$v = \frac{dX_2}{dt} = -\frac{dX_1}{dt} \tag{9}$$

and therefore, the time behavior of X_1 is

$$X_1(t) = X_1(0)e^{-kt} \tag{10}$$

where $X_1(0)$ is the abundance of species X_1 at time $t = t_0$. It follows that

$$\frac{dX_2}{dt} = -X_1(0)e^{-kt} \tag{11}$$

so that the time behavior of X_2 is

$$X_2(t) = -X_1(0)\left(e^{-kt} - 1\right) \tag{12}$$

The convolution between the functions $X_1(t)$ and $X_2(t)$ on the interval $[t_0 = 0, t]$ is

$$(X_1 * X_2)(t) = \int_0^t e^{-k\xi}\left[e^{-k(t-\xi)} - 1\right]d\xi = \frac{1}{k}(1 - e^{-kt})X_1^2(0) \tag{13}$$

and its plot is shown Fig. 1.

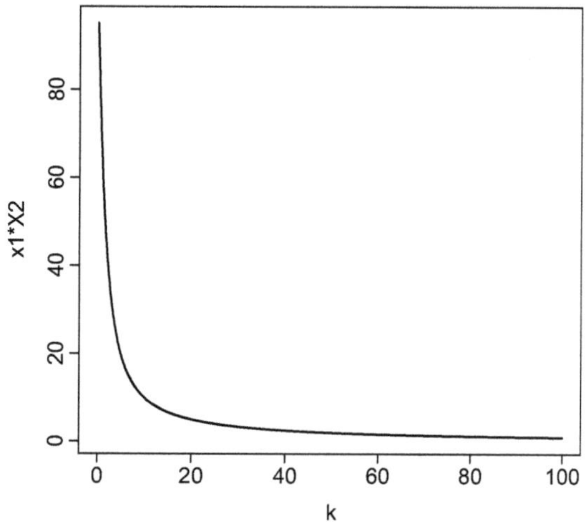

Fig. 1 $(X_1 * X_2)$ as in Eq. 13, at $t = 1$

In the following sections, we delve more deeply into the concept of geometric entropy for a graph and the relationships that bind it to the other two physical quantities on which our clustering algorithm is based, i.e., the correlation between time series (which accounts for the dynamics of the network) and the topological structure of the network.

3 The Topological Meaning of the Entropy

The determinant of the covariance matrix of a set of data points has a very well-known geometrical interpretation, what is not so well known is that it is related to the entropy of the point distribution. Since the covariance matrix is positively defined, all its eigenvalues are not negative and the determinant is the product of these eigenvalues. The square root of this determinant measures the volume of k-dimensional σ-cube. The determinant of the sample covariance matrix Σ measures the differential entropy of the distribution up to constant factors and a logarithm. Indeed, the geometrical entropy of a connected component is

$$E_r = \frac{c}{2}\left(1 + \ln(2\pi)\right) + \frac{1}{2}\ln \det\Sigma \qquad (14)$$

where c is the dimensionality of the space, i.e., the number of nodes.

4 How to Run TD-WGcluster

TD-WGcluster is implemented in R (https://www.r-project.org/), so it is mandatory to install R on the PC in order to use TD-WGcluster. To run the tool, from the command prompt type

```
>Rscript TD-WGCluster.R input_time_series.txt
input_network.txt cross_corr.pdf lags.txt lag.series.txt
```

where

- *input_time_series.txt* is the file of the time series, one for each node in the network. The file is an input and has to be formatted in a text format in which each row is a time series.
- *input_network.txt* is the file of the network to be clustered. The file is an input and has to be formatted in a text format with two columns. The first column is the list of the source nodes, and the second column is the list of target nodes.
- *cross_corr.pdf* is a name (chose by the user) for the output PDF file of the time-lagged correlation plot for each interaction in the network

- *lags.txt* is a name for the output file reporting in text format the largest and the smallest value of time-lagged correlation and their corresponding lags for each couple of source and target node in the network as follows:

Source	Target	Max-Corr	Lag-Max-Corr	Min-Corr	Lag-Min-Corr
X1	X29	0.625810818529982	-10	-0.0442578092832854	16
X1	X33	0.963673598767258	0	0.0852071473990701	16
X1	X39	0.966021379771689	0	0.0861776070049978	16
X1	X81	0.975884821750677	0	0.0190734303202269	-16
X1	X87	0.952034559981157	-1	0.0855938673314824	16
X10	X100	-0.00347531022815283	16	-0.858986223755437	0
X10	X11	0.999831889933307	0	0.0333570215740381	16
X10	X12	0.998970261116114	0	0.0326181370327232	16
...	...	...	...	...	...

- lag.series.txt is a name for the output file reporting in text format the time-lagged correlation at each time lag. Each row reports the time-lagged correlation as a function of the lag for each interaction in the network.

TD-WGcluster outputs the results for the graph clustering in two folders *Subgraphs* and *Connected_Components*, where the sub-graphs and their connected components are stored in GraphML format (http://graphml.graphdrawing.org/).

TD-WGcluster is provided as .ZIP archive including:

- *data*: a subfolder, where the input data concerning the real case study of PPI network is stored
- *TD-WGcluster.R* is the main script to invoke
- *impute.R* is the R script for missing data imputation called by *TD-WGcluster.R* if the input time series has less than 10 time points
- *time-series-analysis.R* is the R script implementing the calculations of the shape feature of the time series; it is called by *TD-WGcluster.R*
- *graph.processing.R* is the R script for the detection of connected components in sub-graphs and for the calculation of the entropies.

5 Generation of Synthetic Use Cases

The exemplification of TD-WGcluster features is based on the in silico generation of random networks consisting of 100 nodes and showing different topological features. The adopted network models include Erdös–Rényi random graphs, random scale-free power

Table 1
R igraph functions and settings of their input parameters for the generation of the graphs

Function generating graph	Function arguments	Assigned value
sample_pa(n, power, directed) Generate scale-free graphs according to the Barabasi–Albert model	*n := number of vertices* *power := power of the preferential attachment (p=1 corresponds to linear preferential attachment)* *directed := logical value (TRUE/ FALSE to create a directed/ undirected network)*	*100* *0.25, 1, 2* *TRU*
sample_gnp(n, p, directed, loops) Generate random graphs according to the G(n,p) Erdos–Renyi model, where every possible edge is created with the same constant probability p	*n := number of vertices* *p := the probability for drawing an edge between two arbitrary vertices* *directed := logical value (TRUE/ FALSE to create a directed/ undirected network)* *loops := logical value (TRUE to add loop edges)*	100 0.1 TRUE FALSE
sample_k_regular(no. of nodes, k, directed, multiple) Generate random graphs, where each vertex has the same degree	*no.of.nodes: number of vertices* *k := the degree of each vertex in the graph, or the out-degree and in-degree in a directed graph* *directed := logical value (TRUE/ FALSE to create a directed/ undirected network)* *multiple := logical value (TRUE to allow multiple edges)*	100 4 TRUE FALSE
graph.formula(... .,. . . .) Generate cortical network via a simple interface	R expressions giving the structure of the graph consisting of vertex names and edge operators. An edge operator is a sequence of "−" and "+ " characters, the former is used for the edges and the latter for arrow heads.	A−+E, B−+F, C−+D, D−+H, E−+G, F−+I or
graph.formula(... .,. . . .) Generate cortical network via a simple interface	R expressions giving the structure of the graph consisting of vertex names and edge operators. An edge operator is a sequence of "−" and "+ " characters, the former is used for the edges and the latter for arrow heads.	N2−+N7, N3−+N8, N4−+N8, N4−+N9, N5−+N9, N5−+N10, N6−+N11, N6−+N12, N7−+N13, N8−+N13, N9−+N13, N9−+N14, N8−+N15,

law graphs, regular graphs, and cortical networks. Instances of networks featuring each kind of topology were generated within the R statistical computing and graphics environment (Table 1) almost entirely by functions defined in the `igraph` package [8].

Erdös–Rényi Random Graph: *The Erdös–Rényi random graph is an important reference model introduced by Erdös and Rényi. Let* n *denote the number of vertices of the graph. Each realization of an Erdös–Rényi random graph is generated by iterating over the* n (n − *1)/2 pairs* (i, j) *of nodes and adding the edge between* i *and* j *with probability* p. *Here, we concentrate on the case* p = *0.1. In this model, the node degree follows a Poisson distribution so that the probability that a vertex has* k *edges is* $P(k) = \frac{e^{-k}\lambda^k}{k!}$, *where* $\lambda = \binom{n-1}{k} p^k (1-p)^{n-1-k}$.

Scale-Free Power Law Graph: *Many large real networks show the common power law scaling, whereby the probability* P (k) *that a vertex in the network interacts with* k *other vertices is free of scale and decays as a power law, following* P (k) ~ k^{-g} *with the exponent* g *between 2.1 and 4. Growth and preferential attachment are the prominent features of real networks responsible for the power law scaling [1]. We generated scale-free power law networks featuring* g = *0.25, 0.5, 1, 2.5.*

Regular Graph: *A regular graph is a graph, where each vertex has the same degree. A regular graph with vertices of degree* k *is called a* k*-dimensional hypercubic lattice. Here we set* k = *4.*

Cortical Graph: *In cortical networks, cortical areas or modules are treated as nodes, each consisting of a network of synaptically coupled and interconnected excitatory and inhibitory neurons, with these nodes joined by associational synaptic connections. Here, we generated two cortical graphs as shown in Table 1.*

We then generated ten weighted versions for each graph's type. We calculated the weights matrix associated with a graph as the matrix M such that

$$M\mathbf{v}_1 = \lambda_1 \mathbf{v}_1 \tag{15}$$

where $\mathbf{v}_1$ is the eigenvector relative to the dominant eigenvalues λ_1, known as spectral radius of the graph. The spectral radius and its eigenvector have a precise meaning in terms of graph connectivity.

λ_1 represents the average distance to traverse across the entire graph and is therefore an approximate estimate of graph connectivity [30]. A low (high) spectral radius suggests high (low) connectivity. Each component of $\mathbf{v}_1$ measures the relative contribution of each node to the overall connectivity of the graph. In this work, we set λ_1 to the value of the dominant eigenvalue of the adjacency matrix of the unweighted variant of the graph. The variance σ of $\mathbf{v}_1$ measures the spread in node contribution across the graph. It is given by the mean squared deviation of node contribution to connectivity. A low variance indicates that each node contributes similarly to network connectivity; a high variance is indicative of a large distribution in node contribution. Shanafelt et al. [30] report the following bound for σ, for a graph of n nodes

$$\sigma \leq \frac{1}{n} + \left[\frac{n-1}{n} \cdot \frac{w_{min}}{\lambda_1}\right]^2 \tag{16}$$

where $w_{\min}$ is user-defined minimum value of the entries of the graph adjacency matrix. For each type of network, we performed ten random generations of the components of $\mathbf{v}_1$ within the range of variation $[0, \sigma]$, and for each of them we calculated a matrix M satisfying Eq. 16. In this way, M is the matrix of the graph weights.

To generate the time series of the node dynamics, we adopted a mass action-law model, where nodes' rate equations are expressed as the product of $M\mathbf{v}_1^{(r)} = \mathbf{b}^{(r)}$, where $\mathbf{v}_1^{(r)}$ is the r-th random realization of the principal eigenvector ($r = 1, \ldots, 10$), and $\mathbf{b}^{(r)}$ is the r-th random realization of the known terms vector.

This method for generating the time series associated with each node proved to be sufficiently accurate for all network types except for cortical networks. For this type of networks, the time series obtained with this method do not have sufficient variance to determine an appreciable dynamic. We believe this is due to the particular topology of these networks, where each layer relays input from the previous layer and is in turn input for the next layer. On this topology, the method of the principal eigenvector then propagates almost unchanged signals that are not processed by the topology, itself invariant.

5.1 Centrality Measures

We examined the distribution of node centrality in the non-weighted graph and in each weighted associated graph. To gauge node centrality, we employed different centrality measures such as degree, entropy, diversity, betweenness, and eigenvector centrality. Centrality measures quantify how central is the position of individual vertices in the graph, and could rely on walks structure such as node degree, betweenness, and eigenvector centrality, or on statistical properties such as structural diversity and entropy. The characteristics of the distribution of these centrality measures are linked to the number of connected components identifiable with the currently available algorithms designed to this purpose, because each of these centrality measures refers to a precise definition of *proximity* of nodes. In order to better understand this important point, we briefly review the definition of all the centrality measures used in this study.

Betweenness Centrality: *The betweenness centrality is based on shortest paths. For every pair of vertices in a connected graph, there exists at least one shortest path between the vertices such that either the number of edges that the path passes through (for unweighted graphs) or the sum of the weights of the edges (for weighted graphs) is minimized. The betweenness centrality for each vertex is the number of shortest paths that pass through the vertex. The* `igraph` *function computing betweenness is inspired by U. Brandes algorithm [5].*

Eigenvector Centrality: *The eigenvector centrality* C_i *of node* i *is given by:*

$$C_i = \frac{1}{\lambda}\sum_k a_{ki}\, x_k \tag{17}$$

where $\lambda \neq 0$ *is a constant. In matrix form Eq. 17 is written as:*

$$\lambda C = CA \tag{18}$$

where A *is the graph adjacency matrix.*

If a directed network is not strongly connected, only vertices belonging to strongly connected components or in the out-component of such components have non-zero eigenvector centrality [35]. The other vertices all have null centrality. This is due to the fact that vertices with no incoming edges have, by definition, a null eigenvector centrality score, and so have nodes that are pointed to only by nodes with a null centrality score. This is scarcely justifiable.

The problem can be solved by assigning to each vertex a small amount of centrality for free *[35], regardless of the position of the vertex in the graph. It follows that each vertex has a minimum positive amount of centrality that it can transfer to its target vertices. The centrality of vertices that are never referred to is this minimum positive amount, while linked vertices have higher centrality. It follows that highly linked vertices have high centrality, regardless of the centrality of the partners. However, vertices that have few links may still have high centrality if the partners have large centrality.*

This method to compute eigenvector centrality was proposed by Katz [22], and refined by Hubbel [15], and it is still widely influencing the study of graph topological properties [12]. The use of Katz's centrality is recommended to calculate eigencentrality for directed graph for which the classical eigencentrality is meaningless. For this reason, in our analysis we used Katz's centrality. The Katz's centrality of a node i *is defined as:*

$$C_{Katz}(i) = \alpha\sum_k a_{ki}\, x_k + \beta \tag{19}$$

or in matrix form

$$C_{Katz} = \alpha C_{Katz} A + \beta \tag{20}$$

where α *and* β *are constants. In particular,* α *is called attenuation factor and can take only values less than the reciprocal of the dominant eigenvalue of* A, β *is a vector whose elements are all equal a user-defined positive constant.*

Therefore from the definition (Eq. 20), we obtain

$$C_{Katz} = \beta(I - \alpha A)^{-1}. \tag{21}$$

Diversity: *The diversity* D_i *of node* i *index was defined in Eagle et al. [10]. It is defined as the normalized Shannon entropy (here named simply* entropy*) of the incident edges' weights:*

$$D_i = -\frac{\sum_j (p_{ij} \log p_{ij})}{k_i} \tag{22}$$

where

$$p_{ij} = \frac{w_{ij}}{\sum_i w_{il}}, \quad l = 1, \ldots, k_i \tag{23}$$

and k_i *is the total degree of vertex* i, w_{ij} *is the weight of the edge (s) between vertices* i *and* j.

The centrality measures are indicative of the presence of connected components in a graph. For example, if a directed graph is not strongly connected, only nodes that are in strongly connected components or in the out-components of such components can have high Katz's eigenvector centrality. The other nodes, such as those in the in-components of strongly connected components, all have very low Katz's eigenvector centrality. Consequently, we can state that the distribution of eigenvector centrality can indicate the presence of connected components, as illustrated in Fig. 2. Since eigenvector centrality is a generalization of the degree and the degree-derived centrality measures, such as entropy, diversity, and betweenness, these observations hold also for these centrality measures.

Moreover, since centrality measures reflect the amount of participation of nodes in the connectedness between different parts of the network, they inspired algorithms to explore the internal organization of a network. For instance, node degree has been employed in several algorithms to explore local community structure [7, 23, 26, 36]. Additionally, a number of algorithms were designed on the basis of shortest-path distance [3] and/or random

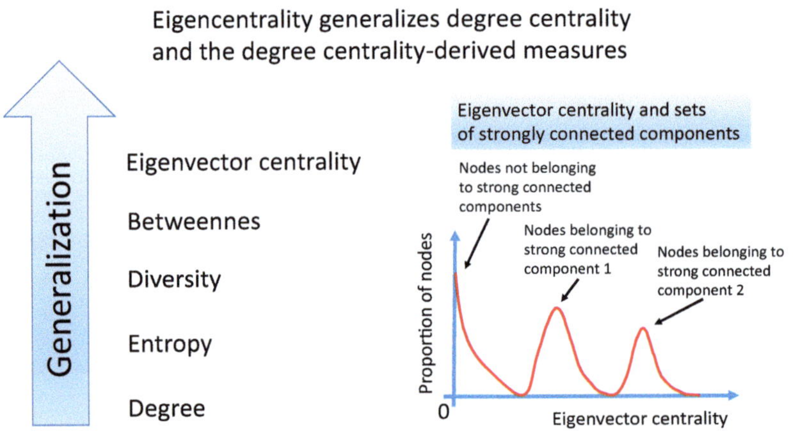

Fig. 2 Eigenvector centrality generalizes degree, entropy, diversity, and betweenness. The analysis of its distribution in a graph allows to approximately estimate the number of connected components

walk on the graph [16, 28]. In this study, we used community detection methods drawn from statistical mechanics that still refer to pathwise connectedness [29]. Indeed, in these approaches the community structure of the network is interpreted as the spin configuration that minimizes the energy of the spin glass with the spin states being the community indices. This interpretation of a network allows a considerable gain in computational efficiency. In this study, just for the sake of efficiency, we used the walk-trap algorithm and the spin-glass clustering whose implementations are available in the `igraph` R package through the functions `walktrap.community` [20] and `cluster_spinglass` [17]. We also used the `igraph` function `decompose`, that, in its current version, unlike `walktrap.community` and `cluster_spinglass` is able to detect only weakly connected components.

5.2 Graph Kernels

Since TD-WGcluster accounts for spatial and temporal dimensions in performing the graph decomposition, we expect to obtain connected components in number and in structure different from those obtained by algorithms that identify connected components on the basis of topological classical centrality (i.e., measures assuming a representation of the graph as a simplicial 1-complex). In order to quantify the dissimilarity between the connected components obtained with TD-WGcluster with those obtained by other algorithms (in particular, here we chose the `decompose` function algorithm in `igraph` R package [18]) we calculate the graph kernels and, from them, we derived the Euclidean distance between the compared connected components.

A Kernel κ is a function that maps two vectors $\mathbf{w}$ and $\mathbf{w}'$ of the vector space W via mapping ϕ into a feature space, that in this study we consider to $\mathbb{R}$, i.e.,

$$\kappa : W \times W \rightarrow \mathbb{R} \tag{24}$$

$$(\mathbf{w}, \mathbf{w}') \rightarrow \langle \phi(\mathbf{w}), \phi(\mathbf{w}') \rangle \tag{25}$$

where $\langle \cdot, \cdot \rangle$ denotes the inner product. Kernels are bilinear forms represented by Gram matrices. A Gram matrix G of a set a vectors $\mathbf{w} \in W$ is a positive semidefinite matrix with indices corresponding to the nodes and entries g_{ij} corresponding to all possible inner products of the vectors $\mathbf{w} \in W$:

$$g_{ij} = \mathbf{w}_i^T \mathbf{w}_j. \tag{26}$$

A Gram matrix K associated with a kernel κ is thus a matrix of the form

$$K = \begin{bmatrix} k(\mathrm{w_i}, \mathrm{w_i}) k(\mathrm{w_i}, \mathrm{w_j}) \\ k(\mathrm{w_j}, \mathrm{w_i}) k(\mathrm{w_j}, \mathrm{w_j}) \end{bmatrix}$$

Given two graphs G_1 and G_2 from the space $\mathcal{G}$, the similarity between G_1 and G_2 can be estimated by a kernel that computes an inner product on graphs:

$$\kappa(G_1, G_2) = \langle \phi(G_1), \phi(G_2) \rangle \tag{27}$$

In literature, we find different models for graph kernels according to different definitions of the inner product and of ϕ. In this study, we considered four models of graph kernels: linear kernel between edge histograms [33], Weisfeiler–Lehman subtree kernel [32], connected graphlet kernel [31], and linear kernel between vertex-edge label histograms [33], all implemented in the `graphkernels` library [19].

Finally, we point out that from the proof of Proposition 3 in [6], it follows that every kernel κ generates a distance measure d

$$d(x, y) = \frac{1}{2}\left(\kappa(x, x) + \kappa(y, y)\right) - \kappa(x, y), \tag{28}$$

that can quantify the dissimilarity between the objects x and y.

6 Results

In Figs. 3, 5, 7, 9, 11, 13, we report the distributions of the centrality measures examined in this study. As expected from what we presented in the previous chapter, a Katz's eigenvector centrality distribution peaked at a non-zero value is reflected into the existence of connected components, as they are detected by the random walks-based and the spin-glass model-based algorithms. This tendency is also true for the other measures of centrality of which the eigenvector is a generalization. We could notice that the lower variability shown by the Katz's eigenvector centrality in the Erdös–Rényi graph compared to scale-free graphs reflects into lower variability in the size of the detected connected components. It is worth mentioning that, whether based on random walks or on spin-glass model, the reconstruction of the community organization produced by those algorithms is strongly dependent on the choice of internal parameters (such as the length of the random walks to perform or the start/stop temperatures in the simulations) which in fact influence the number and size of the returned communities. The parameter space on our artificial random graphs would be exceedingly high to explore without additional information at hand. Therefore, we opted for limiting the comparison of the connected components detected by TD-WGcluster with the decomposition produced by the sole function `decompose` in the `igraph` R package as it can be applied without arbitrary parameters tuning. The drawback in the latter case lies in the algorithm

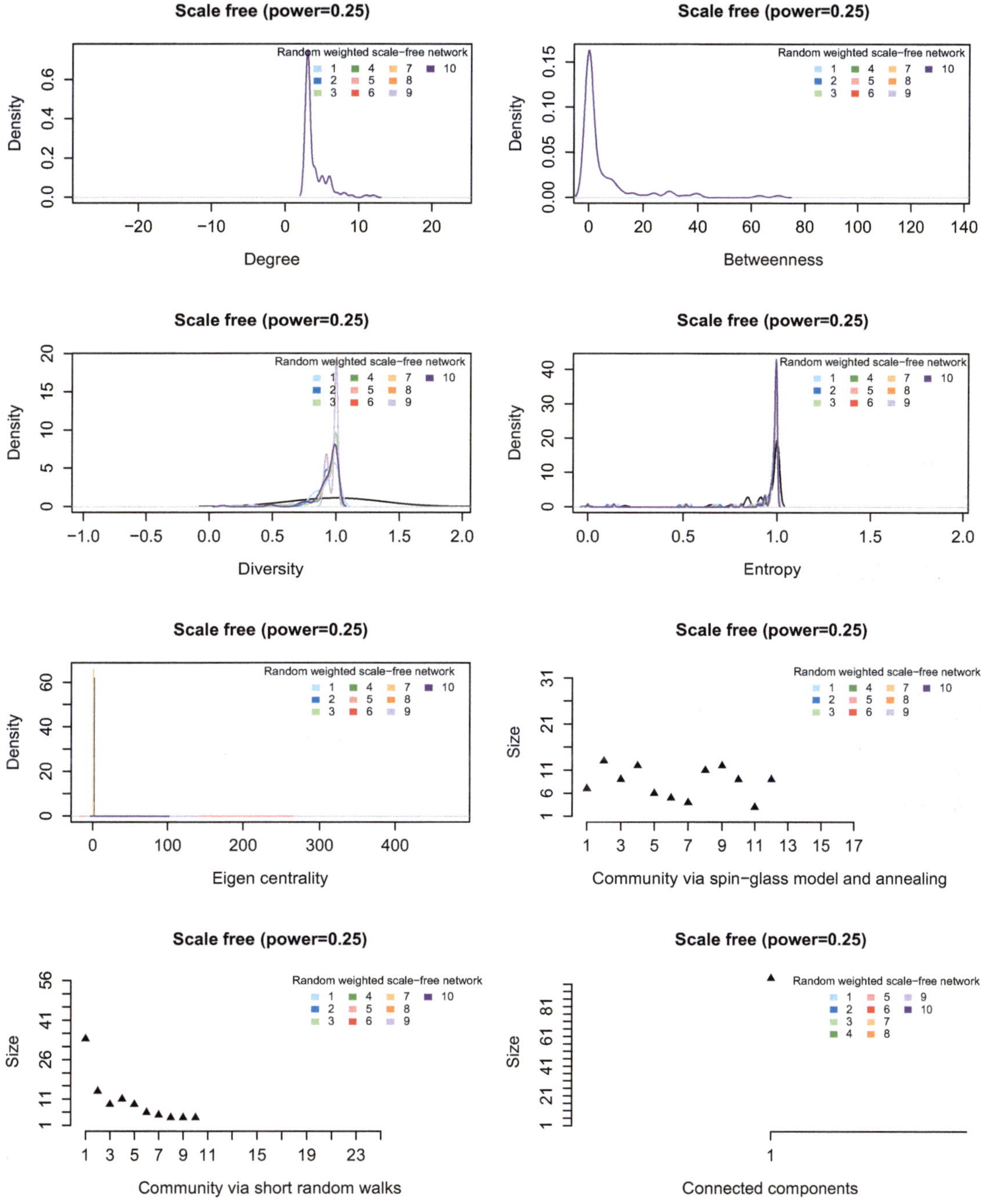

Fig. 3 Figure displaying the distribution of the degree, betweenness, diversity, entropy, and eigenvector centrality measures for the non-weighted scale-free power law graph ($g = 0.25$) and its weighted variants. The distribution for the unweighted variant is shown in black. Furthermore, the figure displays the number and size of communities detected either by a spin-glass model and simulated annealing or by random walks. Finally, the last plot displays the unique connected component detected by the decompose function

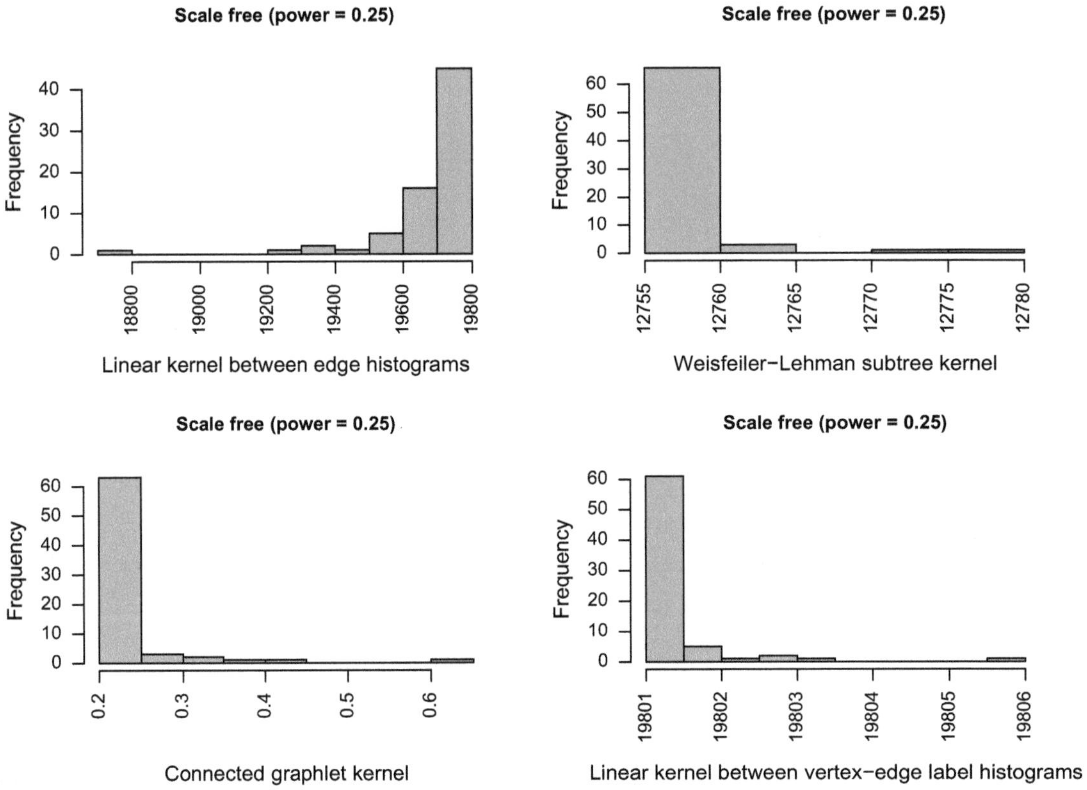

Fig. 4 Euclidean distances between TD-WGcluster connected components and the unique connected components (identical to the entire graph) returned by `igraph decompose` function for scale-free graphs ($g = 0.25$)

implementation which is currently able to decompose the graph of interest into weakly rather than strongly connected components. TD-WGcluster provides the user with a completely different scenario as proven by the histograms of the Euclidean distances between TD-WGcluster detected connected components and the unique connected component (identical to the entire graph) detected by decompose, as it is shown in Figs. 4, 6, 8, 10, 12 (for the cortical networks there is not an analogous analysis, as the time series generated with the method of the principal eigenvector did not produce dynamics with an appreciable variance, so that the use of TD-WGcluster in this case is not recommended). Indeed, the low values of the connected graphlet kernel distances as well as the indefinitely high distances obtained when using the remaining three kernels point at high dissimilarity in the graphs decompositions, irrespective of the kind of graph analyzed. These results are only partially explainable with the size of the two compared graphs, as it is proven in Fig. 14, where we compare graph of 26 nodes and 28 edges with 100 sub-graphs of 5 nodes randomly extracted. In Fig. 14, we see that the distances are lower than those in the

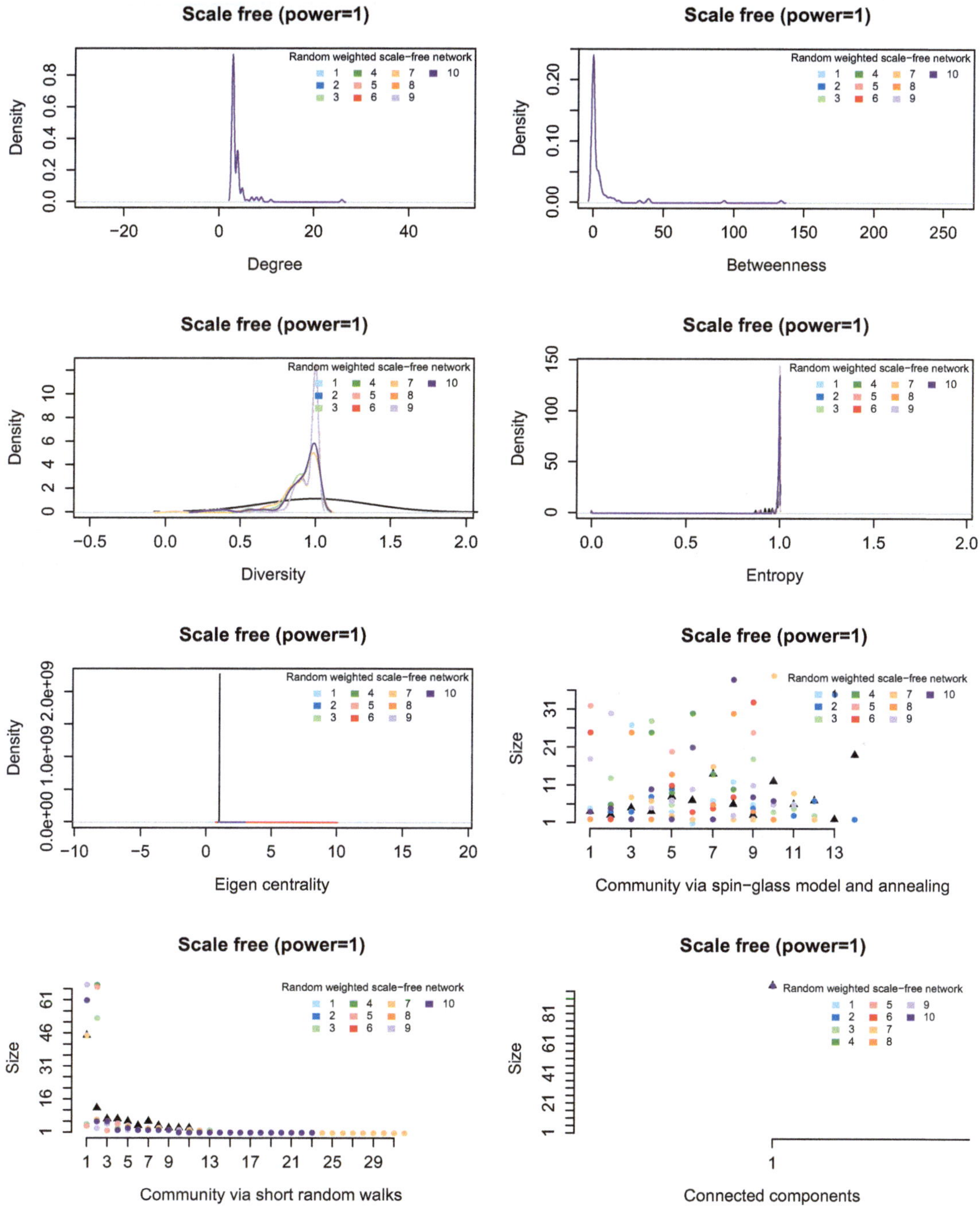

Fig. 5 Figure displaying the distribution of the degree, betweenness, diversity, entropy, and eigenvector centrality measures for the non-weighted scale-free power law graph ($g = 1$) and its weighted variants. The distribution for the unweighted variant is shown in black. Furthermore, the figure displays the number and size of communities detected either by a spin-glass model and simulated annealing or by random walks. Finally, the last plot displays the unique connected component detected by the decompose function

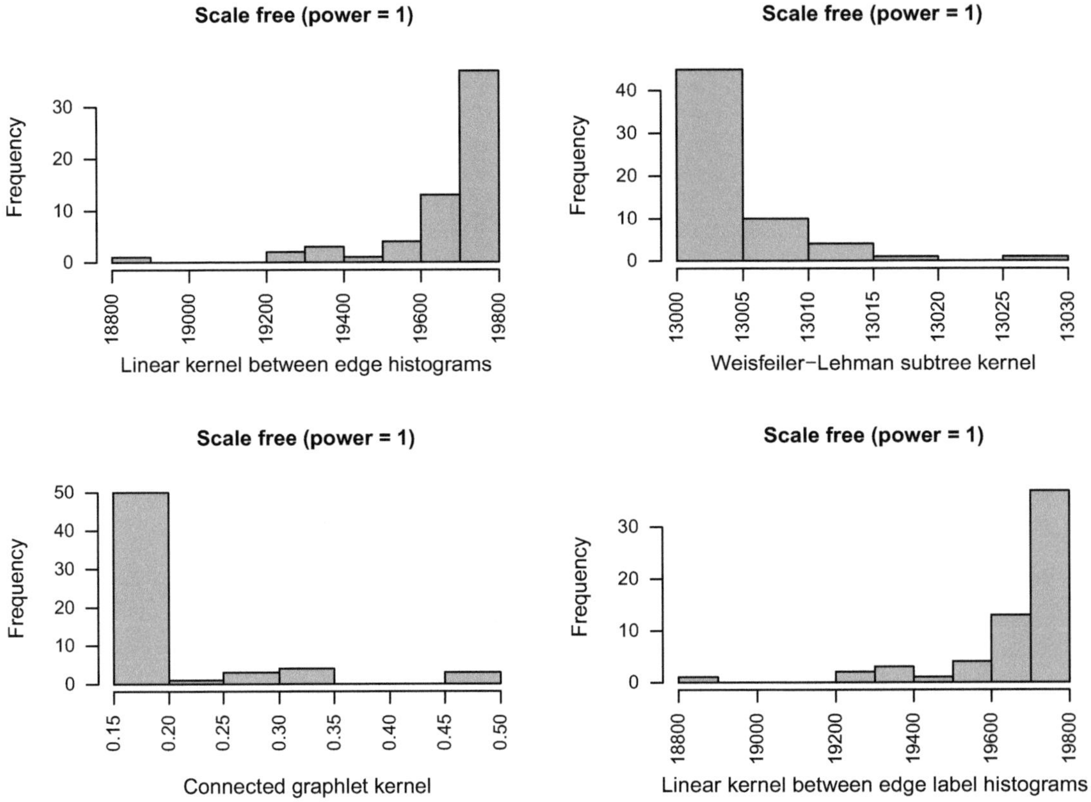

Fig. 6 Euclidean distances between TD-WGcluster connected components and the unique connected components (identical to the entire graph) returned by igraph `decompose` function for scale-free graphs ($g = 1$)

previous examples even if the disproportion of the size of compared graphs is high. Finally, it is worth noting this remarkable difference in rendering graph internal organization is due to the fact that the concept of node proximity underlying the graph decomposition performed by TD-WGcluster is not applied to a simplicial 1-complex but to a multidimensional topological space which includes the temporal dimension besides the spatial ones. As a matter of fact, the geometrical entropy used in TD-WGcluster affords the quantification of the statistical significance by accounting both for the topological structure and for the dynamical nature (deterministic or stochastic) the nodes in the connected components are endowed with. We wish to conclude by suggesting any reader to select a community structure detection algorithm on a case-by-case basis according to the available data types and the desired amount of informativeness on the resulting graph decomposition.

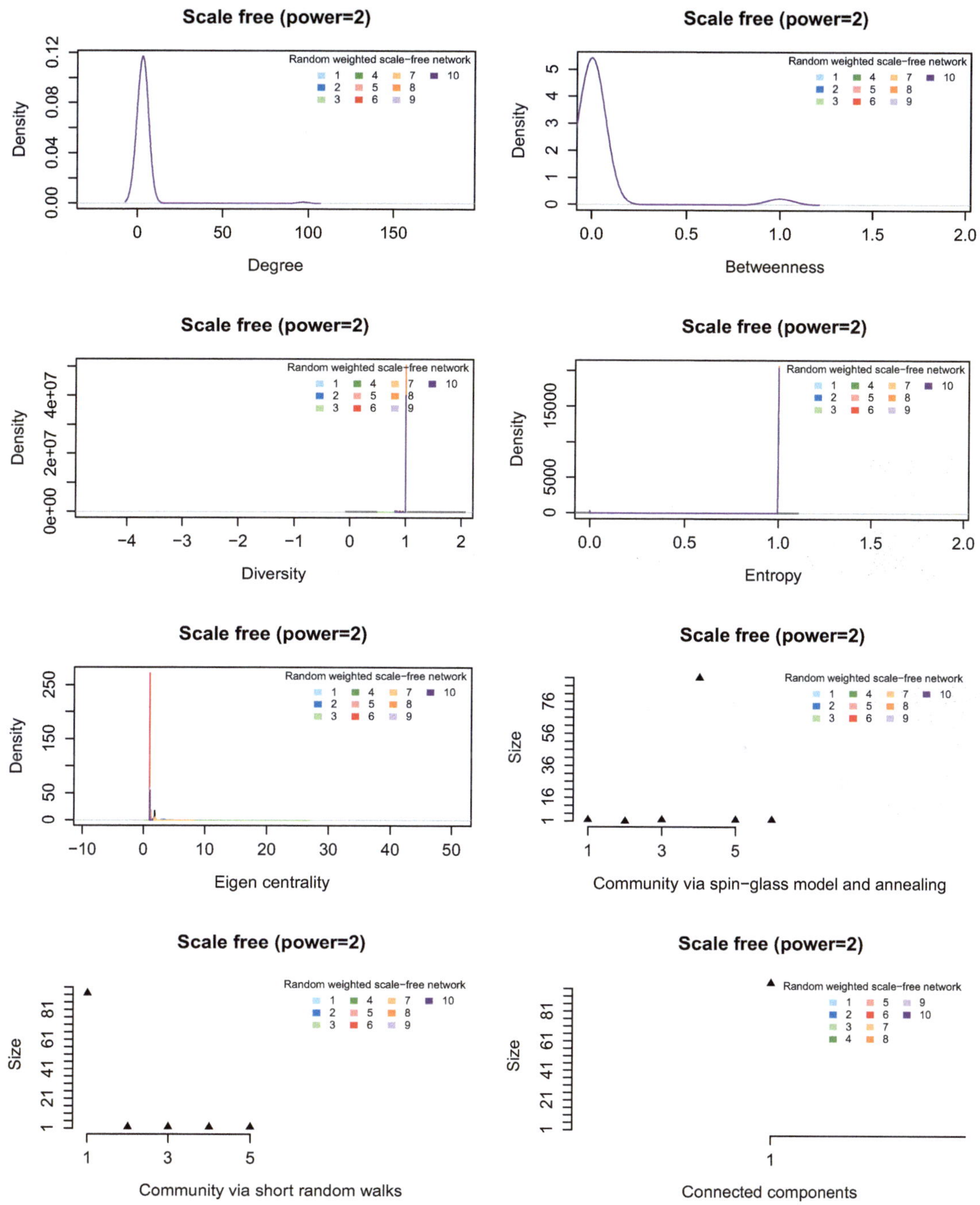

Fig. 7 Figure displaying the distribution of the degree, betweenness, diversity, entropy, and eigenvector centrality measures for the non-weighted scale-free power law ($g = 2$) network and its weighted variants. The distribution for the unweighted variant is shown in black. Furthermore, the figure displays the number and size of communities detected either by a spin-glass model and simulated annealing or by random walks. Finally, the last plot displays the unique connected component detected by the decompose function

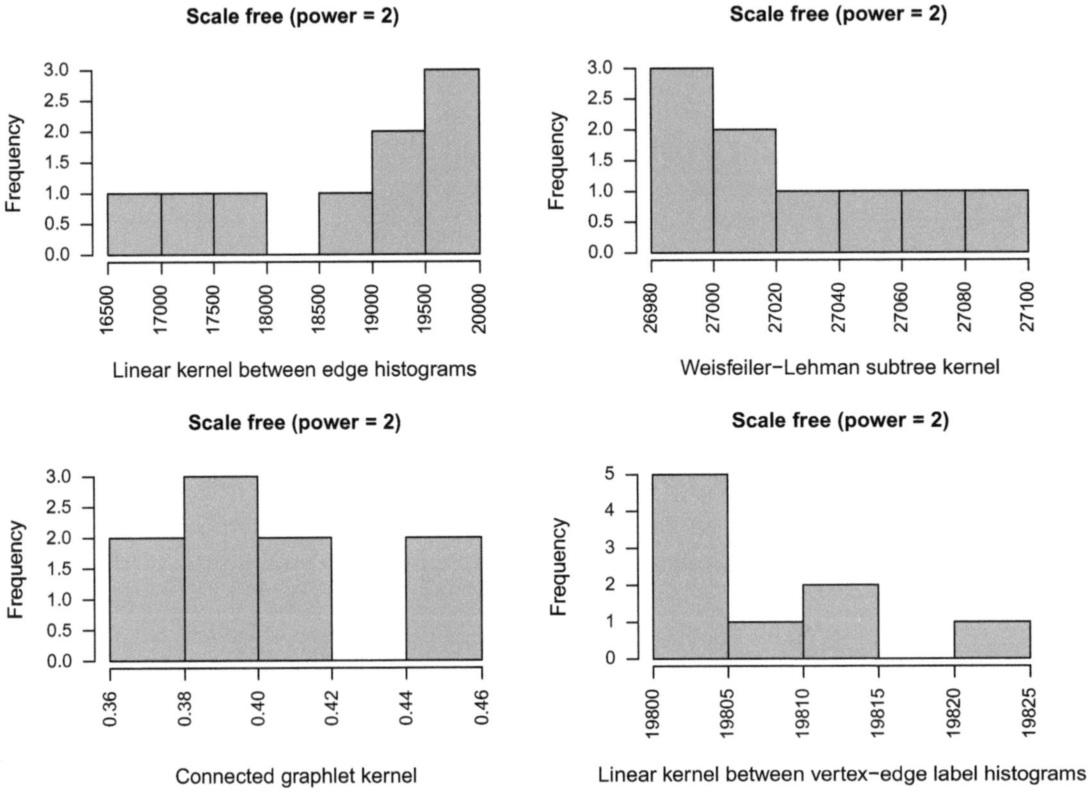

Fig. 8 Euclidean distances between TD-WGcluster connected components and the unique connected components (identical to the entire graph) returned by `igraph decompose` function for scale-free graphs ($g = 2$)

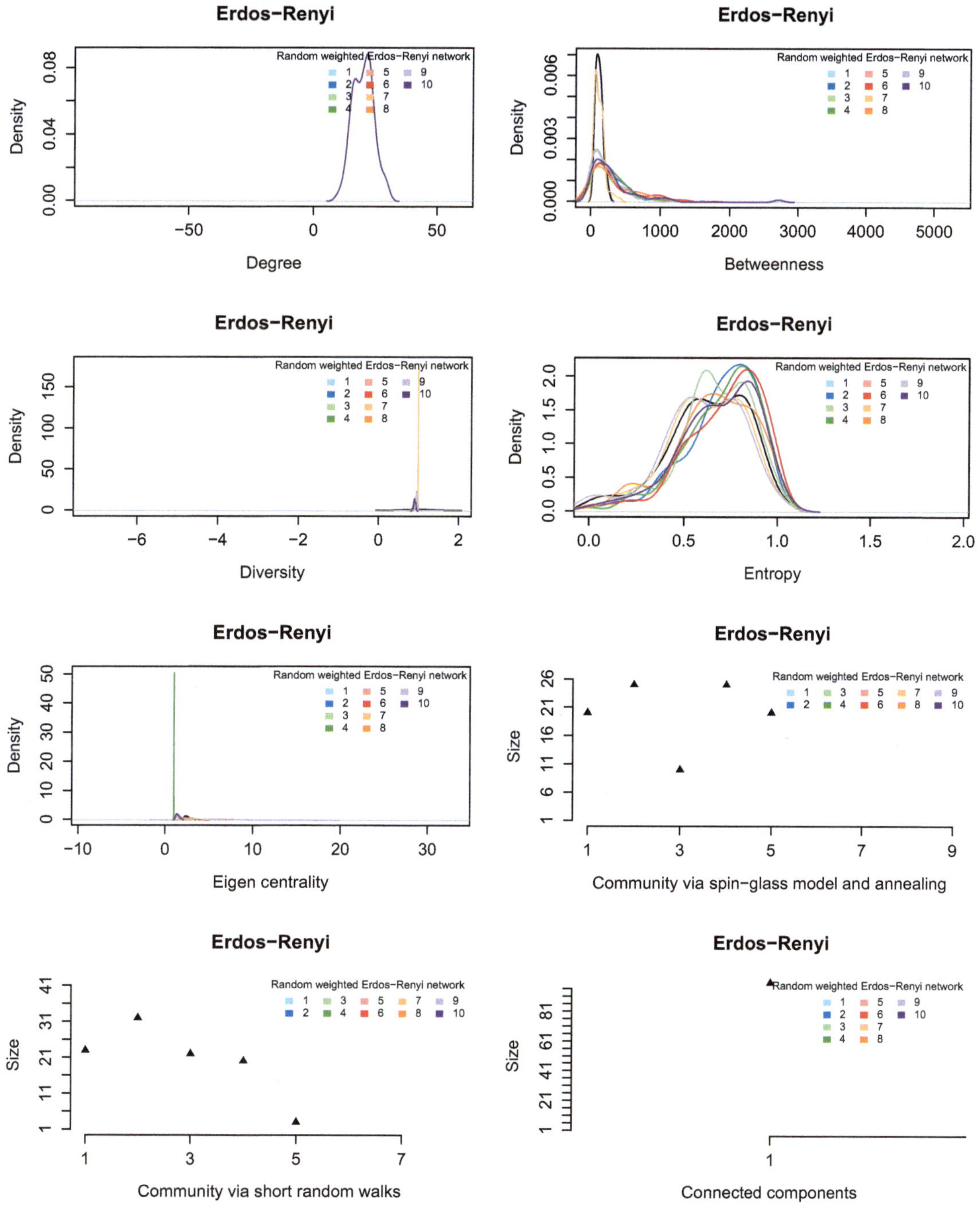

Fig. 9 Figure displaying the distribution of the degree, betweenness, diversity, entropy, and eigenvector centrality measures as well as summarizing statistics on the decomposition of non-weighted Erdös–Rényi network and its weighted variants into connected components. The distribution for the unweighted variant is shown in black

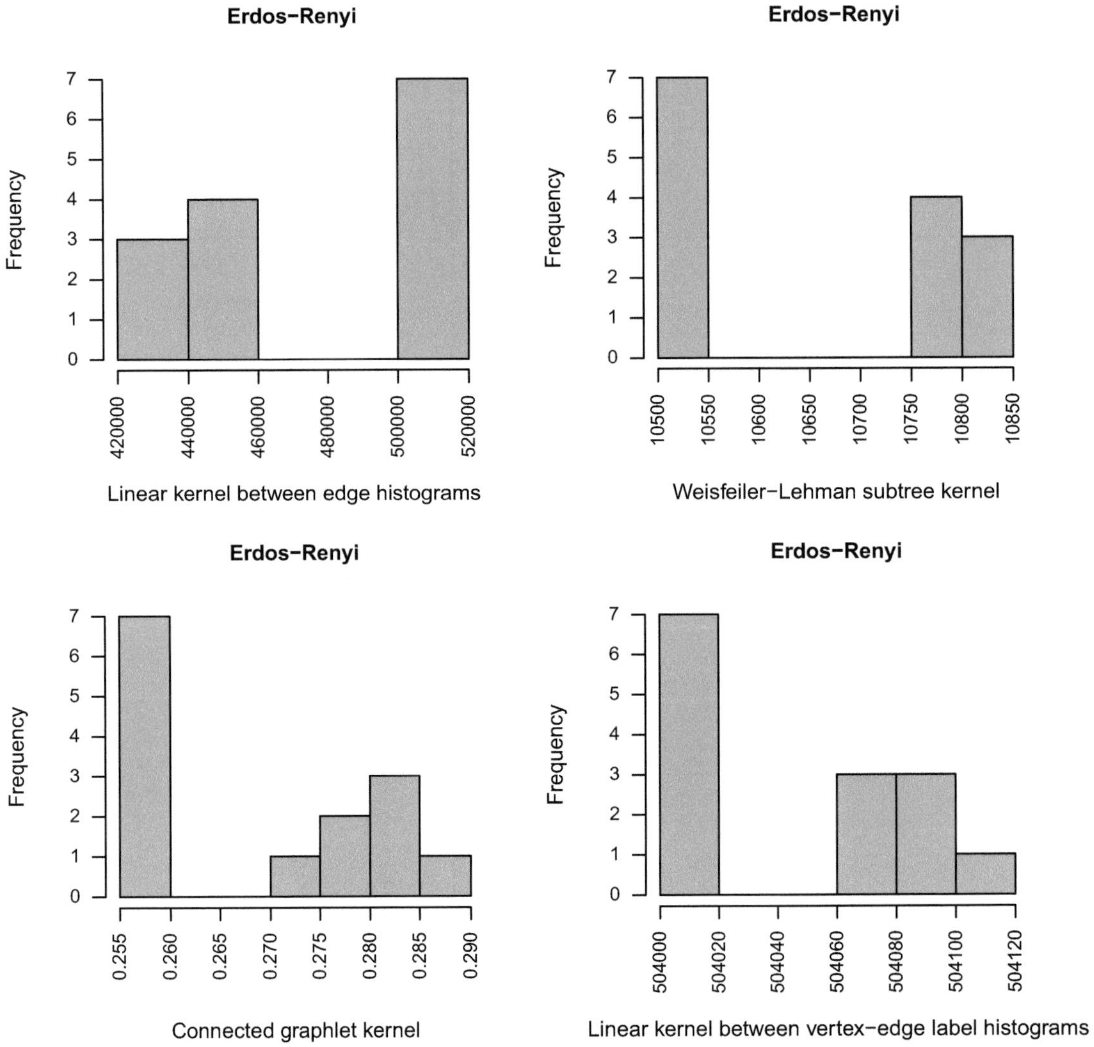

Fig. 10 Euclidean distances between TD-WGcluster connected components and the unique connected components (identical to the entire graph) returned by `igraph decompose` function for Erdös–Rényi graph

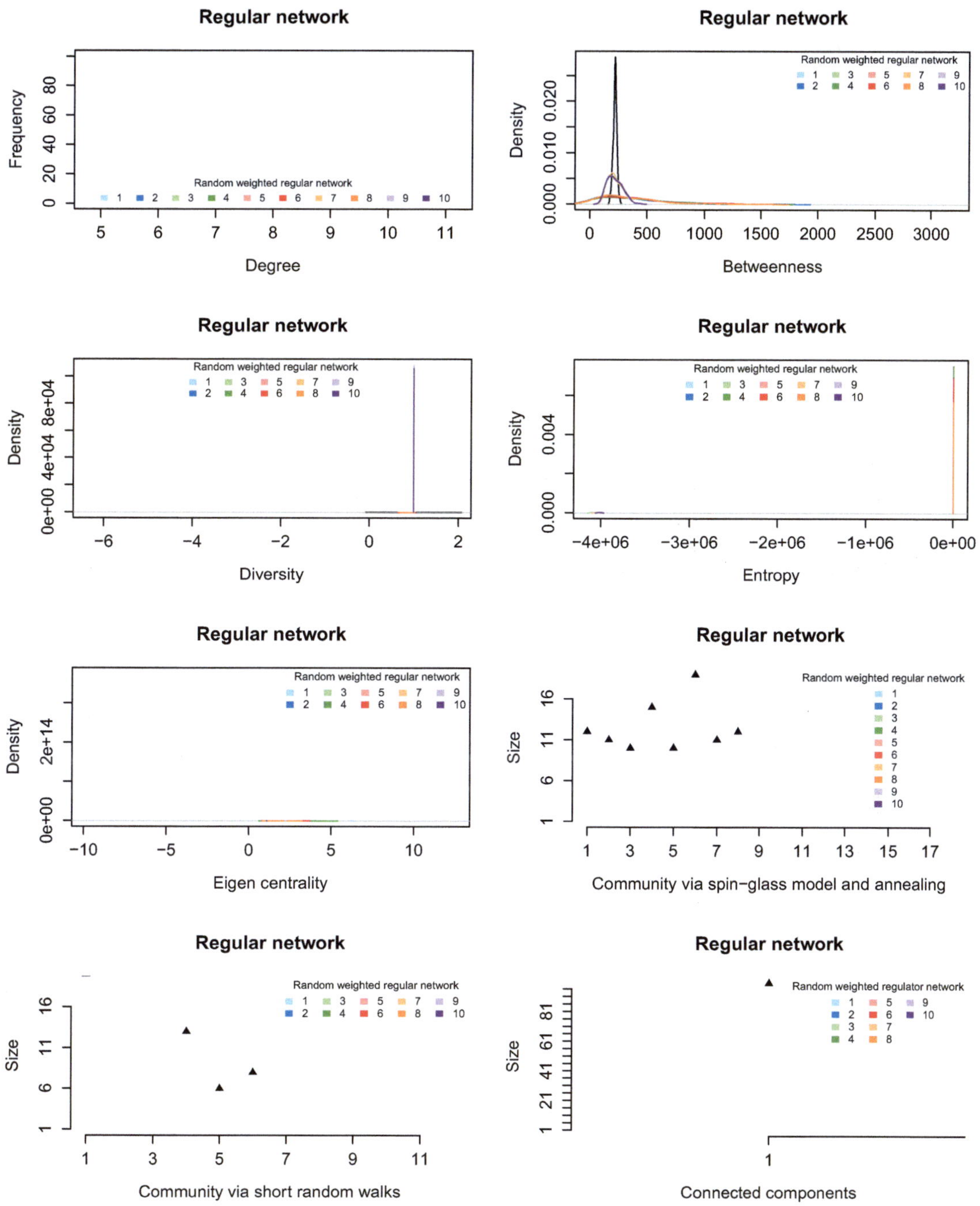

Fig. 11 Figure displaying the distribution of the degree, betweenness, diversity, entropy, and eigenvector centrality measures as well as summarizing statistics on the decomposition of non-weighted Erdös–Rényi network and its weighted variants into connected components. Note that the degree for regular directed networks of $k = 4$ is equal to 8 overall nodes. The distribution for the unweighted variant is shown in black

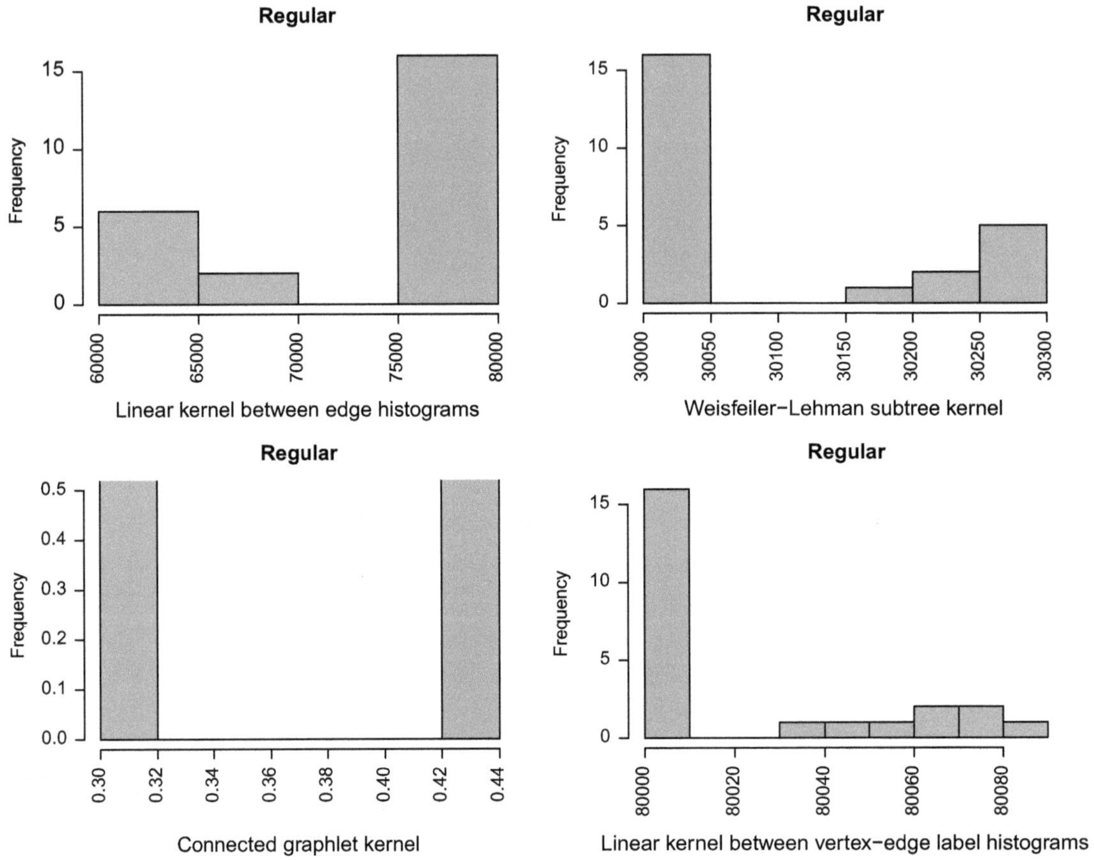

Fig. 12 Euclidean distances between TD-WGcluster connected components and the unique connected components (identical to the entire graph) returned by `igraph decompose` function for a regular graph

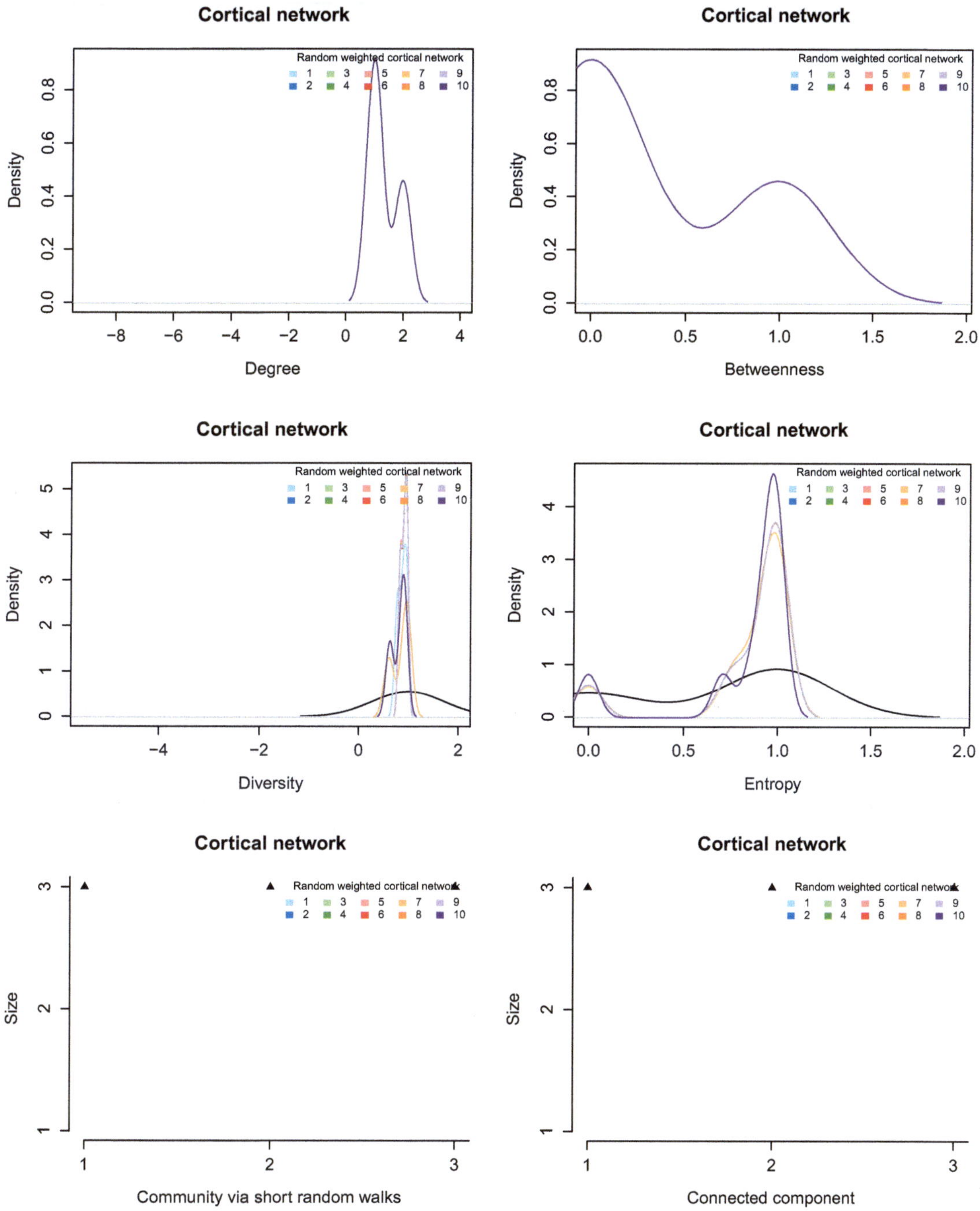

Fig. 13 Figure displaying the distribution of the degree, betweenness, diversity, entropy, and eigenvector centrality measures as well as summarizing statistics on the decomposition of non-weighted cortical network and its weighted variants into connected components. The distribution for the unweighted variant is shown in black

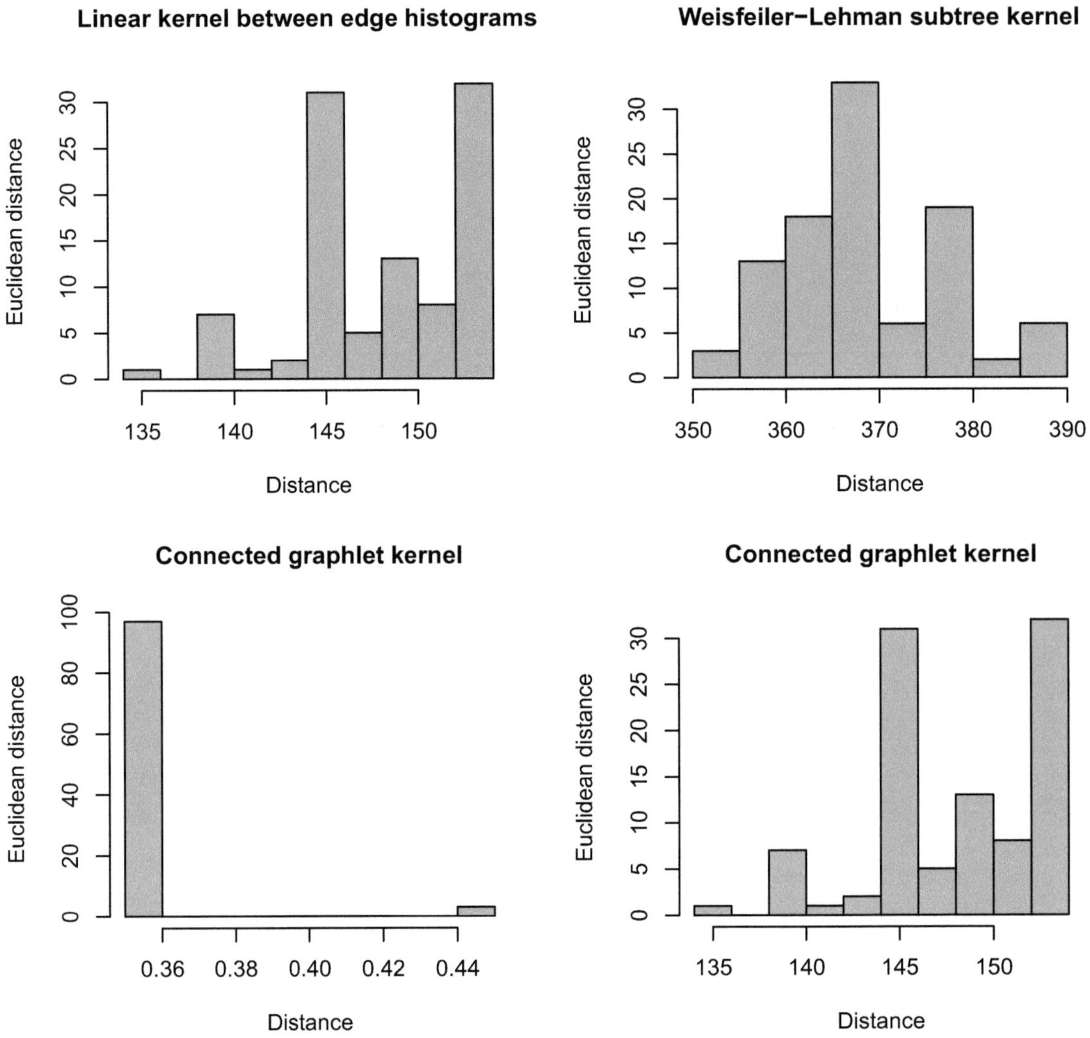

Fig. 14 Distributions of the distances of 100 randomly sub-graphs extracted from a random graph of 28 nodes and 26 edges from the graph itself

References

1. Barabási AL, Albert R (1999) Emergence of scaling in random networks. Science 286 (5439):509–512. https://doi.org/10.1126/science.286.5439.509. http://science.sciencemag.org/content/286/5439/509
2. Barahona M, Poon CS (1996) Detection of nonlinear dynamics in short, noisy time series. Nature 381:215–217
3. Berenhaut KS, Barr PS, Kogel AM, Melvin RL (2018) Cluster-based network proximities for arbitrary nodal subsets. Sci Rep 8(1). https://doi.org/10.1038/s41598-018-32172-0
4. Bilgin C, Yener B (2012) Dynamic network evolution: models, clustering, anomaly detection. https://www.cs.rpi.edu/research/pdf/08-08.pdf. Accessed 27 Nov 2015
5. Brandes U (2001) A faster algorithm for betweenness centrality. J Math Sociol 25 (2):163–177. https://doi.org/10.1080/0022250X.2001.9990249
6. Chebotarev PY, Shamis EV (1998) On the duality of metrics and σ-neighborhoods. Autom Remote Control 59:608–612

7. Clauset A (2005) Finding local community structure in networks. Phys Rev E Stat Nonlin Soft Matter Phys 72(2 Pt 2):026132
8. Csardi G, Nepusz T (2006) The igraph software package for complex network research. InterJournal Complex Syst 1695. http://igraph.org
9. Du Q, Emelianenkom M, Ju L (2006) Convergence of the Lloyd algorithm for computing centroidal Voronoi tessellation. SIAM J Numer Anal 44(1):102–119. http://www.personal.psu.edu/qud2/Res/Pre/dej06sinum.pdf
10. Eagle N, Claxton NMR (2010) Network diversity and economic development. Science 328:1029–1031. https://doi.org/10.1126/science.1186605. http://science.sciencemag.org/content/328/5981/1029
11. Felice D, Mancini S (2015) Gaussian network's dynamics reflected into geometric entropy. Entropy 17(8):5660–5672. https://doi.org/10.3390/e17085660
12. Grindrod P, Parsons MC, Higham DJ, Estrada E (2011) Communicability across evolving networks. Phys Rev E 83:046120. https://doi.org/10.1103/PhysRevE.83.046120. https://link.aps.org/doi/10.1103/PhysRevE.83.046120
13. Harvey AC (1993) Time series models. Harvester Wheatsheaf, New York
14. Hazewinkel M (ed) (2001) "Asymmetry coefficient", Encyclopedia of mathematics. Springer, Berlin
15. Hubbell CH (1965) An input-output approach to clique identification. Sociometry 28:377–399
16. igraph R. Reference manual of igraph. https://igraph.org/c/doc/igraph-Community.html. Accessed 02 Jan 2019
17. igraph R. Web page of manual of (igraph) cluster_spinglass function. https://igraph.org/r/doc/cluster_spinglass.html. Accessed 15 Dec 2018
18. igraph R. Web page of manual of (igraph) decompose function. https://igraph.org/r/doc/decompose.html. Accessed 01 Nov 2018
19. igraph R. Web page of manual of (igraph) graphkernels function. https://www.rdocumentation.org/packages/graphkernels/versions/1.6. Accessed 15 Nov 2018
20. igraph R. Web page of manual of (igraph) walktrap.community function. https://www.rdocumentation.org/packages/igraph/versions/0.5.1/topics/walktrap.community. Accessed 01 Dec 2018
21. Johannesson T, Bjornsson H (2012) Stineman, a consistently well behaved method of interpolation. http://rpackages.ianhowson.com/cran/stinepack/. Accessed 07 Jan 2015
22. Katz L (1953) A new status index derived from sociometric analysis. Psychometrika 18:30–43. https://doi.org/10.1007/BF02289026. https://link.springer.com/article/10.1007/BF02289026#citeas
23. Lancichinetti A, Fortunato S, Kertész J (2009) Detecting the overlapping and hierarchical community structure in complex networks. New J Phys 11(3):033015. https://doi.org/10.1088/1367-2630/11/3/033015
24. Lecca P (2009) Deterministic chemical chaos identification in models and experiments. In: International conference on bioinformatics & computational biology, BIOCOMP 2009, July 13–16, 2009, Las Vegas Nevada, 2 Volumes, pp 307–312
25. Lecca P, Re A (2016) Module detection in dynamic networks by temporal edge weight clustering. In: Angelini SRC, Rancoita P (ed) Computational intelligence methods for bioinformatics and biostatistics, CIBB 2015. Lecture notes in computer science, vol 9874. Springer, Cham, pp 54–70
26. Luo F, Wang J, Promislow E (2006) Exploring local community structures in large networks. In: Proceedings of the 2006 IEEE/WIC/ACM international conference on web intelligence, pp 233–239. https://doi.org/10.1109/WI.2006.72
27. Makridakis SG, Wheelwright SC, Hyndman RJ (1998) Forecasting: methods and applications. Wiley, New York
28. Pons P, Latapy M (2005) Computing communities in large networks using random walks. In: Computer and information sciences - ISCIS 2005. Springer, Berlin, pp 284–293. https://doi.org/10.1007/11569596_31
29. Reichardt J, Bornholdt S (2006) Statistical mechanics of community detection. Phys Rev E 74:016110. https://doi.org/10.1103/PhysRevE.74.016110. https://link.aps.org/doi/10.1103/PhysRevE.74.016110
30. Shanafelt DW, Salau KR, Baggio JA (2017) Do-it-yourself networks: a novel method of generating weighted networks. R Soc Open Sci 4(11). https://doi.org/10.1098/rsos.171227
31. Shervashidze N, Vishwanathan SVN, Petri T, Mehlhorn K, Borgwardt KM (2009) Efficient graphlet kernels for large graph comparison. In: Proceedings of the 12th international conference on artificial intelligence and statistics (AISTATS), vol 5, pp 488–495. http://proceedings.mlr.press/v5/shervashidze09a.html

32. Shervashidze N, Schweitzer P, van Leeuwen EJ, Mehlhorn K, Borgwardt KM (2011) Weisfeiler-Lehman graph kernels. J Mach Learn 12:2539–2561. http://www.jmlr.org/
33. Sugiyama M, Borgwardt KM (2015) Halting in random walk kernels. In: Advances in neural information processing systems (NIPS 2015), vol 28, pp 1630–1638. https://papers.nips.cc/paper/5688-halting-in-random-walk-kernels.pdf
34. Teraesvirta T, Lin CF, Granger CWJ (1993) Power of the neural network linearity test. J Time Ser Anal 14:209–220. https://doi.org/10.1103/PhysRevE.70.066111
35. www.sci.unich.it: Katz centrality. https://www.sci.unich.it/~francesc/teaching/network/katz.html. Accessed 15 Nov 2018
36. Yang L, Xin-Sheng J, Caixia L, Ding W (2014) Detecting local community structures in networks based on boundary identification. Math Probl Eng 2014:1–8. https://doi.org/10.1155/2014/682015

Chapter 18

An Introductory Guide to Aligning Networks Using SANA, the Simulated Annealing Network Aligner

Wayne B. Hayes

Abstract

Sequence alignment has had an enormous impact on our understanding of biology, evolution, and disease. The alignment of biological *networks* holds similar promise. Biological networks generally model interactions between biomolecules such as proteins, genes, metabolites, or mRNAs. There is strong evidence that the network topology—the "structure" of the network—is correlated with the functions performed, so that network topology can be used to help predict or understand function. However, unlike sequence comparison and alignment—which is an essentially solved problem—network comparison and alignment is an NP-complete problem for which heuristic algorithms must be used.

Here we introduce SANA, the *Simulated Annealing Network Aligner*. SANA is one of many algorithms proposed for the arena of biological network alignment. In the context of global network alignment, SANA stands out for its speed, memory efficiency, ease-of-use, and flexibility in the arena of producing alignments between two or more networks. SANA produces better alignments in minutes on a laptop than most other algorithms can produce in hours or days of CPU time on large server-class machines. We walk the user through how to use SANA for several types of biomolecular networks.

Key words Network alignment, Biological networks, Simulated annealing

1 Introduction

A biological network consists of a set of nodes representing entities, with edges connecting entities that are related in some way. They come in many varieties, such as protein–protein interaction (PPI) networks [1, 2], gene regulatory networks [3, 4], gene-μRNA networks [5–9], metabolic networks [10], brain connectomes [11], and many others [12]. It is believed that the *structure* of the networks, in the form of the network topology, is related to the function of the entities [3, 13, 14]. The *alignment* of such networks aims to use connectivity between nodes—the *topology* of the network—to aid extraction of information about the nodes and their function. Network alignments can be used to build taxonomic trees and find highly conserved pathways across distant species

Stefan Canzar and Francisca Rojas Ringeling (eds.), *Protein-Protein Interaction Networks: Methods and Protocols*, Methods in Molecular Biology, vol. 2074, https://doi.org/10.1007/978-1-4939-9873-9_18,

[15]; and by extension finding such topological similarities may aid in transferring functional knowledge from better-understood species to less well-understood ones, much like how sequence alignment has been doing so for sequence for decades. Networks are even starting to have an influence on individual human health [16].

Network alignment is a fundamentally difficult problem: it is a generalization of the NP-complete subgraph isomorphism problem [17, 18]; and adding to the difficulty is that current data sets are very noisy [19]. Therefore, modern alignment algorithms try to approximate solutions using heuristic approaches.

There are several sub-classes of network alignment. *Global Network Alignment* (GNA) is the task of attempting to completely align entire networks to each other; GNA applied to just two networks is called *pairwise* GNA [15, 20–25], while aligning more than two whole networks is called *multiple* GNA. In contrast, *Local Network Alignment* (LNA) attempts to find similarity in the local wiring patterns among small groups of nodes, either in the same network or across many networks. In all of these cases, alignments can map nodes 1-to-1 or many-to-many; the latter is more biologically realistic since, for example, one gene in yeast may have multiple homologs in mammals. However, the 1-to-1 assumption makes programming simpler and so the majority of aligners take the 1-to-1 mapping as a simplifying assumption. A more recent version of network alignment looks into modeling dynamic networks (see, for example, [26]). An excellent comprehensive survey of all these types of alignments is provided by Faisal et al. [27]. SANA was originally a 1-to-1 pairwise global network alignment algorithm, although we here also introduce a prototype multiple network alignment version.

1.1 User/System Requirements

Source code to SANA is available on GitHub at

http://github.com/waynebhayes/SANA, and is best cloned from GitHub on the Unix command line using

`git clone` http://github.com/waynebhayes/SANA

SANA is written in C++ and runs best on the Unix command line. It has been tested with gcc 4.8, 4.9, 5.2, and 5.4, and runs on Unix, Linux, Mac OS/X, and under the Windows-based Unix emulator Cygwin (http://cygwin.com), 32-bit or 64-bit. SANA has a rudimentary Web interface at http://sana.ics.uci.edu, and a rudimentary SANA app is available in the Cytoscape app store. SANA expects its input networks to be in a two-column ASCII format we call *edge list format*: each line is one edge, specified by listing the two nodes at each end of the edge in arbitrary order (unless `-bipartite` is specified, see below). Duplicate edges and self-loops are not allowed. We also supply a program called `createEdgeList` that can convert the following types of formats into SANA's edgeList format: XML, GML, LEDA, .gw, CSV, LGF.

1.2 Alignment Measures and Objective Functions

An *alignment measure* is any quantity designed to evaluate the quality of a network alignment. Alignment measures can be classified along many axes.

1.2.1 Objective vs. Non-objective Measures

The first axis is the distinction between *objectives* and what we call *post-hoc* measures. While both can be evaluated on any given alignment, any measure used to *guide* an alignment *as it is being created* is called an *objective function*; any measure not used to guide the alignment is generally applied after-the-fact as an independent measure of quality. A good alignment algorithm should be able to use virtually any measure as an objective, and also evaluate the alignment after-the-fact using any other measures which were not used as objectives.

1.2.2 Graph Topology vs. Biological Measures

Another axis along which measures can be classified is *topological* vs *biological* measures.

A Topological Measure quantifies a network alignment based solely on graph-theoretic grounds. Several such measures exist: *EC* [15], *ICS* [25], and S^3 [21] quantify the number of edges in one network that are mapped to edges in the other network(s); they are all described in more detail below. Other topological measures use graphlets to quantify local structure [20, 28–30], while still others use graph measures such as spectral analysis [25] and degree similarity-based measures such as Importance [23].

Biological Measures: In contrast, biological measures are usually used to compare the nodes from different networks that have been paired together by the alignment. For genes or proteins, a common measure is the sequence similarity or BLAST score between the aligned nodes [31]; sequence similarity is also frequently combined with topology to produce a hybrid objective function (see, for example, [20–22, 32], among many others). Another biology-based measure is the *functional* similarity between pairs of aligned proteins as expressed by Gene Ontology (GO) terms [33]. While many authors quantify the functional similarity exposed by an alignment using the mean value of various pairwise GO similarity measures across the alignment, such mean-of-pairwise-scores assume each pair of aligned proteins is independent of all others, which is not true in an alignment since every pair is implicitly related to every other pair via the alignment itself. This problem is alleviated by the NetGO score as implemented in SANA [34], which is a global rather than local scoring mechanism (see below for the meaning of local vs. global measures).

1.2.3 Local vs. Global Measures

The final axis along which network alignment measures can be classified is what we refer as *local* vs. *global* measures.

A Local Measure is one that involves evaluating node pairs that are aligned to each other, and has no explicit dependence on the alignment edges and thus has no explicit dependence on network topology. Examples of local measures include sequence similarity and pairwise GO term similarity as described above; some local measures such as graphlet similarity [15, 20, 21] and Importance [23] include topology indirectly by pre-computing all-by-all pairwise local topological similarities between all pairs of nodes in one network and all pairs of nodes in the other.

Global Measures are ones that implicitly or explicitly can be computed only on the entire alignment and have nothing to do with pairwise node similarities. The most common global measures are *EC*, *ICS*, and S^3, described in more detail below.

1.3 Major Topological Measures

1.3.1 A Useful Analogy for Topological Measures

In order to more easily understand and discuss topological measures, we introduce an analogy between pairwise network alignment and the old board game of *Battleship*. A Battleship game consists of many holes in a board, and some pegs that are placed into the holes. In our analogy, assume G_1 is a "smaller" network with n_1 nodes and m_1 edges, and G_2 is a "larger" network with n_2 nodes and m_2 edges, and we assume that $n_1 \leq n_2$—that is, G_1 is the smaller network in terms of number of nodes. We will furthermore depict G_1 as blue and G_2 as red. Consider Fig. 1: this board has $n_2 = 6$ red holes with red edges painted between two holes if there is an edge between the two corresponding nodes in G_2. The smaller network G_1 is represented by $n_1 = 4$ blue pegs; edges between the pegs are represented by blue "laser beams" between the corresponding pegs (because laser beams don't get tangled as pegs are moved from hole to hole). Any placement of the n_1 pegs

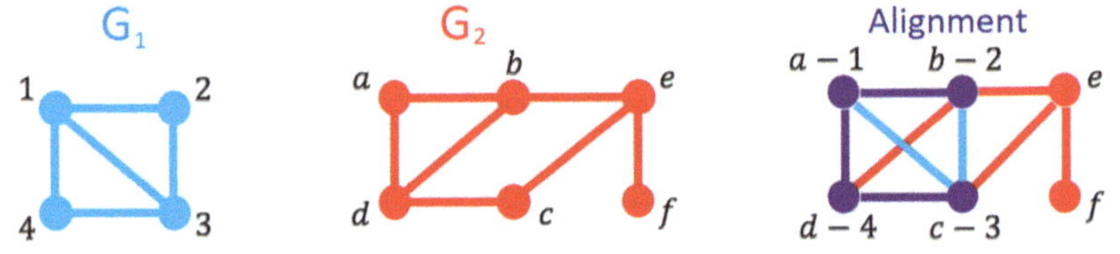

Fig. 1 A simple example of a network alignment. The smaller network G_1 (far left) has its pegs, numbered 1–4, and edges ("laser beams") depicted in blue; the larger network G_2 (middle) has its holes and painted edges depicted in red. One possible alignment (in this case the "visually obvious" one) is depicted at the far right. Here, aligned nodes and edges are depicted as purple; unaligned laser beams from G_1 are still blue, and unaligned holes and edges from G_2 are still red. As stated in the text, in an alignment figure like the one on the right, the number of edges in G_1 is always $m_1 =$ (purple + blue edges), and the number of edges in G_2 is always $m_2 =$ (purple + red edges). Thus, from the figure, it can be easily seen that $EC = 3/5$, and $S^3 = 3/6$ (where 6 is the total number of edges visible across all colors on the subgraph induced by the alignment); also $ICS = 3/4$, since there are 4 edges induced in G_2 by the alignment (i.e., by purple nodes). The purple network is called the *Common Subgraph*, and it can consist of several connected components. In this case, there is only one *Common Connected Subgraph* consisting of 4 nodes and 3 edges

into the n_2 holes represents an alignment between G_1 and G_2; for now we assume that each peg is placed into exactly one hole, so that there are exactly $n_2 - n_1$ empty holes. Furthermore, since mixing red and blue creates purple, we depict the alignment (far right of Fig. 1) in purple: a blue peg in a red hole is purple, and a blue edge lying on top of a red one is also depicted as purple.

1.3.2 Edge-Based Measures: EC, ICS, S^3

We can now define some edge-based topological measures based on this analogy. The fraction of laser beams that lie on top of painted edges is called the *EC* (Variously called Edge Coverage, Edge Correspondence, or Edge Correctness by various authors) [15]. The numerator of *EC* is the number of (purple) edges that are aligned between the two networks, call it *AE* (an integer), while the denominator is m_1; note that since at most m_1 edges can be aligned, the value $EC = AE/m_1$ is always less than or equal to 1. The authors of MAGNA [21] noted that *EC* is asymmetric: in particular, if $n_1 = n_2$, then we can "turn the board upside down," swapping the roles of pegs and holes. In that case, the *EC* changes because G_1 and G_2 are swapped: in particular, the numerator is always the number of aligned edges *AE*, but the denominator switches from m_1 to m_2.

The authors of MAGNA fixed the asymmetry of *EC* by introducing the *Symmetric Substructure Score* or S^3. Consider the rightmost section of Fig. 1, which depicts a proposed alignment. In our analogy, if we "look down" on the alignment from above, we can see four different types of edges. They are: (1) *AE* aligned (purple) edges; (2) UE_1 unaligned (blue) edges from G_1; (3) UE_{2in} unaligned (red) edges in G_2 induced between purple nodes; and (4) UE_{2out} unaligned (red) edges outside the alignment (i.e., not induced between purple nodes). Note that the following equations always hold: $m_1 = AE + UE_1$ and $m_2 = AE + UE_{2in} + UE_{2out}$. Whereas $EC = AE/m_1$, S^3 is defined as $AE/(AE + UE_1 + UE_{2in})$, and is thus symmetric with respect to the interchange of G_1 and G_2. Another way of saying this is that both *EC* and S^3 are rewarded for purple edges in the numerator, but *EC*'s denominator is penalized only for blue edges in its denominator, whereas S^3 is penalized in its denominator for both blue and red edges induced by the alignment.

Another measure called ICS *Induced Conserved Substructure* [25] measures *AE* divided by the number of painted edges that exist only between holes that have pegs in them. ICS has the significant disadvantage that it can be maximized by finding a network alignment that *minimizes* the number of edges between filled holes [21, 22, 35], which can hardly be said to be a good alignment. Consider again Fig. 1. The reason ICS is a bad measure is because we could make it equal to 2/2, i.e. 1, by moving node 2 to align with *e* and 3 to align with *f*; then there would be 2 purple edges (*a*-1 to *d*-4, and *e*-2 to *f*-3) and no red edges *induced by the alignment* on G_2, even though there would be 3 blue edges (1–2, 4–3, and 1–3) unaligned from G_1. Thus there exists an alignment with $ICS = 1$ even though it only exposes 2 edges of common topology, which is less common topology discovered by maximizing either EC or S^3. This demonstrates the general principle that *choosing the right objective function* ***is crucial*** *to getting good alignments.*

1.3.3 Graphlet-Based Measures

Graphlets [28, 36] are small, connected, induced subgraphs on a larger graph. They have myriad uses, such as quantifying global topological structure [28, 30]. Enumerating graphlets in a large graph is an *NP*-hard problem and much work has gone into heuristics to make their enumeration more efficient. SANA uses ORCA [37] to exhaustively enumerate graphlets in a network. By computing an *orbit degree vector* [29], one can create a local measure that compares the orbit degree vectors of two nodes (one from each network); that local measure can then be used as an objective to guide the alignment. GRAAL [15] was the first to use orbit degree vectors. (In the GRAAL paper, we used the term "graphlet degree vector" but it's more correctly called an "orbit degree vector" because it's a vector of orbit counts, not graphlet counts.) SANA uses the exact same mechanism; however, as networks grow larger, the exhaustive enumeration of its graphlets is becoming very expensive. For example, ORCA takes more than 24 h to compute the orbit degree vectors when aligning the 2018 BioGRID [38] networks of *H. sapiens* and *S. cerevisiae*. Instead, we intend to move SANA towards statistical sampling of graphlets which can be accomplished *far* faster and produce results with low frequency error and high confidence (see, for example, [39–41]).

1.3.4 Which Topological Score to Use?

We believe that one of the major outstanding questions in network alignment is the design of good topological objective functions. While most measures that currently exist have been shown to correlate with interesting biological information, none have been shown to be substantially better than any other in terms of recovering relevant biology. For example, while S^3 is symmetric and can thus be considered a more aesthetically pleasing measure from a mathematical standpoint, it's by no means clear that it actually produces better correlations with biology than EC. And while graphlets have been shown to correlate with biological information [13, 15, 20], it is not clear that we know the best way to *use* them to recover the greatest amount of relevant biological information (cf. Subheading 3.1, especially Table 4). In general, the design of good topological objective functions is a wide-open area of research that deserves to be explored. SANA, with its speed and accuracy, is an ideal playground for exploring objective functions.

The *Software* Development Cycle

1. Edit source of program *P* to implement ideas/changes/fix bugs to so it implements your science goal.
2. Compile *P*: create correct, efficient executable *E(P)* implementing *P* at machine level.
3. Run *E(P)*, producing output.
4. Evaluate output, decide if *P* did what you wanted or expected.
5. Think how to modify *P* to better obtain your science goal.
6. Go back to step 1 (or possibly change science goal).

Proposed *Alignment Objective* Development Cycle

1. Edit objective function *F* to implement ideas/changes/fix bugs to so it implements your science goal.
2. Create an efficient algorithm *S* that optimizes the objective *F(A, G1,G2)* across all possible alignments *A*.
3. Run *S(F, G1, G2)*, producing alignment *A*.
4. Evaluate alignment *A*, decide if *F* did what you wanted or expected
5. Think how to modify *F* to better obtain your science goal.
6. Go back to step 1 (or possibly change science goal).

Fig. 2 Comparison of the standard software development cycle (left) and proposed cycle for developing new objective functions for alignment (right). Red highlights the step that should be entirely automated and requiring no effort on the user's part

To explain what we mean by experimenting with objective functions, consider Fig. 2. There are three orthogonal components to network alignment: (1) a (possibly vague) scientific or informational goal *G*; (2) an objective function *M* created by the user that attempts to formally encode *G*; and (3) an alignment algorithm *S* that builds an alignment trying to optimize *M*. In sequence alignment, the three orthogonal components are clearly delimited: the substitution/indel cost matrix encodes the goal the user wants, and tools like BLAST [31] quickly find (near-)optimal solutions. Practitioners can *use* BLAST without having to understand the details of how it works. It is a trusted tool, like a C++ compiler is to a developer, or a linear solver to a scientist solving a linear system; practitioners iterate the familiar edit-compile-debug loop, gaining knowledge from the feedback process until they are satisfied that they have achieved their goal. Unfortunately, this edit-compile-debug loop is virtually impossible in the network alignment arena, due to (a) the lack of an algorithm fast enough to perform effective edit-compile-debug loops, (b) the lack of a generally accepted "gold standard" of network alignment, and (c) the lack of a clear separation of the *goal*, its formalized *objective*, and the *alignment tool*. SANA fixes the first two; the third is a matter of scientific culture in the network alignment community that we hope to influence by spreading the use of SANA in conjunction with the process depicted in Fig. 2.

1.3.5 Using Sequence Similarity as an Objective: A Necessary but Hopefully Temporary Evil

It may help here to (re-)state the obvious: the whole point of *network* alignment is to align networks based upon their network topology. This is a desirable goal because there is a strong belief that the topology of a network is somehow related to its function. For example, we believe that humans and chimpanzees are very close relatives, taxonomically speaking. If there is a particular protein h_0 in humans that performs a certain function by interacting with seven other proteins $h_1, h_2, \ldots, h_7$, then it is quite likely that there is a very similar protein c_0 in chimpanzees that also interacts with (close to) seven proteins $c_1, c_2, \ldots, c_7$ to perform virtually the same function. Another way of saying this is that the network topology of the protein–protein interaction networks of human and chimp is likely to be very similar in the vicinity of h_0 and c_0, respectively. As such, a natural network alignment between human and chimp should contain the ordered pairs $(h_0, c_0), (h_1, c_1), \ldots, (h_7, c_7)$. If the network of interactions around h_0 and c_0 are in fact similar, then any network alignment algorithm worth its mettle, optimizing an objective that highlights such network similarities, should include the above pairs with high likelihood.

The problem, at least in the research area of protein–protein interaction (PPI) networks, is that the data on current PPI networks is *extremely* incomplete in terms of enumerating the edges in the PPI networks. For example, as of 2018, the most complete PPI network is that of *S. cerevisiae*, and it may be only about 50% complete; the human PPI network is probably less than 10% complete [42]; other species are even far less complete. For instance, we'd expect most mammals to have about the same number of interactions in their PPI networks, and yet the 2018 BioGRID Human network has almost 300,000 interactions, but mouse and rat have only 38,000 and 5000 interactions listed, respectively. If Human is only 10% complete and currently contains 300,000 interactions, then we may expect the complete interactome to have over one million interactions. By this measure, mouse and rat are at most a few percent, and well less than 1% complete, respectively. Here's the crux: if we are missing 90% or more of the edges in most mammal PPI networks, *no network alignment algorithm based solely upon network topology has any hope of providing good alignments.* This is the state of affairs in PPI network alignment.

Thus, it is no surprise that virtually every network alignment algorithm currently in existence must rely on using sequence similarity information to help give network alignments that show decent functional similarity. However, *if network alignment is of any worth whatsoever, the use of sequence similarity should be viewed only as a* **temporary** *crutch—a necessary evil—until such time as the interactions in PPI networks are more completely enumerated.*

On the other hand, since protein function is defined by the shape of the folded protein, and disrupting the function of a protein can be lethal, the folded structure of a protein tends to be better conserved than its sequence [43]. This in turn suggests that the network of interactions may also be better conserved than sequence. If this is the case, then network alignment may ultimately be at least as useful as sequence alignment in terms of learning about protein function. Alas, we must wait until PPI networks are far more complete than they are today to test this hypothesis.

1.4 Search Algorithms

Given two networks with $n_1 \leq n_2$ nodes, respectively, the number of possible 1-to-1 pairwise global network alignments between them is exactly $\frac{n_2!}{(n_2-n_1)!}$. This is an *enormous* number; for example, if the two networks each have thousands of nodes (not uncommon for protein–protein interaction networks), the number of possible alignments can easily exceed $10^{100,000}$. This is an enormous search space, *far* larger, for example, than the number of elementary particles in the known universe—which according to Wikipedia is a paltry 10^{100}.

The task of a network alignment algorithm is to search through this enormous space of possible alignments, looking for ones that score well according to one or more of the measures described in Subheading 1.2. Since network alignment is an NP-complete problem (for those who are inclined to graph theory, the proof is trivial: finding a network alignment with an EC of exactly 1 is equivalent to solving the subgraph isomorphism problem), all such algorithms must use heuristics to navigate this enormous search space. Search methods abound; several good review papers exist [11, 44–46]; for an extensive comparison specifically showing that SANA outperforms about a dozen of the best existing algorithms, *see* ref. 22. SANA is virtually unique in that it was designed from the start to be able to optimize *any* objective function, including the objective functions introduced by other researchers; a preliminary report shows that SANA outperforms over a dozen other algorithms at optimizing their own objective functions [47].

1.5 Requirements of a Good Alignment Algorithm

We believe that, in order to be of general use, a network alignment algorithm must satisfy the following properties:

Speed, If So Desired: SANA can produce better alignments in minutes that most other aligners can in hours. This is useful for many reasons: to perform test alignments; to experiment with objective functions; to perform multiple alignments of the same pair of networks in order to see which parts of the alignment, if any, come out the same each time (more on this later).

High Quality of Results, If So Desired: SANA's primary user-tunable parameter is the amount of time the user wishes to wait. While SANA can produce better alignments in 1 min on a laptop than many existing algorithms can do given hours of CPU, users can also tell SANA to spend any amount of time improving the alignment, such as 5 min, 3 h, or a week. SANA generally produces better scoring alignments with longer runtimes, although we generally see a point of diminishing returns beyond a few hours.

It Should be Simple to Use: By this we mean that if there are any algorithmic parameters that crucially control the quality of the result, those parameters should be tuned automatically without user input—in other words, the user should not need be an expert on the algorithm in order to understand how to use it. The primary parameter controlling the anneal is how long to spend annealing. By default SANA spends a minute or two automatically finding near-optimal starting and ending temperature extremes, before annealing for the amount of time specified by the user. (Another algorithm called SailMCS [48] also uses simulated annealing but fails to automatically determine good temperature extremes, and so SANA produces alignments that are far superior to those of SailMCS [47].)

Providing Confidence Estimates on the Quality of the Alignment: For example, if some set of pegs P_1 always end up in the same holes every time SANA is run and another set of pegs P_2 end up in different holes each time SANA is run, this suggests the set P_1 is *confidently* aligned, whereas we should be suspicious about the alignment of pegs in P_2. Few algorithms are capable of this sort of confidence testing of the alignment; SANA, on the other hand, is so fast that it is easy to look for such *core* alignments [49]—cf. Subheading 3.1.

Flexible with Objective Functions: SANA has over a dozen pre-programmed objective functions that users can experiment with. In addition, users can supply SANA with externally computed similarity matrices, either node-to-node or edge-to-edge. Finally, we have tried to make the code base of SANA clear so that anybody familiar with C++ can program new objective functions easily.

Able to Handle Nodes That Have ASCII Names Rather Than Only Allowing Integers as Node Identifiers: To a programmer, creating a mapping between ASCII names and integers is easy. To non-programmers this is not so easy, and many aligners have the inexcusable fault of insisting that nodes are named by sequential integers. SANA does this internally but allows users to use whatever names they want to identify nodes.

Available to Plug in to Existing Popular Tools such as Cytoscape: SANA is available in the Cytoscape App store.

Able to Handle Multiple Input Graph Formats: Currently, SANA only natively accepts networks in edge list format and LEDA.gw format. The former is a line-by-line list of edges (two nodes from the same network listed on one line), while the latter is a rather deprecated format used by an old version of LEDA [50]. However, we do provide a converter called `createEdgeList` that outputs our edge list format given any of the following input formats: GML, XML, graphML, LEDA, CSV, and LGF.

Able to Perform Multi-objective Optimization, If So Desired: SANA has the ability to create a Pareto front across all measures the user chooses to optimize. So far as we are aware, the only other network alignment tool that does this is OptNetAlign [51]. Essentially, a Pareto front consists of a family of alignments that explore the trade-off between all the objective functions being optimized.

In general, we can only increase the value of one objective at the expense of one or more others. The Pareto front approximates the "frontier" of this trade-off; we use the method outlined in [52].

1.6 The Value of Randomness: Core Alignments

SANA shares one important aspect with a few other aligners including MAGNA [21, 35] and OptNetAlign [51]: it is a randomized search algorithm. Like these other algorithms, SANA starts with a random alignment and then starts to move pegs around between holes; each time it tries to swap or move pegs around, it asks if the objective function has gotten better or not. As time progresses, the alignment gets better according to the objective function. If the objective function is an easy one to optimize, SANA will quickly find the optimal or near-optimal alignment [22, 47]; in harder cases, it will simply find better-and-better solutions as it is given more time.

The fact that SANA intentionally injects randomness has some surprising positive aspects. In particular, if there exist highly similar regions between the two networks G_1 and G_2, SANA is likely to find them and align them identically every time, despite starting with a different random alignment each time. If there are other parts of the networks that are dissimilar and there is no obvious way to align them correctly, those regions are likely to get aligned differently each time SANA is run. Given two regions R_1 in G_1 and R_2 in G_2, the more topologically similar R_1 is to R_2, the more likely it is that SANA will align them the same way every time it is run, independent of the randomness. Since SANA is extremely fast, and since it has this random aspect, it is relatively painless to run SANA many times on the same pair of networks and look for pairs of nodes that are aligned together frequently. We use the term *core alignment* to refer to pairs of nodes that are stable across many runs of SANA; the more frequently a pair of nodes is aligned together, the more confident we are that they truly *belong* together according to the objective function being optimized. So for example, if we run SANA 10 times on the same 2 networks and produce output files out0.align, out1.align, out2.align, ..., out9.align, then we can trivially measure the core frequencies on the Unix command line as follows:

```
$ sort out?.align | uniq -c | sort -nr
```

The first `sort` puts identical lines from all 10 files side-by-side; the `uniq -c` counts how many unique lines are side-by-side (thus measuring core frequency), and the final `sort -nr` then sorts the aligned pairs of nodes by frequency, most frequent pairs of nodes first—that is, the most confident parts of the alignment are listed first. Note that the output of the above command line is a list of pairs preceded by their frequency. Note in particular that, even though SANA is a 1-to-1 aligner *per run*, with multiple runs we can produce non-1-to-1 mappings between the two networks, along with a confidence level for each particular pair. We are also working on functionality to produce core alignments in one run of SANA; that functionality may exist by the time this article goes to press and accessible via the command-line option "`-cores`".

1.7 Limitations of SANA

Currently, SANA aligns only two networks at a time. Each time, it produces a 1-to-1 mapping between the nodes of the smaller network to the nodes of the larger one (i.e., an arrangement of pegs into holes). So technically, SANA is a global, pairwise, 1-to-1 alignment algorithm—the simplest type of global alignment algorithm. However, as we described above, SANA produces *good* alignments so quickly that it can be run many times on the same pair of networks in the same time it takes to run most other algorithms just once; by running SANA many times we effectively produce not only a non-1-to-1 mapping, but also a *confidence estimate* of each pair of nodes we output. So far as we are aware, no other algorithm produces such confidence estimates.

Furthermore, even though SANA technically aligns only 2 networks at a time, in **Note 1** we describe a prototype version of multi-SANA that uses pairwise alignments to construct a multiple network alignment.

Thus, although SANA is technically only a 1-to-1 pairwise network aligner, it can effectively produce both many-to-many alignments (with confidences) and multiple alignments.

2 Examples of Usage

2.1 Getting Started with SANA

Table 1 contains a sequence of Unix Shell commands that will download the repo from GitHub, compile SANA, and perform your first test of SANA to ensure everything works.

The most basic run of SANA requires the user only to specify which two networks to align; in Table 1 it is the 2018 BioGRID renditions of *Rattus norvegicus* (the common sewer rat, aka lab rat), and the single-celled yeast *Schizosaccharomyces pombe*. SANA defaults to using S^3 as the objective function, and 5 min as the amount of time to perform simulated annealing. Total runtime is about 6–7 min including the initial computation of the temperature schedule, which we now describe.

Simulated annealing only works well if the temperature schedule is chosen carefully. We must start with a temperature high enough that moves are essentially random, so that even bad moves are frequently accepted (this keeps us out of local minima); and then end with a temperature low enough that only good moves are accepted (to hone in on the best local maximum once we've found its general vicinity). Empirically, we are controlling the *probability of accepting a bad move* or `pBad`; it must start close to 1, and end close to zero. Unfortunately there's no analytical method to compute these extremes, so the first 1–2 min of SANA are spent

Table 1
Getting started with SANA on the Unix command line

```
# Lines like this are comments. The Unix/Bash prompt is the dollar sign.
# First use "git" to clone the repo:
$ git clone http://github.com/waynebhayes/SANA
Cloning into 'SANA'... #git output deleted
$ cd SANA; make   # now wait a few minutes...
# Run SANA for the first time on the 2018 BioGRID networks of rat and S. pombe:
$ ./sana -fg1 networks/RNorvegicus18.el -fg2 networks/SPombe18.el
# wait while SANA computes a temperature schedule and then performs the alignment...
$ cat sana.out # look at the output file; first line is an internal
# representation of the alignment and can be ignored.
$ head -3 sana.align # left column is a BioGRID node name from rat, right from S.pombe.
361207  2542195
316265  2541287
499382  2539901
$
```

We first clone the repo from GitHub, then "make" SANA, then run it on the two smallest BioGRID 2018 networks: *R. norvegicus* and *S. pombe*. We then look at the output file `sana.out`, which contains scores and other useful information, as well as the actual alignment file `sana.align`. SANA has *many* command-line options; type "`../sana -h | less`" to see a long list of them

estimating the initial temperature $t_{initial}$, the final temperature t_{final} that gives a pBad starting near 1 and ending near zero, along with the t_{decay}, the temperature decay rate that gets us from one to the other in the allotted time (5 min by default).

Next you will see the statement, `Start execution of SANA_s3` which says SANA is finally starting the anneal, optimizing `s3`. After that, you'll see updates every few seconds as SANA progresses. These updates show the update number, the elapsed time so far, the current score, some statistical theoretical values that don't concern us here, and the sampled pBad, which should start above 0.98 and end somewhere below about 1e−6.

Once SANA is finished running, there are exactly two output files (whose names can be changed with the "-o" option): `sana.out` contains as its first (long) line an internal representation of the alignment, followed by some human-readable statistics; an example is in Table 2. The second file, called `sana.align`, contains the actual alignment in two-column format: on each line, the left column contains a node ("peg") from G_1 and the right column is the aligned node ("hole") from G_2.

The default objective function is S^3; changing the objective function is easy on the command line. For example to have SANA optimize a 50-50 combination of *EC* and S^3, type

```
./sana -ec 0.5 -s3 0.5 -fg1 ...
```

To turn off S^3 entirely and perform an *EC*-only alignment, do

```
./sana -s3 0 -ec 1 -fg1 ...
```

To perform an alignment that optimizes 90% Importance as defined by HubAlign [23] 5% graphlets as used by GRAAL [15], 5% EC, and no S^3, do

Table 2
The `sana.out` file (whose name can be changed using the `-o` command-line option) contains information about the input networks (nodes, edges, connected components) and an analysis of the alignment (various measures applied to the entire alignment and also applied to the common connected subgraphs)

```
2018-06-15 15:21:57

G1: yeast
n    = 2390
m    = 16127
#connectedComponents = 158
Largest connectedComponents (nodes, edges) = (1994, 15819) (10, 32) (6, 11)

G2: human
n    = 9141
m    = 41456
#connectedComponents = 94
Largest connectedComponents (nodes, edges) = (8934, 41341) (5, 4) (4, 3)

Method: SANA_s3
Temperature schedule:
T_initial: 0.000316228
T_decay: 6.61993
Optimize:
weight s3: 1
Requested Execution time: 5 minutes

Actual execution time = 300.976 seconds

Random Seed: 514154230

Scores:
ec: 0.397966
mec: 1
ses: 35381
ics: 0.831563
s3: 0.368279
lccs: 0.248768
sec: 0.222913

Common subgraph:
n    = 2390
m    = 6418
#connectedComponents = 395
Largest connectedComponents (nodes, edges) = (1059, 4805) (53, 263) (48, 69)

Common connected subgraphs:
Graph  n     m      alig-edges  indu-edges  EC        ICS       S3
G1     2390  16127  6418        7718        0.397966  0.831563  0.368279
CCS_0  1059  4805   4805        5790        1.000000  0.829879  0.829879
CCS_1  53    263    263         268         1.000000  0.981343  0.981343
CCS_2  48    69     69          73          1.000000  0.945205  0.945205
CCS_3  34    68     68          70          1.000000  0.971429  0.971429
CCS_4  33    50     50          52          1.000000  0.961538  0.961538
```

Table 3
Measures accepted by SANA on the command line

Name	Description
`s3`	Symmetric Substructure Score [21]
`ec`	Edge Coverage/Correspondence/Correctness [15]
`ics`	Induced Conserved Structure [25]
`graphlet`	Orbit Degree Vector (ODV) Similarity [15, 29]
`graphletlgraal`	LGRAAL-normalization of ODV sim [20]
`go`	Mean ResnikMax GO similarity [62, 63]
`netgo`	Network-alignment-based GO similarity [34]
`wec`	Weighted EC [24]
`esim`	External file defining node-pair similarities
`sequence`	BLANT-based sequence similarities [31]
`lccs`	Largest Common Connected Subgraph [15]
`nc`	Node Correctness (if known, defines the exact alignment)
`spc`	Shortest Path Conservation [22]
`edgeCount`	Degree difference
`edgeDensity`	Relative degree difference
`importance`	HubAlign's Importance [23]
`nodeDensity`	Local node density
`ewec`	External edge-based similarity matrix, e.g., edge-graphlet similarity [56]
`sequence`	BLAST bit scores based on protein sequence similarity [31]

Note that "Name" means "command-line option," so for example to give `ec` a weight of 0.5, use "`-ec 0.5`" on the SANA command line. As more objectives become available, they can be listed by running "./sana -help".

`./sana -s3 0 -importance 0.9 -graphlets 0.05 -ec 0.05` ... Note that one does not need to manually ensure that all the weights specified on the command line add to 1; if they do not, SANA will simply re-normalize them all so that they add to 1.

Similarly, the are many other objective functions defined by SANA; currently implemented ones are listed in Table 3.

Finally, as mentioned previously, SANA is capable of approximating the Pareto Frontier with a family of solutions called the *Pareto Front*. This mode is invoked using `-mode pareto`. See the output of `./sana --help` for more details on usage.

2.2 Direct Comparison with Other Aligners

As a part of our first publication on SANA [22], we wanted to automate the process of directly comparing to many other existing aligners. Thus, the external source code of over a dozen existing aligners was directly incorporated into SANA so that they can be called from the SANA command line. This was done to ensure consistent calling conventions to these other aligners during our comparisons. These other methods can be called from the SANA command line using the `-method` argument. In the SANA repo, these other aligners are in the directory `wrappedAlgorithms`; see the online SANA documentation for more details. (If you are an author of one of these aligners and notice that SANA is not using your algorithm optimally, feel free to contact us with any corrections.) The other aligners currently incorporated into SANA are LGRAAL [20], MAGNA++ [35], HubAlign [23], WAVE [24], NETAL [53], MIGRAAL [32], GHOST [25], PISWAP [54], OptNetAlign [51], SPINAL [55], GREAT [56], NATALIE 2.0 [57], GEDEVO [58], CytoGEDEVO [59], BEAMS [60], HGRAAL [49], PINALOG [61].

3 An Example of Objective Function Experimentation

As shown in Fig. 2, SANA can be used to experiment with objective functions; we believe that such experimentation is one of the most important but apparently under-appreciated aspects of the science of network alignment. Here, we describe one such experiment with a very well-defined scientific goal.

3.1 Gene–microRNA Networks

Consider a set of gene–microRNA (mRNA) networks [9], one network for each species. These networks are bipartite, meaning that genes interact with microRNAs, but neither genes nor microRNAs interact with their own type. Thus, when aligning two gene–mRNA networks, we wish to align genes from one network to genes in the other, and mRNAs in one to mRNAs in the other, but we should never align a gene to an mRNA or vice versa. In essence, the nodes have two *types*, and we must provide a type-specific network alignment.

At first, SANA did not have the functionality to provide a typed-node alignment. (It does now, using the `-bipartite` argument, in which case we assume that the first column in the edge list is one type, and the second column is the other type. Only two types are supported at the moment, but more may become available; run "./sana -help" for more info). The question was: how do the various topological objective functions compare in their ability to *automatically* align types correctly, given that typing is not *enforced* by the alignment algorithm?

Referring to Fig. 2, the scientific goal is clear: **maximize the fraction of nodes that are aligned to like-type nodes in the other network**. The question is now, *which topological objective function best achieves this scientific goal?*

We received 535 networks directly from one of the authors of [9]. We chose 1000 pairs of networks at random out of the $\binom{535}{2} = 142,845$ possible pairs of networks. For each pair of networks, we tested the following objective functions for their ability to correctly align nodes of like type to each other when this was not enforced: *EC*, S^3, Importance [23], GRAAL-type graphlet orbit signatures [15, 29], and LGRAAL-type graphlet orbit signatures [20]. To further test the dependence on runtime, we ran SANA on all the above objectives for all 1000 networks for runtimes of 1 and 4 min. Finally, to look at the frequency of core alignments, we performed each of the above pairs 5 times each. The results are in Table 4.

Table 4
Table of results when testing various objective functions (leftmost column) for their ability to correctly align genes-to-genes, and mRNAs-to-mRNAs, when aligning a pair of gene–mRNA networks [9]

Objective	Pairs	2*Gene	mix	2*RNA	coreFreq (GG)> 1	coreFreq (MG)> 1	coreFreq (MM)> 1
1 minute runs							
ec	30424880	29953074	198792	273014	1268806	570	3169
s3	30424880	29986047	284470	154363	1093307	2947	688
Importance	30241594	25434876	4658345	148373	651969	114137	1386
graphlet-GRAAL	30424880	24109670	6176510	138700	1902554	449738	17331
graphlet-LGRAAL	30424880	23056815	7305611	62454	1718519	584735	7086
4 minute runs							
ec	30424880	30055465	97811	271604	1245103	208	5908
s3	30424880	29953309	283313	188258	1092319	3779	1508
Importance	30292547	25473995	4669942	148610	652830	114815	1621
graphlet-GRAAL	30424880	24104880	6180836	139164	2208583	502806	25308
graphlet-LGRAAL	30424880	23051615	7310109	63156	2090416	692752	10504

Objectives tested were *EC* [15], S^3 [21], Importance [23], graphlet [15, 29], and graphlet-LGRAAL [20]. The columns are as follows. **pairs**: total number of pairs of nodes aligned in all 1000 network pairs that were run 5 times each. **2*Gene**: number of pairs in which a gene was correctly aligned to another gene. **mix**: number of pairs in which a gene in one network was aligned to an mRNA in the other. **2*RNA**: number of pairs in which an mRNA was aligned to another mRNA. **coreFreq(XY)>** 1: the number of aligned pairs that had a core frequency greater than 1 (indicating the objective function strongly prefers to align this pair of nodes together) for type-pairs GG, MG, and MM

One column of great interest is the "mix" column, which counts the number of times, out of the approximately 30 million pairs of aligned nodes, in which a gene from one network was aligned to an mRNA in the other network—which is the kind of mis-typed node alignment we are trying to avoid. The rows are sorted best-to-worst by this measure, in each of the 1-min and 4-min sub-tables. As we can see, the *EC* objective scores best at avoiding this kind of mis-typed alignment. In the 1-min runs, *EC* aligns unlike typed node-pairs in only 0.65% of cases; S^3 is a close second, mis-typing just under 1% of the aligned pairs of nodes. In contrast, HubAlign's Importance measure [23] is almost 20 times worse in terms of incorrectly aligning nodes of different types, doing so in about 15% of aligned pairs of nodes, while both graphlet measures fare the worst, aligning unlike-type nodes in over 20% of cases.

Even more interesting is the 4 min runs, in which *EC* cuts its mis-typed node alignment in half, down to about 0.3% of aligned pairs, while all other measures fail to improve their "mix" column with the longer runtime.

Recall that if SANA aligns the same pair of nodes together in more than one run, we say that pair is in the *core* alignment, because the objective function is unlikely to align two nodes together more than once by chance. Another column of great interest is thus the **coreFreq(MG)>** 1 column, which tells us how frequently the objective function seems to *strongly* prefer mis-aligning a pair of nodes of different types. Again we see that the *EC* measure is by far the best measure by this criterion: in the 1 min runs, only 570 mis-typed pairs appear out of 30 million (about 2 per 100,000 pairs), while the 4 min runs cut that "error rate" in half, suggesting that longer runs will do a better job of correctly aligning types. Meanwhile, S^3 does 10× worse at 1 min and gets *more* bad in the 4 min runs, while importance and both graphlet measures misalign orders of magnitude more typed pairs, presenting a strong preference for misaligning nodes in about 1–2% of pairs.

We conclude that the *EC* measure is, *by far*, the best available objective function for this particular purpose among those we tested. For the moment we do not hypothesize why this is the case, but empirically the result seems iron-clad. While we agree that the S^3 measure is mathematically more aesthetically pleasing and would seem to be a better measure intuitively, for this particular purpose *EC* seems to work better. The author finds the poor performance of graphlet-based measures particularly surprising, since the author is a strong believer that graphlets are a useful tool for network analysis (see, for example, [41])—and graphlets have certainly demonstrated their value in other contexts [13, 30]. However, these results suggest that perhaps orbit degree signatures as they are currently defined [15, 20, 29] may not be the best way to leverage graphlet-based information in the context of global pairwise network alignment.

4 Conclusion

We have described the use of SANA [22], the *Simulated Annealing Network Aligner*, in the context of the pairwise 1-to-1 global alignment of biological networks. SANA provides many advantages over the many other aligners currently available: as a search algorithm, it is lightning fast, producing well-scoring alignments in minutes rather than hours; it provides a large array of objective functions users may wish to experiment with, as well as the facility to add more objectives in the future; it does not require the user to know much about the internal workings of the aligner in order to use it; and it is well on the way toward being fully integrated into popular network analysis tools such as Cytoscape.

We have introduced the concept of *objective function experimentation* (cf. Fig. 2 and Subheading 3.1), which we believe is at the core of future developments in network alignment. SANA's speed and effectiveness makes it the ideal aligner to implement the process depicted in Fig. 2.

5 Note

1. A prototype of a multiple-network-alignment version of SANA is available in the SANA GitHub repo. Simply re-compile SANA with the `-DMULTI_PAIRWISE` option on the command line (see the `Makefile`), and consult the Bourne shell script `multi-pairwise.sh`; running it without any arguments provides a short help message.

References

1. Williamson MP, Sutcliffe MJ (2010) Protein--protein interactions. Portland Press Limited, London
2. Jaenicke R, Helmreich E (2012) Protein-protein interactions, vol 23. Springer, Berlin
3. Davidson EH (2010) The regulatory genome: gene regulatory networks in development and evolution. Academic press, San Diego
4. Karlebach G, Shamir R (2008) Modelling and analysis of gene regulatory networks. Nature reviews. Mol Cell Biol 9(10):770
5. Chen K, Rajewsky N (2007) The evolution of gene regulation by transcription factors and microRNAs. Nat Rev Genet 8(2):93
6. Prescott DM (2012) Cell biology a comprehensive treatise V3: gene expression: the production of RNA's, vol 3. Elsevier, Amsterdam
7. Farazi TA, Hoell JI, Morozov P, Tuschl T (2013) Micrornas in human cancer. In: MicroRNA cancer regulation. Springer, Berlin, pp 1–20
8. Kotlyar M, Pastrello C, Sheahan N, Jurisica I (2015) Integrated interactions database: tissue-specific view of the human and model organism interactomes. Nucleic Acids Res 44 (D1):536–541
9. Tokar T, Pastrello C, Rossos AE, Abovsky M, Hauschild A-C, Tsay M, Lu R, Jurisica I (2017) mirdip 4.1—integrative database of human microRNA target predictions. Nucleic Acids Res 46(D1):360–370
10. Fiehn O (2002) Metabolomics-the link between genotypes and phenotypes. In: Functional genomics. Springer, Berlin, pp 155–171
11. Milano M, Guzzi PH, Tymofieva O, Xu D, Hess C, Veltri P, Cannataro M (2017) An extensive assessment of network alignment

algorithms for comparison of brain connectomes. BMC Bioinf 18(6):235

12. Junker BH, Schreiber F (2011) Analysis of biological networks, vol 2. Wiley, New York
13. Davis D, Yaveroğlu ÖN, Malod-Dognin N, Stojmirovic A, Pržulj N (2015) Topology-function conservation in protein–protein interaction networks. Bioinformatics 31 (10):1632–1639. https://doi.org/10.1093/bioinformatics/btv026
14. Sporns O (2010) Networks of the brain. MIT Press, Cambridge
15. Kuchaiev O, Milenković T, Memišević V, Hayes W, Pržulj N (2010) Topological network alignment uncovers biological function and phylogeny. J R Soc Interface 7 (50):1341–1354. https://doi.org/10.1098/rsif.2010.0063
16. Van El CG, Cornel MC, Borry P, Hastings RJ, Fellmann F, Hodgson SV, Howard HC, Cambon-Thomsen A, Knoppers BM, Meijers-Heijboer H *et al* (2013) Whole-genome sequencing in health care: recommendations of the European society of human genetics. Eur J Hum Genet 21(6):580
17. Cook SA (1971) The complexity of theorem-proving procedures. In: Proceedings of the third annual ACM symposium on theory of computing. ACM, New York, pp 151–158
18. Garey MR, Johnson DS (1979) Computers and intractability: a guide to the theory of NP-completeness. W.H. Freeman, New York
19. Von Mering C, Krause R, Snel B, Cornell M *et al* (2002) Comparative assessment of large-scale data sets of protein-protein interactions. Nature 417(6887):399
20. Malod-Dognin N, Pržulj N (2015) L-GRAAL: Lagrangian graphlet-based network aligner. Bioinformatics 31(13):2182–2189
21. Saraph V, Milenković T (2014) Magna: maximizing accuracy in global network alignment. Bioinformatics 30(20):2931–2940
22. Mamano N, Hayes WB (2017) SANA: simulated annealing far outperforms many other search algorithms for biological network alignment. Bioinformatics. https://doi.org/10.1093/bioinformatics/btx090
23. Hashemifar S, Xu J (2014) HubAlign: an accurate and efficient method for global alignment of protein-protein interaction networks. Bioinformatics 30(17):438–444. https://doi.org/10.1093/bioinformatics/btu450
24. Sun Y, Crawford J, Tang J, Milenković T (2015) Simultaneous optimization of both node and edge conservation in network alignment via WAVE. In: Pop M, Touzet H (eds) Algorithms in bioinformatics. Lecture notes in computer science, vol 9289. Springer, Berlin, pp 16–39. http://dx.doi.org/10.1007/978-3-662-48221-6_2
25. Patro R, Kingsford C (2012) Global network alignment using multiscale spectral signatures. Bioinformatics 28(23):3105–3114. https://doi.org/10.1093/bioinformatics/bts592. http://bioinformatics.oxfordjournals.org/content/28/23/3105.full.pdf+html
26. Vijayan V, Milenković T (2017) Aligning dynamic networks with dynawave. Bioinformatics 34(10):1795–1798
27. Faisal FE, Meng L, Crawford J, Milenković T (2015) The post-genomic era of biological network alignment. EURASIP J Bioinforma Syst Biol 2015(1):3
28. Pržulj N, Corneil DG, Jurisica I (2004) Modeling interactome: scale-free or geometric? Bioinformatics 20(18):3508–3515. https://doi.org/10.1093/bioinformatics/bth436. http://bioinformatics.oxfordjournals.org/content/20/18/3508.full.pdf+html
29. Milenković T, Pržulj N (2008) Uncovering biological network function via graphlet degree signatures. Cancer Inform 6:257–273. (Epub 2008 Apr 14)
30. Yaveroğlu N, Malod-Dognin N, Davis D, Levnajic Z, Janjic V, Stojmirovic RKA, Pržulj N (2014) Revealing the hidden language of complex networks. Sci Rep 4:4547
31. Altschul SF *et al* (1990) Basic local alignment search tool. J Mol Biol 215:403–410
32. Kuchaiev O, Pržulj N (2011) Integrative network alignment reveals large regions of global network similarity in yeast and human. Bioinformatics 27:1390–1396. https://doi.org/bioinformatics/btr127
33. The Gene Ontology Consortium (2008) The gene ontology project in 2008. Nucleic Acids Res 36(Suppl 1):440–444. https://doi.org/10.1093/nar/gkm883. http://nar.oxfordjournals.org/content/36/suppl_1/D440.full.pdf+html
34. Hayes WB, Mamano N (2017) Sana netgo: a combinatorial approach to using gene ontology (go) terms to score network alignments. arXiv preprint, arXiv:1704.01205
35. Vijayan V, Saraph V, Milenković T (2015) Magna++: maximizing accuracy in global network alignment via both node and edge conservation. Bioinformatics. https://doi.org/10.1093/bioinformatics/btv161
36. Pržulj N, Wigle D, Jurisica I (2004) Functional topology in a network of protein interactions. Bioinformatics 20(3):340–348
37. Hočevar T, Demšar J (2014) A combinatorial approach to graphlet counting. Bioinformatics

30(4):559–565. https://doi.org/10.1093/bioinformatics/btt717

38. Chatr-Aryamontri A, Oughtred R, Boucher L, Rust J, Chang C, Kolas NK, O'Donnell L, Oster S, Theesfeld C, Sellam A *et al* (2017) The biogrid interaction database: 2017 update. Nucleic Acids Res 45(D1):369–379
39. Rossi RA, Zhou R, Ahmed NK (2017) Estimation of graphlet statistics. arXiv preprint, arXiv:1701.01772
40. Yang C, Lyu M, Li Y, Zhao Q, Xu Y (2018) SSRW: a scalable algorithm for estimating graphlet statistics based on random walk. In: International conference on database systems for advanced applications. Springer, Berlin, pp 272–288
41. Hasan A, Chung P-C, Hayes W (2017) Graphettes: Constant-time determination of graphlet and orbit identity including (possibly disconnected) graphlets up to size 8. PLoS ONE 12 (8):0181570
42. Vidal M (2016) How much of the human protein interactome remains to be mapped? American Association for the Advancement of Science, Washington
43. Lesk A, Chothia C (1986) The response of protein structures to amino-acid sequence changes. Philos Trans R Soc Lond A 317 (1540):345–356
44. Clark C, Kalita J (2014) A comparison of algorithms for the pairwise alignment of biological networks. Bioinformatics 30(16):2351–2359
45. Faisal FE, Meng L, Crawford J, Milenković T (2015) The post-genomic era of biological network alignment. EURASIP J Bioinforma Syst Biol 2015(1):1
46. Guzzi PH, Milenković T (2017) Survey of local and global biological network alignment: the need to reconcile the two sides of the same coin. Brief Bioinform. https://doi.org/10.1093/bib/bbw132
47. Kanne DP, Hayes WB (2017) SANA: separating the search algorithm from the objective function in biological network alignment, Part 1: Search. arXiv preprint, arXiv:1709.01464
48. Larsen SJ, Alkærsig FG, Ditzel HJ, Jurisica I, Alcaraz N, Baumbach J (2016) A simulated annealing algorithm for maximum common edge subgraph detection in biological networks. In: Proceedings of the 2016 on genetic and evolutionary computation conference. ACM, New York, 341–348
49. Milenković T, Ng WL, Hayes W, Pržulj N (2010) Optimal network alignment with graphlet degree vectors. Cancer Informat 9:121–137. https://doi.org/10.4137/CIN.S4744
50. Mehlhorn K, Naher S (1999) LEDA: a platform for combinatorial and geometric computing. Cambridge University Press, Cambridge
51. Clark C, Kalita J (2015) A multiobjective memetic algorithm for PPI network alignment. Bioinformatics 31(12):1988–1998. https://doi.org/10.1093/bioinformatics/btv063. http://bioinformatics.oxfordjournals.org/content/31/12/1988.full.pdf+html
52. Smith KI, Everson RM, Fieldsend JE, Murphy C, Misra R (2008) Dominance-based multiobjective simulated annealing. IEEE Trans Evol Comput 12(3):323–342
53. Neyshabur B, Khadem A, Hashemifar S, Arab SS (2013) Netal: a new graph-based method for global alignment of protein-protein interaction networks. Bioinformatics 29 (13):1654–1662. https://doi.org/10.1093/bioinformatics/btt202. http://bioinformatics.oxfordjournals.org/content/29/13/1654.full.pdf+html
54. Chindelevitch L, Ma C-Y, Liao C-S, Berger B (2013) Optimizing a global alignment of protein interaction networks. Bioinformatics 29 (21):2765–2773. https://doi.org/10.1093/bioinformatics/btt486. http://bioinformatics.oxfordjournals.org/content/29/21/2765.full.pdf+html
55. Aladağ AE, Erten C (2013) Spinal: scalable protein interaction network alignment. Bioinformatics 29(7):917–924. https://doi.org/10.1093/bioinformatics/btt071. http://bioinformatics.oxfordjournals.org/content/29/7/917.full.pdf+html
56. Crawford J, Milenković T (2015) Great: graphlet edge-based network alignment. In: 2015 IEEE international conference on bioinformatics and biomedicine (BIBM). IEEE, Piscataway, pp 220–227
57. El-Kebir M, Heringa J, Klau GW (2011) Lagrangian relaxation applied to sparse global network alignment. In: IAPR international conference on pattern recognition in bioinformatics. Springer, Berlin, pp 225–236
58. Ibragimov R, Malek M, Guo J, Baumbach J (2013) Gedevo: an evolutionary graph edit distance algorithm for biological network alignment. In: OASIcs-OpenAccess series in informatics, vol 34. Schloss Dagstuhl-Leibniz-Zentrum fuer Informatik
59. Malek M, Ibragimov R, Albrecht M, Baumbach J (2016) Cytogedevo: global alignment of biological networks with cytoscape. Bioinformatics 32(8):1259–1261
60. Alkan F, Erten C (2014) Beams: backbone extraction and merge strategy for the global many-to-many alignment of multiple PPI networks. Bioinformatics 30(4):531–539

61. Phan HT, Sternberg MJ (2012) Pinalog: a novel approach to align protein interaction networks—implications for complex detection and function prediction. Bioinformatics 28 (9):1239–1245
62. Resnik P (1995) Using information content to evaluate semantic similarity in a taxonomy. arXiv preprint cmp-lg/9511007
63. Ashburner M, Ball CA, Blake JA, Botstein D, Butler H, Cherry JM, Davis AP, Dolinski K, Dwight SS, Eppig JT, Harris MA, Hill DP, Issel-Tarver L, Kasarskis A, Lewis S, Matese JC, Richardson JE, Ringwald M, Rubin GM, Sherlock G (2000) Gene Ontology: tool for the unification of biology. Nat Genet 25 (1):25–29. https://doi.org/10.1038/75556

INDEX

Stefan Canzar and Francisca Rojas Ringeling (eds.), *Protein-Protein Interaction Networks: Methods and Protocols*, Methods in Molecular Biology, vol. 2074, https://doi.org/10.1007/978-1-4939-9873-9,

P

R

S

T

V

W

MIX
Papier aus verantwortungsvollen Quellen
Paper from responsible sources
FSC® C105338

If you have any concerns about our products,
you can contact us on
ProductSafety@springernature.com

In case Publisher is established outside the EU,
the EU authorized representative is:
Springer Nature Customer Service Center GmbH
Europaplatz 3, 69115 Heidelberg, Germany

Printed by Libri Plureos GmbH
in Hamburg, Germany